Global Pathways to Water Sustainability

David E. McNabb

Global Pathways to Water Sustainability

palgrave
macmillan

David E. McNabb
Pacific Lutheran University
Tacoma, WA, USA

ISBN 978-3-030-04084-0 ISBN 978-3-030-04085-7 (eBook)
https://doi.org/10.1007/978-3-030-04085-7

Library of Congress Control Number: 2018962904

Cover illustration: Jason Jones Travel Photography/Gettyimages

This Palgrave Macmillan imprint is published by the registered company Springer Nature Switzerland AG
The registered company address is: Gewerbestrasse 11, 6330 Cham, Switzerland

Preface

This is not an easy time for water managers; their world appears to be in what Joshua Busby described in 2018 as "a state of permanent crisis." Not only must they find, treat and deliver freshwater from limited and, in some cases, declining resources to a rapidly growing and migrating population, they have to invest heavily in repairing and replacing aging infrastructure to be sure they remain resilient and sustainable in an increasingly complex and threatening future. In many parts of the world, they are running out of water, while in others, they must cope with too much, or with water that comes at the wrong time of the year for the existing social and economic structure. To meet growing demand, water and wastewater must invest billions in the technology, infrastructure and knowledge needed for desalination facilities while they also invest on the infrastructure needed for coping with more occurring floods and heavier stormwater events. Still others are turning to the tertiary treatment of wastewater to augment over-allocated surface or groundwater resources. Together with these technical and water efficiency requirements, global pressures are forcing water managers to include ecosystem protection along with their sustainability planning.

While water managers are doing their best to provide their critical but too often un-recognized public service, global climate change is affecting water resources in ways far beyond water managers' ability to control. The increasing existence of carbon dioxide and other greenhouse gases in the atmosphere has already changed precipitation patterns and triggered the melting of glaciers and global icecaps. In 2005, world leaders initiated what they called the Global Water Decade. That effort brought water problems onto the political and economic stage but did not slow the increasing trend of water scarcity. It neither reversed the need for eliminating the pollution of existing water resources, nor did it result in more efficient use of the world's supplies of freshwater. Box 1 helps explain the state of the global water problem.

Box 1 The changing water environment

The global demand for water has been increasing at a rate of about 1% per year as a function of population growth, economic development and changing consumption patterns, among other factors, and it will continue to grow significantly over the next two decades. Industrial and domestic demand for water will increase much faster than agricultural demand, although agriculture will remain the largest overall user. The vast majority of the growing demand for water will occur in countries with developing or emerging economies. At the same time, extreme weather events in the water cycle are intensifying due to climate change, with wetter regions generally becoming wetter and drier regions becoming even drier. At present, an estimated 3.6 billion people ... live in areas that are potentially water-scarce at least one month per year, and this could increase to some 4.8–5.7 billion by 2050. Since the 1990s, water pollution has worsened in almost all rivers in Africa, Asia and Latin America. The deterioration of water quality is expected to further escalate over the next decades and this will increase threats to human health, the environment and sustainable development. Globally, the most prevalent water quality challenge is nutrient loading.

The trends in water availability and quality are accompanied by projected changes in flood and drought risks. The number of people at risk from floods is projected to rise from 1.2 billion today to around 1.6 billion in 2050... The population currently affected by land degradation/desertification and drought is estimated at 1.8 billion people, making this the most significant category of 'natural disaster' based on mortality and socio-economic impact relative to gross domestic product (GDP) per capita.

Source UN World Water Development Report (2018, 3).

Despite the great strides in water conservation that have occurred around the world, population growth is expected to continue for most of this century. People are still migrating from rural to urban centers; water use for agriculture in many parts of the world is still inefficient; production of fuels to generate power is still is still water-hungry; water and wastewater infrastructure is still aging faster than it can be repaired or replaced; and climate change is still changing natural water supply and distribution.

To better explain how these factors affect water and how and what water managers are doing to meet these challenges, the book is divided into five parts. The first three chapters in Part I deal with the water and wastewater problem and world leaders' call to action to ensure that the human right to water is met. In the next six chapters I look at the role of each of the mega-scale problems facing water managers, beginning with the related factors of population growth and urbanization. This is followed by a look at how inefficient water use by agriculture is limiting the availability of water for producing the food needed by the growing population, and how water is wasted in the production of primary fuels and power generation. I cap this section with what many consider to be the one of the most salient problems facing the world: climate change and its impact on the availability of water.

Part III of the book shifts direction from a look at water problems to focus on examples of the positive steps many water organizations and nations are taking to alleviate problems and become resilient enough to overcome future crises and achieve sustainability. The section topics range from paths taken to achieve sustainable urban supplies to the global search for alternative water supplies. The final part includes chapters on the path to sustainability and is guided by the more macro view of strategic planning asset management planning.

Tacoma, USA David E. McNabb

Reference

UN (United Nations). 2018. 2018 World Water Development Report. Accessed July 23, 2018 from http://unesdoc.unesco.org/images/0026/002614/261424e.pdf.

Contents

Part I Water for a Sustainable Future

1	The Global Goal of Sustainable Growth	3
2	Water's Role in a Sustainable Future	17
3	The Global Need for Water Sustainability	31

Part II Barriers on the Path to Sustainability

4	The Barrier of Climate Change	53
5	The Population Growth Barrier	67
6	The Urbanization Barrier	83
7	Agriculture and Inefficient Water Use	99
8	Water-Thirsty Energy Production	117
9	The Barrier of Failing Infrastructure	133

Part III Pathways to Water Sustainability

10 Pathways to Water Supply Security 153

11 Pathways to Sustainable Cities 169

12 Pathways to Sustainable Agriculture 185

13 Pathways to Water-Smart Power 201

14 Paths to Alternative and Supplemental Water Supplies 217

15 Pathways to Supplemental Water Resources 233

16 Alternative Sources of Water Supply 251

Part IV Innovative Water Management

17 Managing Assets for Water Sustainability 265

18 Planning for Achieving Sustainability 285

Glossary 305

Bibliography 309

Index 339

About the Author

David E. McNabb, Ph.D. is the elected audit commissioner of a combined water and wastewater district in the Pacific Northwest, and Professor Emeritus of the Pacific Lutheran University School of Business. He has taught business and public administration courses for Oregon State University, the University of Washington Tacoma, and the University of Maryland-Europe. This is his fifteenth book. His previous works include *Water Resource Management*, *Comparative History of Commerce and Industry*, *Nation Building in the Baltic States*, *Research Methods for Public Administration and Nonprofit Organizations*, and *Research Methods for Political Science*.

List of Figures

Fig. 3.1	Top ten countries with disasters and the damage amounts, 2005–2014 (*Source* UNISDR 2015)	48
Fig. 4.1	U.S. coastal areas vulnerable to water pollution (*Source* NASA 2016)	57
Fig. 5.1	World population growth and global growth rate, 1750–2100 (*Source* Roser and Ortiz-Ospina [2017]. From UN and the Netherland Environmental Assessment Agency's HYDE database)	68
Fig. 5.2	Estimated increases (%) in annual water demand, 2005–2030 (*Source* 2010 Water resources global water demand)	71
Fig. 7.1	Overhead irrigation system watering a potato crop in the arid U.S. West (*Source* Washington State Department of Ecology photo)	107
Fig. 8.1	U.S. energy consumption by source	118
Fig. 8.2	U.S. regions with high water use for fracturing, 2011–2014 (*Source* USGS 2015)	126
Fig. 11.1	Abbreviated model of the all-water scope of the IUWM system (*Source* The author)	176
Fig. 13.1	Tennessee Valley Authority natural gas combined-cycle power plant showing steam evaporation during the cooling cycle (*Source* Public domain photo www.tva.gov)	209
Fig. 13.2	Growth of wind power's share of U.S. electricity generation, 1990–2017 (*Source* US EIA)	211
Fig. 13.3	Schematic of a simple heat and power combined-cycle system (*Source* EPA, n.d. www.epa.gov/chp/what-chp)	212

Fig. 17.1 Major elements of a comprehensive asset management system (*Source* The author) 271
Fig. 18.1 Overview of utility general planning processes (*Source* After material in EPA 2016) 286
Fig. 18.2 Assessing and selecting which trait to focus on (*Source* EPA 2008) 296

List of Tables

Table 1.1	Top 15 water-stressed countries, 2015	14
Table 3.1	Increasing use of collaborative innovations by public sector, 1990–1994 and 2010	38
Table 3.2	Water efficiency rules and compliance dates	42
Table 4.1	Major cities running out of drinking water (estimates only)	62
Table 5.1	Selected countries with the high population growth rates in 2017	69
Table 5.2	Population growth estimates for Mexico, 2020–2050	75
Table 5.3	Population growth in China, with forecasts to 2050	78
Table 6.1	The world's 25 largest urban areas in 2017 and projections to 2030	86
Table 6.2	Sustainability problems associated with rampant urbanization	87
Table 6.3	Highly urbanized countries with the highest use of freshwater, 2017 estimates	90
Table 7.1	Categories of major water pollutants from agriculture and impact by source	110
Table 8.1	Consumption of combustible fuels for electricity generation, 2015–2017	119
Table 8.2	Water withdrawals and consumption by sector, billion cubic meters (bcm), 2014	122
Table 8.3	Major producers of electricity from fossil fuels, 2014 data	124
Table 8.4	Fuels used in the USA and globally to generate electricity	129
Table 9.1	Infrastructure concern scores of AWWA 2018 water system survey respondents	135
Table 9.2	Numbers of water supply organizations in the USA (% rounded)	137

Table 9.3	Estimated service lives by materials in Western USA (average years of service)	140
Table 9.4	Total 20-year needs by system size and infrastructure categories, 2015–2035	146
Table 9.5	Shares of 20-year needs by new vs. existing infrastructure and by project category	147
Table 10.1	The ten largest river systems in India	159
Table 11.1	The top ten U.S. cities and sustainability index rating on a scale of 1–100%	173
Table 11.2	Water sustainable city items and activities for Seattle, Washington	173
Table 11.3	Technologies and practices implemented by cities for water sustainability	179
Table 12.1	Adoption of Conservation Agriculture (CA) methods by continent, 2011 data	194
Table 12.2	Indian conservation agriculture experiment results	195
Table 13.1	Cooling methods used in power plants by fuel source, 2014	203
Table 15.1	Major green infrastructure design models and techniques	241
Table 15.2	Example alternative green infrastructure solutions to water quality problems	242
Table 15.3	Example alternative green infrastructure solutions for mitigating flooding	244
Table 15.4	Selected characteristics of the five case studies in green infrastructure	244
Table 16.1	Considerations for selected alternative water sources	252
Table 17.1	Form for core components of an asset management plan	272
Table 17.2	Possible sub-topics that might be included in the asset inventory phase	273
Table 17.3	Example of an asset inventory and asset categories for a small water system	273
Table 17.4	Asset management plan level of service scores	275
Table 17.5	Sample list of level of service statements and performance scoring system	276
Table 17.6	Rating plan for probability of failure for a water system asset	278
Table 17.7	Rating plan for quantifying potential consequences of an asset failure	278
Table 17.8	A multiplication method for establishing a hierarchy score for a critical asset	279
Table 17.9	Examples of major water system asset repair sand replacements, 15-year schedule	280
Table 18.1	Results of a hypothetical SWOT analysis for a small water and sewer utility	294
Table 18.2	Strategic actions and performance measures for water quality	298
Table 18.3	Potential water and/or wastewater utility sustainability goals	300

List of Boxes

Box 1.1	The 1992 Dublin Statement on Water and Sustainable Development	11
Box 2.1	Selected elements of opening paragraphs of 2003 UN General Comment No. 5	20
Box 3.1	Combination of forces threatening the global water supply	33
Box 4.1	Proposals for preparing water management for the impact of climate change	63
Box 6.1	The race to an urbanized world	85
Box 7.1	Sustainable agriculture in U.S. Food, Agriculture, Conservation, and Trade Act of 1990	99
Box 7.2	Agriculture, the environment and water	102
Box 7.3	Landowners fined for allegedly pumping 500 million gallons from Odessa Aquifer	107
Box 8.1	What is hydraulic fracturing?	120
Box 9.1	California high dam nearly collapses	134
Box 10.1	Cape Town faces severe water scarcity	162
Box 11.1	Cities ranked by their preparedness to survive extreme weather events	180
Box 12.1	Water's role in achieving sustainable agriculture	188
Box 13.1	Planning for a water-smart energy future	204
Box 14.1	Water and wastewater projects in Jordan	228
Box 15.1	Rainwater harvesting in a Seattle office building	235

Box 15.2	Washington, DC adds green infrastructure to planned drain overflow control plan	245
Box 15.3	Philadelphia's stormwater control green infrastructure programs	248
Box 17.1	Communicating the need for a change	268
Box 18.1	Draft mission statement, vision and value statements of a small water system	293

Part I

Water for a Sustainable Future

1

The Global Goal of Sustainable Growth

This book examines some of the major water-related challenges that societies around the globe face in their efforts to achieve and maintain sustainable growth. It is also a discussion of some of the paths that many of the most threatened water suppliers are following in their efforts to overcome these barriers. After the first two introductory chapters, the book focuses on the barriers contributing most to water scarcity in regions where sustainable water supplies are paramount, and on the actions now underway to achieve sustainability. My goal has been to find and report answers to such questions about global water sustainability as: What do we mean when we talk about achieving a sustainable future? What is water's role in achieving and maintaining that sustainability? How does population growth affect a sustainable water supply? What effect is the global migration towards cities having on water sustainability? What effect does water sustainability have on the problem of global food scarcity? How and why does water shape the sustainable economic health of societies? What are the world's leaders doing to achieve water, food, and development sustainability?

Defining Sustainability

The way to start answering these questions is to agree on a definition of sustainability. To begin, when researchers use the term 'sustainability' they are usually referring to development that entails the prudent use of natural resources at levels that are unlikely to damage the environment while ensuring

D. E. McNabb, *Global Pathways to Water Sustainability*,
https://doi.org/10.1007/978-3-030-04085-7_1

their availability for future generations as well. The term is often used in concert with the concept of economic growth, as in achieving and maintaining a sustainable quality of life with economic growth. Too often, however, less attention is given to the environmental protection clause in that definition and far more attention is given to development.

Stainability has also been defined in many other ways, depending on the objective or action that is to be sustained. One definition often seen is: sustainability is "the physical development and institutional operating practices that meet the needs of present users without compromising the ability of future generations to meet their own needs, particularly with regard to use and waste of natural resources."

The most commonly accepted official definition of sustainable development today is the definition included in the 1987 report of United Nations' World Commission on Environment and Development. Named after its chairperson, the Brundtland Commission Report states: "Sustainable development is development that meets the needs of the present without compromising the ability of future generations to meet their own needs." Hence, it assumes that a sustainable future is one in which resources are used at a replaceable rate, and one that ensures their availability for future generations.

Looming Water Scarcity

World food production is being threatened by water scarcity problems that in many locations are already cutting into efforts at achieving food sustainability. Changes in precipitation patterns and increasing and longer-lasting droughts associated with climate change are limiting production. Population growth and urbanization are fast bringing an end to family farming as more farmland is taken out of production. The Swiss nonprofit organization World Economic Forum forecasts that half of the world production of grains that have high water use such as wheat and rice will be subject to water stress by 2050. Long-term drought in the American West and an earlier series of droughts in Australia have resulted in forced water reductions in irrigation water allocation and the resulting reduction in agriculture production are costing billions of dollars in agriculture production.

In recognition of the water scarcity predicted for many parts of the globe possibly as soon as 2025, the UN established a program of action to occur from 2005 to 2015, labeling it the *Water for Life Decade*. The following statement on the relationship between water and sustainable agriculture was included in the description of the UN's focus on water and agriculture (UN 2014):

> A sustainable source of fresh water is key to food security. Crops and livestock need water to grow. Agriculture requires large quantities of water for irrigation and of good quality for various production processes. While feeding the world and producing a diverse range of non-food crops such as cotton, rubber and industrial oils in an increasingly productive way, agriculture [is] also … the biggest user of water on the globe. Irrigation now claims close to 70 percent of all freshwater appropriated for human use … There is enough water available for our global future needs, but this world picture hides large areas of absolute water scarcity which affects billions of people, many of whom are poor and disadvantaged. Major changes in policy and management, across the entire agricultural production chain, are needed to ensure best use of available water resources in meeting growing demands for food and other agricultural products.

Water security means ensuring that a supply of clean water is available for human uses and for other living things (Schnoor 2010). Water sustainability also means ensuring that water security can be maintained not only for this generation, but for all future generations as well. However, water sustainability is at risk due to the combined effects of human activity and shifting natural phenomena. Many scientists consider the following activities to be the contributing causes of growing risks to water sustainability:

- Climate change caused by concentration of greenhouse gases in the atmosphere,
- Shifting precipitation patterns with more and greater extreme weather events,
- The melting of polar icepacks and glaciers,
- Excessive groundwater withdrawals,
- Increased water demand from population growth, and
- The critical need for water and sanitary infrastructure associated with urbanization.

Global warming refers to an increase in the Earth's average surface temperature from the burning of greenhouse gas emissions. Changes in the Earth's average surface temperature are resulting in changes in traditional climate and precipitation patterns. Changes in precipitation are resulting in more severe and longer periods of droughts in some areas and flooding in others. Global warming is also resulting in variations in the amounts of snowfall and earlier snowpack thawing. Rising sea levels from sea ice melting are also associated with climate change. The relationship between global warming and the burning of fossil fuels with the resulting greenhouse gas rise was explained in this NASA report:

> In its Fifth Assessment Report, the Intergovernmental Panel on Climate Change, a group of 1,300 independent scientific experts from countries all over the world under the auspices of the United Nations, concluded there is a more than 95 percent probability that human activities over the past 50 years have warmed our planet. The industrial activities that our modern civilization depends upon have raised atmospheric carbon dioxide levels from 280 parts per million to 400 parts per million over the last 150 years. The panel also concluded there's a better than 95 percent probability that human-produced greenhouse gases such as carbon dioxide, methane and nitrous oxide have caused much of the observed increase in Earth's temperatures over the past 50 years.

Three Water Crises

A Columbia University meta-analysis of the then current research on water sustainability identified three types of crises resulting from the ongoing world water supply problem (Lall et al. 2008). The first is a crisis of access to safe drinking water, including governments' inability to supply appropriate infrastructure. The second is a crisis of water pollution, including pollution resulting from changes associated with resource-use changes caused by climate change. The third is a crisis of water supply security.

The Crisis of Access

A World Health Organization (WHO) progress report on achieving sustainability goals published in 2017 included an amended and updated list of goals that included water security. First on the list of water goals was achieving universal and equitable access to safe and affordable drinking water for all (WHO 2017). The WHO report added that in 2015, some 2.1 billion people lacked access to a supply of safe, available water at home. In 2015, 844 million still lacked even the most basic drinking water service; 263 million people (mostly women and children) had to walk more than 30 minuets to collect water from a treated source; and 169 million still collected their untreated water from an open surface water source—some of which were three or more hours from away. However, much progress has taken place toward the goal of ensuring that all persons have access to safe water. The number of people with some access to safe drinking water grew from 77% in 1990 to 83% in 2002 and 89% in 2015. More than 1 billion people gained access to piped water between 2000 and 2015. While WHO and other agencies are having continued success in their goals of achieving global access to water and sanitation facilities for all, achieving 2030 goals requires much more work. The UN adopted 2030 sustainable development goals agenda that included goals for water, sanitation and hygiene (WHO 2017, 7). The 2030 water access goals are:

1. Achieving universal and equitable access to safe and affordable drinking water for all.
2. Substantially increase water-use efficiency across all sectors and ensure sustainable withdrawals and supply of freshwater to address water scarcity and substantially reduce the number of people suffering from water scarcity.
3. Implement integrated water resources management (IWM) at all levels, including through transboundary cooperation as appropriate.
4. Expand international cooperation and capacity-building support to developing countries in water and sanitation-related activities and programs.
5. Support and strengthen participation of local communities in improving water and sanitary management.

The Crisis of Pollution

Drinking water is polluted from both point and non-point sources. The U.S. Environmental Protection Agency (EPA) defines point source pollution from any single identifiable source of pollution from which a pollutant is discharged (into the air, on land, or water). The source can include a pipe, ditch, ship or factory smokestack, or another single source. Examples of point sources of pollution include:

- Discharges from wastewater (sewage) treatment plants;
- Operational wastes from industries; and
- Combined storm and sewer outfalls.

Globally, significant improvement in controlling pollution from point sources has occurred since the late twentieth century. Advancements in technology for identifying point pollution sources and for the collection and treatment of pollutants before they are discharged, together with more and stronger environmental regulations have resulted in major progress in controlling the pollution of water from point sources. Industrial wastewater is often treated in-house and reused, with new industrial and thermoelectric generation water cooling technologies, greater adoption of the tertiary treatment of wastewater and the reuse of wastewater resulting in great strides towards reducing point pollution from industries and municipalities. While that progress has not been universal, it is indeed possible. Pollution from point sources in now under control in most areas of the world. However, many poor countries of the world do not have the political will nor the funds to eliminate neither water pollution nor the ability to invest in the infrastructure needed to treat and deliver clean water.

The EPA defines non-point pollution as pollution that occurs from land runoff and precipitation. As rainfall or snowmelt moves over and through the ground, it picks up natural and human-made pollutants. The polluted water is carried into lakes, rivers, wetlands, coastal and ground waters. Nonpoint sources include the following (EPA 2017):

- Excess fertilizers, herbicides and insecticides from agricultural and residential areas;
- Oil, grease and toxic chemicals from urban runoff and primary energy production;
- Sediment from improperly managed construction sites, crop and forest lands;
- Salt winter road coverage, irrigation practices and acid drainage from abandoned mines;
- Bacteria and nutrients from livestock, pet wastes and faulty septic systems;
- Atmospheric fallout.

These two objectives that combined pollution-related objectives with water access objectives were also included in the 2017 WHO sustainability progress report: (1) Improve water quality by reducing pollution, eliminating dumping and minimizing release of hazardous chemicals and materials, halving the proportion of untreated wastewater and substantially increasing recycling and safe reuse globally; and (2) protect and restore water-related ecosystems, including mountains, forests, wetlands, rivers, aquifers and lakes.

Also addressed in the report was the contribution to water pollution from the widespread lack of access to adequate sanitation facilities. The following findings were among the points addressed: 27% of the global population still used private privies with excreta disposed of on site; 2.3 billion people still lacked even a basic sanitary service; and 892 million people worldwide still practiced open defecation.

The Crisis of Water Supply Security

The declining availability of a secure and sustainable water supply may be the most salient of the three barriers in the path to achieving water sustainability. Yet, despite this importance to civilization, there does not seem to be a universally accepted definition for water scarcity. Different measurements are used for different areas and different water uses. Water scarcity occurs when there is not enough water available to meet the usage demands of all sectors within a region. It is also identified as the lack of access to adequate quantities of water for human and environmental uses. Water scarcity is considered to exist when the supply drops below a total of 1000 cubic meters (264,172 US gallons) per day per person available for use by all sectors. Critical water scarcity occurs at 500 cubic meters (132,086 US gallons) of water per day. As these definitions suggest, water scarcity for a use or sector occurs when there is not enough to meet the needs of that user or sector. Take, for example, agriculture, the largest user of water. Agriculture uses 70% of the world's accessible freshwater. However, some 60% of this water is often wasted because of obsolete or leaky irrigation systems, inefficient application methods, or evapotranspiration (evaporation from the soil or plant life). Another major loss of water is the cultivation of crops that are too thirsty for the environment in which they are grown. Examples include many popular grains. It is estimated that because of traditional farming practices, rapid population growth, and urbanization, fully one third of the developing world will be experiencing sever water scarcity before the end of this century.

As much of the world has discovered, a result of poor water management is the disappearance of springs and streams, and the drying out of rivers, reservoirs, lakes and aquifers. Accordingly, "Many countries that produce large amounts of food—including India, Brazil, China, Australia, Spain and the United States—have reached or are close to reaching their water resource limits. Added to these thirsty crops are the fact that agriculture also generates considerable freshwater pollution—both through fertilizers as well as pesticides—all of which affect both humans and other species" (WWF 2017).

A 2016 survey of a variety of water industry professionals, researchers, and university water scholars found general agreement that many parts of the developing world are already well on the path to severe water stress, both in terms of water quantity and quality (Mayor et al. 2016).

Identified as forces contributing to this water crisis were changes in precipitation patterns due to global warming, the high cost of energy, population growth, urbanization, traditional agriculture methods, poor sanitation, lack of capital for necessary infrastructure enhancement and maintenance, and the lack of political will for addressing the problems. The production of biofuels and shale gas extraction both contribute to the water scarcity problem and the role that energy will play in alleviating the problem.

Achieving security of water supplies has generated a global need for engineering, repairing and replacing aged water-related infrastructure has been estimated by the OECD to require investments of US$1.7 trillion every year for at least the next decade. Infrastructure means all the dams, canals, pipes, collection, treatment and distribution systems needed to ensure a consistent supply clean, safe water is available when and where it is needed by people, municipalities, businesses, and industries and farms. It also includes the complex facilities needed to collect, treat and discharge or reuse as recycled wastewater, and for managing storm water runoff and prevent flooding. Effective and secure water and wastewater infrastructure is critical to public and environmental health worldwide. Developing nations often lack the financial ability to repair or replace damaged systems. Keeping pace with new water and wastewater system demands due to population growth and rapid urbanization is a problem of supplying clean water and sanitation service for developed and emerging economies. The need for new water and wastewater infrastructure is particularly important for achieving and maintaining water sustainability.

Every year, disruptions in or loss of water or wastewater services costs water and wastewater service providers billions of dollars, while also adding to declines in water quantity and quality. Water and wastewater systems often close to a century or older must be repaired or replaced. Additional damage or destruction of infrastructure results from floods and other extreme weather events. Climate change is adding to system failures. Past underinvestment in water and wastewater infrastructure management contributes to public health problems and environmental damage.

Water for Sustainable Development

The UN considers water sustainability status to be a critical condition for achieving sustainable development. Beginning in the 1990s, the UN began a series of sessions that focused on coming up with a plan for achieving global waster sustainability. Participants at those sessions officially recognized the critical importance of water to future development programs. Member governments agreed to work on ensuring that all 17 goals to be achieved by 2030. Box 1.1 includes elements of the water portion of the sustainable development goals (SDGs) developed during an International Conference on Water and the Environment (ICWE) in Dublin, Ireland, on January 26–31, 1992.

During its *Millennium Summit* in September 2000, the UN adopted the *Millennium Declaration.* From that declaration were a set of SDGs, with target dates for their implementation. Three key objectives of the SDGs for sustainable development were the eradication of poverty, eliminating persistent malnutrition, and improving the health and welfare of people everywhere. Ensuring access to water plays a fundamental role in this effort. Among the goals was cutting in half the proportion of people without sustainable access to safe drinking water.

At the UN World Summit for Sustainable Development held during 2002 in Johannesburg, South Africa, the target water goal was expanded to include basic sanitation, and water was considered a critical resource and critical to meeting all the development goals. The UN General Assembly then proclaimed the decade of 2005–2015 the International Decade for Action, the goals of which were to support national efforts to fulfil commitments made on water and water-related issues by 2015.

Box 1.1 The 1992 Dublin Statement on Water and Sustainable Development

"Scarcity and misuse of fresh water pose a serious and growing threat to sustainable development and protection of the environment. Human health and welfare, food security, industrial development and the ecosystems on which they depend, are all at risk, unless water and land resources are managed more effectively in the present decade and beyond than they have been in the past.

The experts saw the emerging global water resources picture as critical. ... The problems highlighted are not speculative in nature; nor are they likely to affect our planet only in the distant future. They are here, and they affect humanity now. The future survival of many millions of people demands immediate and effective action.

The Conference participants [called for] fundamental new approaches to the assessment, development and management of freshwater resources, which can only be brought about through political commitment and involvement from the highest levels of government to the smallest communities." Four guiding principles were adopted:

Principle No. 1: Fresh water is a finite and vulnerable resource, essential to sustain life, development and the environment.

Principle No. 2: Water development and management should be based on a participatory approach, involving users, planners and policy-makers at all levels.
Principle No. 3: Women are central in the provision, management and safeguarding of water.
Principle No. 4: Water has an economic value in all its competing uses and should be recognized as an economic good. Within this principle, it is vital to recognize first the basic right of all human beings to have access to clean water and sanitation at an affordable price.

Source UN (1992)

The Water Sustainability Goal

The underlying objective of the Water for Life decade was to encourage and assist members in implementing and continuing commitment to the sustainability goals so that the water-related goals of the Millennium Declaration and the Johannesburg Plan of Implementation of the World Summit for Sustainable Development, and Agenda 21 would be achieved.

The challenge of the Water Decade was to keep everyone's focus of attention on activities and policies that would advance the long-term goal of sustainable management of water resources, in terms of both quantity and quality, and include measures to improve sanitation. Achieving the goals of the Water for Life Decade called for a sustained commitment from 2005 to 2015 on the cooperation and investment on the part of all stakeholders.

The Water Goal

The complete list of development goals included one target, Goal Number 6, that called for global action to ensure the availability of water for all and the sustainable management of water. The focus was on individual nations' commitment to ensuring the availability and sustainable management of fresh water for all sectors and all citizens. This goal includes eight targets. One was to be achieved by 2020, six by 2030, and one has no target year. The targets and appropriate achievement target dates are (UNDP 2017):

1. By 2030, achieve universal and equitable access to safe and affordable drinking water for all.
2. By 2030, achieve access to adequate and equitable sanitation and hygiene for all and end open defecation, paying special attention to the needs of women and girls and those in vulnerable situations.

3. By 2030, improve water quality by reducing pollution, eliminating dumping and minimizing release of hazardous chemicals and materials, halving the proportion of untreated wastewater and substantially increasing recycling and safe reuse globally.
4. By 2030, substantially increase water-use efficiency across all sectors and ensure sustainable withdrawals and supply of freshwater to address water scarcity and substantially reduce the number of people suffering from water scarcity.
5. By 2030, implement integrated water resources management at all levels, including through transboundary cooperation as appropriate.
6. By 2020, protect and restore water-related ecosystems, including mountains, forests, wetlands, rivers, aquifers and lakes.
7. By 2030, expand international cooperation and capacity-building support to developing countries in water- and sanitation-related activities and programmes, including water harvesting, desalination, water efficiency, wastewater treatment, recycling and reuse technologies.
8. Support and strengthen the participation of local communities in improving water and sanitation management.

SDGs 4 and 5 pertain to water access and water management; SDG 6 calls for protecting and restoring damage ecosystems, including all sources of surface water; SDG 7 is a call for international and cross-border cooperation in water management; and SDG 8 pertains to facilitating better cooperation among local groups in managing water resources and avoiding water pollution.

Measuring National Water Stress

Water stress is ranked as a single, five-point scale, representing the extent of water availability in countries where more than 80% of the freshwater available for agriculture, domestic and industrial (including energy production) uses was already being withdrawn, thereby leaving the nation vulnerable to scarcity. To come up with the final score, the Washington, DC-based Water Resources Institute (WRI) evaluated countries on the following five indicators (Reig et al. 2013):

- Baseline water stress: A ratio of total water withdrawals to total available water for annual resupply.
- Inter-annual variability: Year to year water supply variation.

Table 1.1 Top 15 water-stressed countries, 2015

Rank	Country	Stress score
1	Bahrain	5.0
1	Kuwait	5.0
1	Qatar	5.0
1	San Marino	5.0
1	Singapore	5.0
1	United Arab Emirates	5.0
1	Palestine	5.0
8	Israel	5.0
9	Saudi Arabia	4.99
10	Oman	4.97
11	Lebanon	4.97
12	Kyrgyzstan	4.93
13	Iran	4.91
14	Jordan	4.86
15	Libya	4.77

Source Maddocks et al. (2015)

- Seasonal variability: Within-year water supply variation by month.
- Flood propensity: The number of recorded floods from 1985 to 2011.
- Drought severity: The average length of droughts multiplied by the drought dryness from 1901 to 2008.

Each country was rated on the five indicators using a five-point scale, with a score of five representing the greatest exposure to that stress indicator. Countries were then ranked according to a composite five-point scale. Twelve of the top 15 countries with the greatest water stress potential are located in the Middle East, seven countries were tied with top stress scores; five Middle Eastern countries of the top eight received maximum risk scores of 5.0 (Table 1.1).

Summary

Stainability has been defined in many different ways, depending on the objective or action that is to be sustained. Examples include: (1) Sustainability is "the physical development and institutional operating practices that meet the needs of present users without compromising the ability of future generations to meet their own needs, particularly with regard to use and waste of natural resources"; and (2) the 1987 UN definition of sustainable development, "Sustainable development is development that meets the needs of the present without compromising the ability of future generations to meet their own needs."

Access to a supply of clean, safe water is a basic human right, and therefore adequate supplies of water as an absolute necessity for sustainability, appeared in important UN conferences the 1980s and 1990s; a definition for sustainable development was adopted in 1983. Examples of why adequate water is necessary for sustainability appeared within these UN sustainability-objective categories: Improvement of human health through the elimination of water-borne disease and malnutrition; gender inequality; sustainable energy; control of urbanization; more efficient agriculture and food security; peace and security; and climate change.

Much of the world now faces three types of crises resulting from a growing world water crisis (Lall et al. 2008). The first is a crisis of access to safe drinking water, including governments' inability to supply appropriate infrastructure. The second is a crisis of water pollution, including pollution resulting from changes associated with resource-use changes caused by climate change. The third is a crisis of declining water supplies. World leaders can implement either one of two policy approaches to alleviating water scarcity crises: They can focus on the short-term option of strategies for increasing water supply, or they can adopt the long-term approach of managing demand.

World leaders have agreed upon a series of goals to ensure that we are able to meet and surmount future barriers to achieving sustainability. This book is about the drive for achieving sustainability of the global supply of freshwater for people, agriculture and industry.

References

EPA (Environmental Protection Agency). 2017. Climate impacts on coastal areas. Environmental Protection Agency. Accessed November 28, 2017 from https://19january2017snapshot.epa.gov/climate-impacts/climate-impacts-coastal-areas_.html.

ICWE (International Conference on Water and the Environment). 1992. The Dublin statement on water and the environment. Accessed August 4, 2018 from https://www.gdrc.org/uem/water/dublin-statement.html.

Lall, Upmanu, Tanya Heikkila, Casey Brown, and Tobias Siegried. 2008. Water in the 21st century: Defining the elements of global crises and potential solutions. *Journal of International Affairs* 61 (2): 1–17.

Maddocks, Andrew, Robert S. Young and Paul Reig. 2015. Ranking the world's most water stressed countries in 2040. World Resources Institute, Washington, DC. Accessed November 4, 2017 from http://www.wri.org/blog/2015/08/ranking-world%E2%80%99s-most-water-stressed-countries-2040.

Mayor, Beatriz, R. Rodríguez Casado, J. Landeta, E. López-Gunn, and F. Villarroya. 2016. An expert outlook on water security and water for energy trends to 2030–2050. *Water Policy* 18 (1): 1–18.
Reig, Paul, Andrew Maddocks and Francis Gassert. 2013. *World's most water stressed countries*. Washington, DC: World resources Institute. Accessed November 4, 2017 from www.wri.org/blog/2013/12/world%E2%80%99s-36-most-water-stressed-countries.
Schnoor, Jerald L. 2010. Water sustainability in a changing world. 2010 Clarke Prize Lecture, National Water Research Institute. Accessed October 11, 2017 from www.nwri-usa.org/pdf/2010/ClarkPrizelecture.pdf.
UN (United Nations). 1992. The Dublin Statement on water and sustainable development. Accessed December 7, 2018 from www.un-documents.net/h2o-dub.htm.
UN (United Nations). 2014. International decade for action 'Water for Life' 2005–2015. Accessed November 5, 2017 from http://www.un.org/waterforlifedecade/food_security.shtml.
UNDP (United Nations Development Program). 2017. Goal six targets. Accessed December 7, 2017 from www.undp.org/content/undp/en/home/sustainable-development-goals/goal-6-clean-water-and-sanitation/targets/.
WHO (World Health Organization). 2017. Progress on drinking water, sanitation, and hygiene, update and sustainable development goals. World Health Organization (WHO) and the United Nations Children's Fund (UNICEF) Joint Monitoring Program (JMP). Accessed November 7, 2017 from http://apps.who.int/iris/bitstream/10665/258617/1/9789241512893-eng.pdf?ua=1.
WWF (World Wildlife Fund). 2017. Water scarcity. Accessed November 7, 2017 from https://www.worldwildlife.org/threats/water-scarcity.

2

Water's Role in a Sustainable Future

The United Nations and other global leaders agree that water plays a significant role in achieving and maintaining sustainability. Clean, safe water is a vital resource, without which there cannot be a sustainable future. However, water is a finite resource. We must all learn to do a better job of dealing with this limited supply. Surface or underground water supplies and the organizations that collect, process and distribute water for homes, farms and industries make up the systems that supply water in adequate quantity and quality that meets appropriate needs. They must do this without compromising the ability of future generations to access and make use of the same adequate and safe supply. Excessive withdrawals from groundwater aquifers, discharge of polluted water into surface water courses and groundwater aquifers, and wasteful domestic and industrial use are enemies of sustainability, just as are urbanization, development that harms the environment, and energy use that contributes to global warming.

Clearly, without sufficient supplies of clean, safe water, humanity as we know it will not have a future on this or any other planet. We all know that water is essential for the life of all other living beings, and until 100 years ago, there was general agreement that fresh water will also be there when and where it is needed. Ours was, and indeed remains, a water-rich planet. No water is lost from the planet, but neither is any new water is being created. The water sustainability problem is that almost all of earth's water is unfit for human consumption without expensive treatment. Nearly 97% of our water is in the world's oceans. Moreover, the little freshwater that does exist is not evenly distributed; much is still frozen in glaciers and icecaps,

D. E. McNabb, *Global Pathways to Water Sustainability*,
https://doi.org/10.1007/978-3-030-04085-7_2

although those sources appear to be melting rapidly, resulting in rising oceans and related problems.

Only 29% of the globe is land above the ocean surface. All the rest is covered with salt water. Without extensive desalination, ocean water is unfit to drink or to be used to irrigate crops. Freshwater is, therefore, limited. What we see is what we get. Some may evaporate, but it usually falls shortly thereafter as precipitation. Some is absorbed into the soil and more is taken up by vegetation. We pump it back to the surface and plant life transpiration returns more into the atmosphere, where it again returns to earth as mist, rain or snow. The scientific consensus is the world has enough fresh water to meet the needs of 11 or 12 billion inhabitants. We are well on the way to reaching somewhere near those figures if not by 2050, certainly by 2100.

Our freshwater comes and goes through a hydrologic cycle. As water passes through the cycle, it is always in movement above and below the surface of the Earth. Water is always changing, from liquid, then evaporating to vapor or becoming ice, then returning to a liquid. These processes are continuous and have occurred over millions of years. During this cycle, there is often either too little water for people's needs during a drought in some parts of the world, and too much during a flood in other areas. To determine water's role in a sustainable future requires knowing more about that distribution. This is what world leaders are doing in their efforts to ensure a sustainable supply of freshwater is available now and in the future for everyone.

Clearly, we must manage the existing freshwater resource as well as we can if we hope to have a sustainable future. Two possible solutions for many arid regions are the reuse of treated wastewater and the large-scale desalination of sea and brackish seawater. Recycled water is increasingly in use in the USA's Southwest. Highly treated wastewater is used to recharge underground aquifers and for irrigation of non-human consumption crops and for landscaping in increasing numbers of nations.

As more processes are developed and the costs to produce it continues to drop, desalination is closer to being widely adopted in coastal areas. In 2017, there were more than 1900 desalination plants in operation around the globe. However, it is still very expensive, as this report from the State of Texas indicates: "In 2007, $87 million was spent to build the Kay Bailey Hutchison brackish groundwater desalination plant in Texas. Design capacity of the KBH plant is 27.5 million gallons per day (MGD). The 2010 biennial report on seawater desalination projected that it will cost approximately $32 million to build a 2.5 MGD seawater desalination plant, and approximately $658 million to build a 100 MGD seawater desalination plant in Texas" (TEDB, n.d.).

The Universal Right to Water

Everyone has the fundamental right to access clean water. For nearly 50 years, the United Nations has been promoting the need for development that includes a concern for maintaining a sustainable environment in all development planning. As environmental preservation came to a global concern, the responsibility of governments to provide sustainable access to clean water in all development programs came to be associated with development. The connection between these themes came to the fore in the 1972 Conference on the Human Environment. Little was done to halt global environment degradation after that conference. Over the next ten years, the role of water management as a key element in environmental protection and economic development became a focus of governments around the globe. Environmental degradation, such as ozone depletion, global warming and water pollution, were recognized as serious perils that had to be addressed. That recognition led the UN to establish the first World Commission on Environment and Development in 1983. The Commission, led by Gro Harlem Brundtland, the former prime minister of Norway, produced a report that recommended adoption of the concept of sustainable development as an alternative approach to development simply based on economic growth. Development was to "[meet] the needs of the present without compromising the ability of future generations to meet their own needs" (UN 1997). This definition has been incorporated into all United Nations development programs and has been the theme of many subsequent international conferences. The core concept encapsulated in the Brundtland definition is "intergenerational solidarity or justice" that is founded upon a commitment to maintaining a sustainable future for all future generations (Grimaldos 2015, 69).

Water's essential role in sustainable development appeared in important conferences that followed in 1992 and 1993. The Dublin Conference on Water and Development and the Rio Summit on Environment and Development took place in 1992. Attendees of these conferences adopted the concept that it is a basic right of all human beings to have access to clean water and sanitation. Providing freshwater for all people then became a global imperative (Soubbotina 2004).

In addition to others, three provisions were agreed upon at the Rio Summit: (1) that humans are at the center of concerns for sustainable development, therefore humans have the right to a healthy and productive life in harmony with nature; (2) scientific uncertainty should not delay measures to prevent environmental degradation where there are threats of serious or irreversible damage; and (3) that states have a sovereign right to their own resources but not to cause damage to the environment of other states (UN 1997). Inherent in these principles was the recognition that access to sustainable sources of clean water was a basic human right. The 1993 Vienna World Conference on Human Rights included the following excerpt in Article 47 of the final Declaration and Program of Action dealing with the rights of children the right of access to clean, safe water:

> The Conference calls on States to integrate the Convention on the Rights of the Child into their national action plans. By means of these national action plans and through international efforts, particular priority should be placed on reducing infant and maternal mortality rates, reducing malnutrition and illiteracy rates and providing access to safe drinking water and to basic education. (OHCHR 2017)

The role of water in sustainable development was reaffirmed in subsequent United Nations conferences and summits. The following World Bank excerpt from study on the legal right of water study stated that: "Recognition of the centrality of water to the realization of the rights enshrined in the Universal Declaration of Human Rights, and the International Covenant on Economic, Social and Cultural Rights, and the Declaration on the Right to Development. This was confirmed some 10 years after Dublin, Rio, and in Vienna [where in 2002] General Comment No. 15, [was] issued. The Comment set forth in unequivocal terms, the direct relationship between human rights and water, and provided explicit recognition of the human right to water itself" (Soubbotina 2004, vii).

General Comment 15 explained and explicitly affirmed the right of all humans to clean, safe water, along with the legal foundation for that right. The first paragraphs of Comment 5 are included in Box 2.1.

Box 2.1 Selected elements of opening paragraphs of 2003 UN General Comment No. 5

1. Water is a limited natural resource and a public good fundamental for life and health. The human right to water is indispensable for leading a life in human dignity. It is a prerequisite for the realization of other human rights. The Committee has been confronted continually with the widespread denial of the right to water in developing as well as developed countries. Over 1 billion persons lack access to a basic water supply, while several billion do

not have access to adequate sanitation, which is the primary cause of water contamination and diseases linked to water.

2. The continuing contamination, depletion and unequal distribution of water is exacerbating existing poverty. States parties have to adopt effective measures to realize, without discrimination, the right to water, as set out in this general comment. The legal bases of the right to water.
   a. The human right to water entitles everyone to sufficient, safe, acceptable, physically accessible and affordable water for personal and domestic uses. An adequate amount of safe water is necessary to prevent death from dehydration, to reduce the risk of water-related disease and to provide for consumption, cooking, personal and domestic hygienic requirements.
   b. Article 11, paragraph 1, of the Covenant specifies a number of rights emanating from, and indispensable for, the realization of the right to an adequate standard of living "including adequate food, clothing and housing". The use of the word "including" indicates that this catalogue of rights was not intended to be exhaustive. The right to water clearly falls within the category of guarantees essential for securing an adequate standard of living, particularly since it is one of the most fundamental conditions for survival. Moreover, the Committee has previously recognized that water is a human right...In 2000, the World Health Organization estimated that 1.1 billion persons did not have access to an improved water supply (80% of them rural dwellers) able to provide at least 20 liters of safe water per person a day; 4 billion persons were estimated to be without sanitation...
   c. Further, 2.3 billion persons each year suffer from diseases linked to water. The right to water is also inextricably related to the right to the highest attainable standard of health ... and the rights to adequate housing and adequate food. The right should also be seen in conjunction with other rights enshrined in the International Bill of Human Rights, foremost amongst them the right to life and human dignity.
3. The right to water has been recognized in a wide range of international documents, including treaties, declarations and other standards. For instance, Article 14, paragraph 2, of the Convention on the Elimination of All Forms of Discrimination Against Women stipulates that States parties shall ensure to women the right to "enjoy adequate living conditions, particularly in relation to...water supply". Article 24, paragraph 2, of the Convention on the Rights of the Child requires States parties to combat disease and malnutrition "through the provision of adequate nutritious foods and clean drinking water".

Article 11, paragraph 1, of the Covenant specifies a number of rights emanating from, and indispensable for, the realization of the right to an adequate standard of living "including adequate food, clothing and housing". The use of the word "including" indicates that this catalogue of rights was not intended to be exhaustive. The right to water clearly falls within the category of guarantees essential for securing an adequate standard of living, particularly since it is one of the most fundamental conditions for survival.

Source OHCHR (2003) (United Nations)

Water's Essential Benefits

Water is essential for sustainable development for many reasons. Alex Mung, an editor and director of the Water Initiative at the World Economic Forum emphasized water's importance to human health and security through the following seven roles, beginning with the elimination of water-borne disease: malnutrition; gender inequality; sustainable energy; urbanization; agriculture and food security; and peace; security and climate change.

Eliminating Waterborne Disease

Water-related diseases and poor sanitation are at the top of the list of global health problems. One of the reasons for this is that a billion or so people live in regions where water pollution and poor sanitation practices are endemic, despite the efforts that are underway to improve living conditions and to prevent these diseases. We also know how to prevent the disasters that affect children from exposure to lead and other minerals and toxic chemicals in their water. But the diseases and illnesses accruing from these adverse conditions continue to plague children and adults in both developing and developed nations. An estimated 850,000 people around the globe lack a basic drinking water supply, and nearly 150 million of them depend upon an unprotected surface water source for their basic water needs. Moreover, at least 2 billion people get their drinking water from a source contaminated with human and/or animal feces. This contaminated water can carry such diseases as diarrhea, cholera, dysentery, hepatitis-A, typhoid, and polio.

More than 360,000 children under 5 years of age die from diarrhea every year; as much as 88% of those deaths are caused by exposure to unsafe drinking water, missing or inadequate sanitation, or inadequate hygiene. Each year, another nearly 240 million people suffer from schistosomiasis, a chronic disease cause by parasitic worms in infested water. In many parts of the world, people affected by dengue fever, a disease caused by insects that breed in clean rather than dirty water that they find in open household drinking water containers (WHO 2017).

Finally, population growth and the lack of adequate water treatment facilities and sanitary infrastructure associated with urbanization adds to the impact these conditions are having on the ability of local governments on achieving sustainable water and wastewater services. By 2025, the World Health Organization predicts that fully half of the world's population will be living in water-stressed environments.

Water's Role in Child Development

Malnutrition is a chronic cause of underdevelopment in children in many parts of the world. Although we tend to think that malnutrition is a consequence of not having enough food, scientists now believe that malnutrition is more likely the result of water pollution and poor sanitation. The Water and Sanitation Program (WSP), a multinational partnership of the World Bank Group, reports that chronic diarrhea resulting from unsafe water and poor sanitation and hygiene each year results in the deaths of 1.8 million children under five years of age. The problem is most severe in Africa, Southeast Asian countries and much of Central Latin America. The children that survive their early years often suffer from stunted growth and poor mental development. The World Health Organization reported that the "lack of access to safe, clean drinking water and basic sanitation, as well as poor hygiene cause nearly 90 percent of all deaths from diarrhea, mainly in children. While 87 percent of the world's population now have access to improved water sources, 39 percent still lack access to improved sanitation. Moreover, in developing countries, 1.1 billion people still defecate in the open, and hand washing with soap is practiced, on average, only after 17 percent of toilet uses. Diarrhea most often results from the ingestion of pathogens from feces that have not been disposed of properly, or from the lack of hygiene" (WHO 2011).

The sustainable prevention of water-transmitted diarrhea, improved sanitation and hygiene requires eliminating the fecal–oral transmission of pathogens. It also includes the provision of an improved water supply, water safety planning, household water treatment and safe storage, improved sanitation facilities, and hygiene education. Improved water supplies refer to technologies such as piped household water connections, public taps, standpipes, or protected dug wells, springs or rainwater collection.

Water's Role in Gender Equality

In much of the developing world, collection and transporting of household water is a chore carried out almost exclusively by women and children. Reports from many nations reveal that in nearly two-thirds of households without a drinking water source in the home, women and children collect all the water. The United Nations and World Bank, among many other organizations, are convinced that development in those regions suffers as a result of this gender inequality. The World Bank's 2012 World Development

Report on Gender Equality and Development describes sustainable small-scale development activities that have occurred as a result of greater female participation in economic activity.

The connection between gender inequality in water has long been considered a barrier to sustainability development. The United Nations and other agencies cite these barriers that women face in many developing societies:

- Lack of access to a water supply;
- Lack of access to land ownership and access to water for productive use;
- Lack of access to sanitation facilities;
- Lack of participation and equality in resource decision making; and
- Lack of participation in planning capacity development.

Water's Role in Energy Production

The sustainability of water and energy is interwoven. Water is needed to produce energy and large amounts of energy are needed to withdraw and distribute water for all its uses. A reliable supply of surface water is needed for hydropower. Water is also used for steam-driven turbine generation of thermoelectric power. It is also used for cooling in thermal power plants. Water is also needed for releasing hydrocarbon fuels from underground beds through fracking, extracting the released oil and gas, processing and transporting the products long distances by pipeline. Climate change, excessive withdrawal of groundwater, and protracted drought in low-rainfall environments are having a significant impact on the water supply.

Conversely, the withdrawal, treatment, storage, transportation, and distribution of water requires large amounts of energy. Both energy and water are used in concert for growing and processing both food crops and biofuels. In 2015, more than 780 million people still lacked access to clean, safe water, and more than 1.3 billion people lacked access to electricity.

Population growth, urbanization and expanding economies are placing additional demands on both water and energy. Global consumption of energy is expected to increase 35% by 2015, with water consumption increasing by 85% over the same period. Climate change is resulting in variation in rainfall amounts, more extreme weather events such as floods and droughts, together with changes in surface water flow patterns caused by reduced snowfall-related storage of surface water, are expected to have significant adverse effects on both water and energy (World Bank 2013, 2015; Shuster-Wallace et al. 2015).

Water and Sustainable Cities

Urbanization limits the achievement of water sustainability in many ways. The first limitation is the increased need for both water and sanitation for the people flocking to urban areas at unprecedent rates. A second limitation is the greater incidence of extreme water event flooding caused by the loss of soils to absorb rainfall and recharge underground aquifers. Construction of roads and buildings brought about by planned and unplanned development results in runoff of contaminated water flowing into nearby lakes and streams.

Urbanization is a challenge for both developed and developing regions. In cities in many regions, much of the decayed and decaying water and sanitation infrastructure has long passed the point of a replacement being needed. Water mains and sewer systems in many cities were put into use 100 years ago or longer. As a result, many utility operations are incapable of meeting the needs of a rapidly growing urban population while also repairing and replacing their aging infrastructure. Repairing water and wastewater infrastructure is one of the most urgent needs in many cities, large and small. Senn and Spuhler (2015) identified four demographic and economic factors that they believed are responsible for the urbanization trend:

1. The natural population growth occurring in the already large and growing cities. Urbanizationis fast stripping cities of their capacity to operate and maintain existing water supply and sanitation treatment systems.
2. The absorption into the cities of rural settlements existing on the nearby edges of the cities. Bringing modern water and wastewater facilities to these rural settlements is a slow process, resulting in some of the earlier rural settlements transitioning into desirable suburbs while others fall into unsafe and unsanitary slums.
3. Transformation of rural towns that rapidly become urban centers. These newly transformed urbanized centers have been forced to adopt urban planning and infrastructure expansion as they attempt to keep pace with their growth. Those that followed planned expansion may be closest of the four city growth models to achieving sustainability.
4. Migratory movements from rural areas to cities. This model results in the highest degree of unplanned growth. Many become slums, with their residents lacking access to basic water and sanitation services, lacking affordable transportation, and having little opportunity for reduction of the poverty that characterizes many.

Water's Role in Sustainable Agriculture

A state of food scarcity is defined by the UN (2014) as existing when all people, at all times, do not have physical, social and economic access to sufficient, safe and nutritious food to meet the dietary needs for a productive and healthy life. It exists in many developing nations today and is expected to be even more prevalent in many regions in the next several decades. By 2030, food demand is predicted to increase by 50% and by 70% by 2050. The main challenge facing the agricultural sector is not as much growing 70% more food in 40 years but in making 70% more food available where and when it is needed.

Achieving Waster Sustainability

Steven Cohen provided this brief, inclusive definition of sustainability management (Cohen 2011, 1–2): "Sustainability management is simply the organizational management practices than result in sustainable development. In the modern industrial world, sustainability management is the practice of economic production and consumption that minimizes environmental impact and maximizes resource conservation and reuse." This definition clearly applies to the many costly management decisions that will have to be made, beginning as soon as possible if the world is achieve true water sustainability in light of greater and greater water scarcity.

Sustainable Water Supplies

There is no indication that the amount of freshwater available on the Earth may not be able to meet the needs of its growing population by the end of this century. The 2–3% of water that is drinkable freshwater will probably not decline. However, the way it is spread throughout the hydrological cycle above and below the surface of the earth is unlikely to stay the same as it is now.

The role water plays in global sustainability is changing due to natural and human-caused activities. Globally, population growth, agriculture needed to feed more people, more pollution, more people moving to cities, and production of the energy to move them are resulting in more water withdrawals from existing sources of supply. Meanwhile, a fact of live is that the supply of freshwater is finite. Except for limited efforts at desalination, no new freshwater is being created. The answer to water scarcity is better management of existing water supplies and sources.

Water sustainability begins with management of the surface and groundwater resource for the sustainable provision of quality freshwater when and where it is needed and ends with reliable wastewater treatment systems for ensuring adequate water quantity protection of the environment, and ultimately for the reuse of water for all appropriate needs for this and future generations.

Sustainable water management requires balancing the requirement for meeting the needs of sustainable economic growth with commitments to maintaining social responsibility and environmental integrity. A manager or administrator of a water and wastewater district might be inclined to adopt this modified version of the water sustainability definition used by Jerald Schnoor in 2008 and 2010: "*Water management sustainability is the design, implementation and administration of activities and systems necessary for achieving and maintaining a continual supply of clean, safe water for human uses and for other living things* (emphasis in the original)." Although this definition focuses on drinking water, it may also be applied to all classes of water: freshwater, wastewater and recycled water.

Sustainable Development

Sustainability and sustainable development are often assumed to mean the same thing. Therefore, there is general agreement among researchers that the definition of sustainable development adopted by the Brundtland Commission in 1987 best describes what sustainable development infers. The Brundtland report defined sustainable development as: "Development that meets the needs of the present without compromising the ability of future generations to meet their own needs." (World Commission on Environment and Development 1987; Randall 2007; Cohen 2011). The United Nations has adopted this definition of sustainable development as the underlying aim of long-term economic and social development for all societies. Thus, the objective of sustainable development is government actions for following a path of balanced economic development, social development and environmental protection.

In many parts of the world, surface water courses are still considered a convenient means for disposing of untreated sewage, manufacturing wastes, stormwater and irrigated agricultural runoff. Much of that discharged water is polluted with pesticides, human or animal waste agriculture nutrients, and sediment. A business-as-usual attitude still persists among many local leaders, primarily because they lack the physical or technical means for changing behaviors. Even where progressing water management attitudes do exist, the money to pay for the necessary infrastructure is not available.

Midway through the twentieth century, it became clear to world leaders that the visible decline in the state of the world's ecosystems was being exacerbated by a series of anthropogenic actions. Contributing to the problem were unchecked population growth, urbanization, climate change, inefficient or destructive irrigated agricultural methods, harmful fossil fuel extraction methods and their growing use for transportation and the production of electrical energy, and the high cost of mitigating these problems.

Today, it is apparent that we cannot continue to follow a business-as-usual path with water. Population growth and urbanization, the need to become more efficient with water use for agriculture and energy production, depletion of groundwater, and the discharge of polluted wastewater are all putting extreme pressure on our finite water resources. This is occurring while millions of people still do not have access the clean drinking water and billions do not have access to basic sanitary facilities.

Summary

As water passes through the hydrologic cycle, it is always in movement above and below the surface of the Earth. Water is always changing, from liquid, then evaporating to vapor or becoming ice, then returning to a liquid. The process is continuous and has occurred for millions of years. During this cycle, however, the supply of freshwater is not always the same. This means that there is often either too little water for some people's needs in some parts of the world, and too much water in other areas. A global effort to gain equal and sustainable water supplies is, therefore, extremely difficult.

Water and wastewater utilities everywhere confront formidable challenges as they maintain and expand infrastructure systems that supply clean and safe freshwater to their customers, collect and treat all wastewater, and, for many, manage stormwater runoff. These problems are intensifying existing stresses from increases in population growth, urbanization and deteriorating infrastructure. These barriers are exacerbated by operational conditions that are difficult to control or change. These include missing or inadequate water and sanitary infrastructure, shifting trends in traditional precipitation patterns, and demands for new energy choices.

New challenges to utility operations have also emerged. The effects of climate change and the high costs of providing adequate supplies of clean, safe water and appropriate levels of sanitary services to a rapidly growing population and urban-centered migration are among the latest challenges many nations face in complying with their civic responsibility. As these challenges converge, it is becoming increasingly clear to many leaders that their once-stable operating conditions of supplying their services and investing in replacement infrastructure only when it failed no longer meets current and rapidly changing conditions. Today's water system requirements are such that managers must innovate to ensure sustainability for meeting current and future demands. While they are often absorbed in meeting the needs for clean water to their current customers, water managers must also plan for and invest in the infrastructure needed for meeting future growth in demand. By their sustainability actions they help to ensure economic growth while maintaining their commitment as stewards of a clean and healthy environment for present and future generations (Herrick and Pratt 2012).

Seven of what water professionals and utility managers believe are among the major barriers to achieving and sustaining stable and secure water supplies include climate change, population growth, urbanization, inefficient agriculture, energy choices, sustainable development, and unavailable funding for infrastructure repair and replacement. Each will be discussed in the individual chapters that follow.

References

Cohen, Steven. 2011. *Sustainability Management*. New York: Columbia University Press.

Grimaldos, Angel Simón. 2015. Water for a sustainable future: New paradigm, new vision. In *Water and Sustainable Development*, ed. Ramiro A. Lopera, 68–73. United Nations Water for Life. Accessed December 7, 2017 from www.un.org/waterforlifedecade/pdf/wm-iii-eng.pdf.

Herrick, Charles, and Joanna Pratt. 2012. Sustainability in the water sector: Enabling lasting change through leadership and cultural transformation. *Nature and Culture* 7 (3): 285–313.

OHCHR (Office of the High Commissioner). 2003. *General Comment No. 15*: The right to water. United Nations, January 20. Accessed October 23, 2017 from http://www.refworld.org/pdfid/4538838d11.pdf.

OHCHR (Office of the High Commissioner). 2017. Vienna declaration and program of action. Vienna World Conference on Human Rights, United Nations. Accessed October 23, 2017 from http://www.ohchr.org/EN/ProfessionalInterest/Pages/Vienna.aspx.

Randall, Alan. 2007. Benefit cost analysis and a save minimum standard of conservation. In *Handbook of Sustainable Development*, ed. Giles Atkinson, Simon Dietz, and Eric Neumayer, 91–105. Cheltenham: Edward Elgar.

Schnoor, Jerald L. 2008. Living with a changing water environment. *The Bridge* 38 (2): 46–54.

Schnoor, Jerald L. 2010. Water sustainability in a changing world. 2010 Clarke Prize Lecture, National Water Research Institute. Accessed October 11, 2017 from nwri-usa.org/pdf/2010/ClarkPrizelecture.pdf.

Senn, Dominique, and Dorothee Spuhler. 2015. Water, sanitation and urbanization. Sustainable Sanitation and Water Management (SWWM). Accessed October 25, 2017 from https://www.sswm.info/content/water-sanitation-and-urbanisation.

Shuster-Wallace, C.J., M. Qadir, Z. Adeel, F. Renaud, and S.K. Dickin. 2015. Putting water and energy at the heart of sustainable development. United Nations University (UNU). Accessed October 24, 2017 from http://inweh.unu.edu/wp-content/uploads/2015/03/Putting-Water-and-Energy-at-the-Heart-of-Sustainable-Development-Web.pdf.

Soubbotina, Tatyana P. 2004. *Beyond economic growth: An introduction to sustainable development*, 2nd ed. World Bank. Accessed December 4, 2018 from http://documents.worldbank.org/curated/en/454041468780615049/pdf/2489402nd0edition0Beyond0economic0growth.pdf.

UN (United Nations). 1997. Earth summit. United Nations. Accessed October 22, 2017 from http://www.un.org/geninfo/bp/envirp2.html.

UN (United Nations). 2014. International decade for action 'Water for Life' 2005–2015. Accessed November 5, 2017 from http://www.un.org/waterforlifedecade/food_security.shtml.

WCED (World Commission on Environment and Development). 1987. *Our Common Future*. New York: Oxford University Press.

WHO (World Health Organization). 2011. Water, sanitation and hygiene interventions and the prevention of diarrhea. Accessed October 24, 2017 from http://www.who.int/elena/titles/bbc/wsh_diarrhoea/en/.

WHO (World Health Organization). 2017. Progress on drinking water, sanitation, and hygiene, update and sustainable development goals. World Health Organization (WHO) and the United Nations Children's Fund (UNICEF) Joint Monitoring Program (JMP). Accessed November 7, 2017 from http://apps.who.int/iris/bitstream/10665/258617/1/9789241512893-eng.pdf?ua=1.

World Bank. 2012. Gender equality and development. World Development Report 2012. Accessed October 24, 2017 from http://documents.worldbank.org/curated/en/492221468136792185/pdf/646650WDR0201200Box364543B00PUBLIC0.pdf.

World Bank. 2013. Thirsty energy: Securing energy in a water-constrained world. Accessed October 24, 2017 from http://www.worldbank.org/en/topic/sustainabledevelopment/brief/water-energy-nexus.

World Bank. 2015. Thirsty energy update. Accessed October 24, 2017 from www.un.org/waterforlifedecade/images/newsletter/Thirsty_Energy_general_update_February_2015.pdf.

3

The Global Need for Water Sustainability

Achieving an adequate and sustainable supply of clean, safe water and access to sanitation is not just a problem for the developing world. Water supply scarcity and drinking water tainted by chemicals, pharmaceuticals and metals such as lead is a problem that is common across much of the globe. Population growth and urbanization is generating a global need for more efficient agriculture for an adequate supply of healthful foodstuffs, the installation of the infrastructure needed to ensure access to clean water, and for sanitation. The effects of these barriers to sustainability causes a critical need to deal with climate change and the resiliency to bounce back from more and more extreme weather events.

Water service suppliers have adopted many innovative approaches to overcoming these barriers to sustainability. Water managers recognize that finding new and untested approaches will be necessary in both water treatment technologies and business models for supporting the adoption of these innovations. This chapter looks at some of the ways water suppliers are reacting to the growing scarcity of clean, safe water. The focus is on descriptions of the innovative water management practices and the process implemented by municipalities and other public water suppliers. It begins with a review of changes in the organization's innovation policy and closes with a look at some of the management and governance changes taking place in the industry.

D. E. McNabb, *Global Pathways to Water Sustainability*,
https://doi.org/10.1007/978-3-030-04085-7_3

Problems in Gaining Sustainability

The nature of the many problems blocking the global drive to attaining sustainable water supplies has shown that achieving this goal will be difficult and costly. Many natural and human factors affect the world's supply of fresh water, beginning with climate change and uncontrollable population growth, and including the struggle to find the capital needed for new water retention facilities and for repairing and replacing aging water treatment and distribution infrastructure. Climate change is altering historic precipitation patterns, resulting in more and longer droughts in some areas and highly destructive floods in others. More people mean greater stress on existing facilities and the need for more water and more sanitation services. Growth and the desire of people to live in locations where the climate is fair and where work opportunities are abundant is difficult to alter, let alone halt. Where water systems exist, people pay their water bill and expect clean, safe water to be there when they turn the tap or flush the toilet. Few people think about what it takes to deliver clean water so that it is available whenever and wherever it is needed.

In some areas, water management objectives include a continuing search for better ways to allocate water for agriculture and energy production while also reversing the practice of discarding water after a single use. The search for innovative ways to fund the repair, replacement and augmenting of aging water infrastructure is a common water system objective.

Developing regions typically have neither the financial ability, technical skill, nor the political will to manage their water supply problems. There are no taps to turn and no toilets to flush. Daily, women and children must walk miles to a natural, untreated source to collect enough water for that day. Waterborne diseases are particularly devastating to children under five years of age.

One of the main reasons for the global water sustainability problems is that people have considered that access to water is a human right and that access to water should be free or affordable to everyone. Water is considered a nature-given commodity that should be available for everyone. After all, water just falls from the sky; it has always been there, unattended in rivers and lakes. The United Nations assures people that everyone has a right to water. Because of this, political leaders have often been reluctant to ask for the funds needed to invest in water and wastewater sustainability needs.

In their efforts to achieve sustainable water supply resources and systems, water-system managers are adopting new and innovative water and wastewater management practices and procedures (Lall et al. 2008; Cohen 2011). This chapter examines the some of the many ways that water suppliers are working to achieve and maintain a sustainable supply of clean, safe and affordable water.

1. Developing and implementing long-term policies for ensuring the sustainable supply and distribution of services.
2. Adopting new and innovative water source governance and management models.
3. Seeking and adopting open and cooperative innovation.
4. Adapting to new and innovative sustainability goals.
5. Mandating water conservation practices.
6. Securing alternative water resources.
7. Applying asset management practices to ensure a resilient water infrastructure.

Box 3.1 Combination of forces threatening the global water supply

The combined effects of continued global pullulation growth, urbanization, pollution, and climate change are contributing to a rapidly approaching global water problem. "Feeding 9 billion people by 2050 will require a 60 percent increase in agricultural production (which in 2017 used 70 percent of the available resource), and a 15 percent increase in water withdrawals. Besides this increasing demand, the resource is already scarce in many parts of the world. Estimates indicate that 40 percent of the world population live in water scarce areas... By 2025, about 1.8 billion people will be living in regions or countries with absolute water scarcity. Water security is a major—and often growing—challenge for many countries today.

The fragmentation of (the world's water) resource also constrains water security. There are 276 transboundary basins, shared by 148 countries, which account for 60% of the global freshwater flow. Similarly, 300 aquifers systems are transboundary in nature, meaning 2 billion people worldwide are dependent on groundwater. The challenges of fragmentation are often replicated at the national scale, meaning cooperation is needed to achieve optimal water resources management and development solutions for all riparians.

To deal with these complex and interlinked water challenges, countries will need to improve the way they manage their water resources and associated services."

Source World Bank (2017)

The World Bank, UN and other regional and global organizations have clearly described the water resource crisis that looms ahead in the near future. Box 3.1 is a representative consensus evaluation of the world's water resource problems and the need for better water management.

Organizational Governance and Management Change

Small urban and rural public utility water and wastewater systems continue to struggle with a variety of the same sustainable problems facing large systems. However, small systems usually lack the financial, material and human resource assets needed to address the litany of problems on an integrated basis. High on the list of the most pressing challenges facing administrators of these small but important utilities is implementing the necessary changes in the organization due to difficulties in recruiting or retaining qualified staff, repairing or replacing aging or inadequate infrastructure, insufficient financial reserves to meet emergency conditions, and excessive pressures on product and service resources. Managers and administrators of these systems are receiving managerial guidance and assistance from federal and state offices of the EPA and other federal and state agencies. However, missing from that assistance is instruction in the art and science of adopting open and cooperative innovation as a tool for implementing the necessary changes. Small- and medium-sized water systems are unlikely to have the resources needed to adopt full-fledged Total Water Resource Management (TWRM) operations. To achieve their sustainability goals, they have had to address the need for adopting changes in their organizational structure by adopting new or non-traditional management models.

Organizational Management Change

Public utilities are similar in their search for innovative solutions to all types of management and operational problems. System operators may lack the personnel or management skills needed to cope with these and other problems. Requirements that water and wastewater operators must hold one or more certifications and have worked in their respective industry for several years add to their staffing difficulties. The utilities may be departments of local or state governments, municipal organizations, quasi-government agencies that exist as independent nonprofit organizations, or investor-owned companies. Regardless of the governance model adopted, they are also alike in the managerial and operational challenges they face. Particularly hard hit are small water and wastewater service providers.

Water providers in the USA and elsewhere in the world are dealing with supply problems brought on by the effects of climate change, population growth, urbanization, and aged infrastructure. Collectively, these problems are contributing to critical stress to available freshwater supplies. Excessive source withdrawals, long-term drought, more and more severe extreme weather events, and continuing pollution all add to the difficulty of achieving long-term sustainability by water and wastewater service providers. Additionally, many immediate operating challenges that may not be directly associated with the current warnings about global warming have implications for immediate operations. Increasing population growth and urbanization in the South and Western USA are taxing the ability of many utilities to meet demand. Additional problems are competition over limited water supplies and failing infrastructure to meet increased demand with their present organizational and financial resources. A limited supply of trained water system professionals is restricting utilities' ability to replace retirees and meet new staffing requirements from service growth demands. Like all public organizations, to meet these challenges, utilities need to adopt innovative ways of delivering their services and funding the repair and replacement of infrastructure. To reach their long-term sustainability goals, they must also implement changes in the way they manage the organizations (Sroufe 2017).

Acceptance of and adaptation to changes in external and operating environments in organizations is necessary for an organization's survival. "Organizations have to adjust both their operations according to the changing environment and their organizational structures according to new operational models" (Král and Králová 2016, 1). Change solutions may include replacing traditional organizational structures; for example, to achieve and maintain sustainable operations, public water and wastewater utilities have found many different and innovative ways of organizing and managing their operations. Among public governance options are the following structural models:

- Municipal ownership of both a water collection, treatment and distribution system and a wastewater collection, treatment and effluent discharge system;
- Municipal water system, with a separate, usually regional, publicly owned and operated wastewater system;

- Regional public water system and separate municipal wastewater system or systems;
- Public organization oversight of a group of small, independently managed water and/or wastewater treatment systems.

In addition to the public ownership models, systems may also be privately owned and managed as non-municipal businesses, private non-profit cooperatives, private services supplied to small recreational facilities, supply systems for factories or commercial developments, branch operations of domestic or international corporations, or mutual domestic associations. Different operational management structures for both water and wastewater organizations and major infrastructure construction projects include:

1. Municipal or special district utility ownership with local government unit employee operation.
2. Municipal ownership with both contract management and function operations.
3. A combination of municipal staff management with some but not all water or wastewater operation activities contracted.
4. Municipal ownership with private-commercial partnerships on individual infrastructure developments.
5. Private commercial management and operation.
6. Combination public ownership and private construction and operation (public–private projects).
7. A public system growing in popularity is the joint municipality ownership of an independent system by two or more municipalities or special districts, with operations supervised by a utility manager reporting to a separate utility board of commissioners elected from all represented communities. This model entails following an open, cooperative, innovation policy.

Open and Collaborative Innovation

Many, if not most, water and combined water and wastewater sectors of the public utility industry are facing a complex and challenging future with limited access to the funds needed to add or replace system personnel. Before replacing or adding infrastructure, most must seek grants or borrow large sums that are amortized over decades and repaid from subscriber fees. However, this is not possible for resolving personnel problems. Utilities require funding for the higher salaries that are required to attract personnel from a declining number of administrators and operators with the required water and wastewater certification. Meeting government-mandated health, efficiency and other requirements add to the difficulty small systems face in maintaining their service commitments. System managers must turn to rate payers for the funds needed to meet their mandate of serving their publics and keeping current with the issues and regulatory trends that affect all aspects of their operations. The smallest utilities are stretching their available resources for meeting these challenges to breaking point. They must innovate in order to survive.

Organizational Collaboration

This search for innovative solutions to infrastructure funding problems demands new and better ways of operating. This openness to innovation is not new; it has long been a characteristic of public service operators with limited resources. Moreover, openness to others' best practices has been found to be a common mindset in many societies, a culture borne of a fundamental "belief in government as a solution," not as a problem. It is a belief that government workers can always find better ways to deliver services. Therefore, government "innovators share a conviction that creative problem solving, inspired improvisation, experimentation, and risk-taking within their organizations" are not only possible, they are desired (Borins 2014, 6).

As the need for new and better ways of meeting their commitment to serve continues to grow, so does the willingness to embrace innovation through inter-organizational collaboration. It is happening in all sectors of the utility industry and in all sizes of organizations. The trend is not only happening in the USA; innovation at all levels of government is now a global phenomenon. Collaboration in seeking and applying innovative solutions to problems has increasingly become the norm among organizations providing public services. A wide shift over a 15-year period in the sources and characteristics of innovative programs was found among the many public service organizations applicants to the Harvard University Kennedy School's Innovations in American Government awards. The number of organizations reporting that they collaborated with external organizations increased from 28% of the 1990–1994 pool of applicants to 65% of the semifinalists in 2010 (Table 3.1). Collaboration with other government organizations similarly grew from 21% in the early 1990s to 58% of

Table 3.1 Increasing use of collaborative innovations by public sector, 1990–1994 and 2010

Focus or characteristic	1990–1994 semifinalists (%)	All 2010 applicants (%)
External collaboration	28	65
Collaboration within government	21	58
Information technology	28	36
Process improvement	34	29
Citizen empowerment	26	16
Use of volunteers	7	13
Use of market incentives	8	10
Changes in public attitudes	13	11
Totals	165	238

Source Data from S. Borins (2014, 10)

the applicants in 2010. Decreases were reported in just two of the eight categories of characteristics mentioned: process improvement, 34% in the early 1990s pool to 29% in 2010; and in citizen empowerment, from 26% in 1990–1994 to 16% in 2010 (Borins 2014, 10).

Public water service organizations found it increasingly difficult to secure the funds needed to operate, repair weather damage and aging infrastructure, and expand to meet the mandate to serve new consumers. Innovative solutions to old problems have become a necessity. Innovation begets transformational change in all types of organizations.

Jonathan Breul described the need for greater inter-organizational cooperation in innovation this way (emphasis in the original): "Rising public expectations for demonstrable results and enhanced responsiveness … require fundamental *transformation* of government—where roles and even continued existence of some organizations and functions will be at stake … Government organizations need to pick up the pace to become less hierarchical, process-oriented, stovepiped [i.e., single-purpose], and inwardly focused. They will need to become more partnership-based, results-oriented, integrated and externally focused" (Abramson et al. 2006, 7).

Abramson et al. (2006) outlined six broad categories of innovative solutions that utility managers must develop and adopt in response to the pressures for change they face. Collectively, responses to these and other demands for change are spurring greater and inter-organizational transformation. They include:

1. Changes in the management of government operations through implementation of performance management practices.
2. Changes from bureaucratic to market-based governance.
3. Changes that enable agencies to provide services on demand.
4. Changes from tolerating citizen participation to encouraging and re-engaging the electorate and encouraging greater volunteer involvement.
5. Changes in the structure of services delivery to include collaboration, public/private teamwork, networks, partnerships and coalitions.
6. Changes in the rules of government, in the way public work takes place.

Organizational Change Case Study

A small Washington state water and wastewater district with 532 connections found itself forced by personnel departures to begin a search for replacements. The search quickly evolved into a decision to change from a contract system to a different governance and management model. The district had embraced the concept of open innovation since its formation by seeking assistance from state and regional water and wastewater associations. The manager of a somewhat larger water system that had just undergone a similar problem and solution provided assistance in planning for the change.

Harstine Island, approximately ten miles long and three miles wide, is located at the southern end of Puget Sound, Washington state, USA. The northern portion of the island is a gated community of 532 dwellings. Originally developed in the late 1960s as a recreational community for company employees, individual lots and summer cabins were sold to the public as recreation homes. In 2018, approximately a third of the structures had become primary residences. Fresh water is pumped from the large Puget Sound aquifer and purified in a new US$1,300,000 system and stored in a steel on-site reservoir. Prior to discharge, all wastewater is treated and sanitized to federal and state standards in an onsite treatment facility.

When initially developed, the installed water and wastewater s treatment systems were maintained and managed by the county public utility department. In 2005, residents became concerned over indications that the systems were not receiving mandated upgrades, while locally generated reserves were used to support other small systems managed by the county. They decided to begin action to take over the ownership and operation of both systems. The water and wastewater district became an independent special district water and wastewater service provider in 2009.

The utility is governed by a board of three part-time commissioners elected for staggered six-year terms. While the commissioners have had extensive business operation experience, they are somewhat less familiar with the complications associated with operating sustainable water or wastewater treatment systems. They recognize that they cannot solve their many management problems alone; utility managers and administrators must look outside of their organizations for answers. A new governance model search began with assistance by the state water district association and a nearby public water district that had completed a similar governance change several years earlier.

Examples of the district's ongoing cooperative arrangements include membership in the Washington Association of Sewer and Water Districts (WASWD), the Evergreen Rural Water of Washington (ERWoW), and the Municipal Research and Services Center (MRSC). The WASWD association's membership includes general and associate members. General members are special purpose sewer, water, or combined sewer/water districts. As of January 2014, MRSC listed a total of 2015 water and sewer districts in the state. Associate members are individuals, companies, state or local agencies, non-district local governments, and consultants or associations that are not a special purpose water and/or sewer districts, but in which the association's board of directors have a legitimate interest in the organization's mission. Like MARSC, the association is a source of legal, computer technology, procurement and other operations problems resolved by other member districts.

Guiding the Proposed Change

The research examined in producing this review has indicated that many water and wastewater global, local government, private civil society organizations, and industry trade associations exist for providing water and wastewater service providers get through the process of adopting change. The global water and wastewater sector consists of thousands of local government and private utilities, special municipal service providers, engineering and laboratory service providers, equipment manufacturers and suppliers, contractors, regulatory agencies and many nonprofit associations and aid organizations. In many U.S. states, all these contribute to provide clean, safe water and consistant wastewater treatment services. The makeup of the industry in Washington state is a typical example. Evergreen Rural Water

(ERW) is a nonprofit organization established in 1994 to aid rural water-service providers in managing and planning their operations. ERW is now the Washington state affiliate of the national Rural Water Association. The official mission of Evergreen is to provide and promote the highest level of service, technical assistance, training and advocacy for Washington state utilities. Evergreen provides training and technical assistance to districts directly to help solve operational, regulatory, administrative and financial problems faced by local water systems. Although not a state agency, its staff cooperates with the Washington state departments of health, ecology, and such federal agencies as the U.S. Department of Agriculture, EPA and local health offices and other agencies interested in solving rural water districts. ERW is funded by the EPA and the Department of Agriculture and Rural Development agency in the form of grants and contracts.

The MRSC is a nonprofit center formed in 1934 to support effective local government in Washington state. The organization continues to function as the official research and services agency for cities, towns and utility districts, serving fewer than 10,000 citizens in the state. The center provides operations consultation, research, training and collaboration to all local governments, including separate water and sewer districts. Most of MRSC's services are provided free of charge to local government employees and public officials of cities, counties, and some special purpose districts in the state. MRSC's services to utility administrators include important management responsibilities such as municipal law and finance, urban planning, public works and public management. As a relatively independent public service organization, the district particularly values the center's vetting of private water and sewer operations and equipment vendors.

Adjusting to External Change

Many water suppliers in arid and semi-arid regions of the world such as the U.S. southwest and south are coping with population growth and urban development that are placing increasing demands on finite water resources and limited wastewater treatment facilities. Responding to these problems, many legislatures are adopting water-use efficiency requirement laws. Entrenched water rights holders and real estate developers often challenge these laws. In developing regions, urban slums without access to freshwater or sanitary facilities compound the regions' sustainability challenges. Municipal water suppliers must prove how they are using water more

Table 3.2 Water efficiency rules and compliance dates

Rule requirement	Deadline for systems under 1000 connections
Install water supply production meters	January 22, 2007
Collect water consumption and production data	January 1, 2008
Include efficiency (WUE) program in utility planning documents	January 22, 2008
Set WUE goals for your own utility	January 22, 2009
Submit service meter installation schedule	July 1, 2009
Submit first annual performance report	July 1, 2009
Service meters installed for all connections	January 22, 2017
Meet at least a 10% leakage standard (based on a three-year average)	Three years after all service meters are installed

Source Washington State Department of Health (WSDOH)

efficiently in order to receive certainty of water rights for current and future demand.

State legislatures and local health departments throughout the USA are adopting water efficiency programs for municipal water suppliers. Water utilities are required to meet compliance standards; deadlines for compliance only differ depending on whether the system serves less or more than 1000 connections. Table 3.2 is an example of the requirements and the compliance dates.

Change Adoption Cooperation

Water providers in Washington state worked closely with other small water districts faced with implementing the water efficiency rules, a point particularly valuable when acquiring and installing water meters on all connections in the district. A larger district was in the process of replacing its nearly new traditional meters with smart meters. A smaller district was not only able to acquire the meters that were being replaced at an exceptionally favorable price, they received free guidance and installation assistance from the sister utility. In wastewater effluent compliance processes, the district has benefited from larger utilities' experience in installing new and updated automatic treatment processing and other technology. Personnel at the Evergreen office were a valuable source of innovations designed and implemented by other small utilities in Washington state and across the country.

Cooperative Security Management Changes

The Pacific Northwest Partnership for Regional Infrastructure Security (PNPRIS) is another example of a successful cooperative arrangement. One of the training exercises sponsored by PNPRIS involved a disaster scenario devised by representatives from the Bonneville Power Center; telecommunications companies Telus, Verizon, and Qwest; the Federal Emergency Management Agency (FEMA); the British Columbia Provisional Emergency Program; and the Canadian Office of Critical Infrastructure Protection and Emergency Preparedness (OCIPEP).

The scenario theme was a hypothetical disruption to the regional electric power grid. It also included terrorist and non-terrorist disruptions of natural gas transmission and distribution systems, municipal water systems, regional ports, and telecommunications systems. Disruptions of those critical public services affected other independent infrastructures, including transportation systems, emergency services, public safety services, hospitals, and cross-border cooperation. A chief result of the training session was that many participants discovered that their organizations' contingency plans were negated by the cross-border interdependencies that exist among the region's public services.

Need for New Sustainability Goals

A report proposing extension and revisions of earlier recommended water goals was released in 2014 by UN-Water, the UN agency for coordinating and publishing water-related data. The report presented suggestions for adopting a post-2015 water goal that incorporated all three dimensions of sustainable development: social, economic and environmental stakeholder interests. The purpose was to "support protection of water resources from over exploitation and pollution while meeting drinking water and sanitation needs, energy, agriculture and other uses … [and protect] communities from water-related disasters" (UN-Water 2014, 3). The new goal was to be "securing sustainable water for all."

The proposal included targets and indicators (metrics) for aiding and guiding countries in reaching their 2015 goals. When adopted, the proposed global goal for water was to be applicable to all regions and nations, taking into consideration local contexts and priorities. The following framework of five measurable and associated targets relevant to all countries were included for facilitating and achieving the goal; each of the five targets is then discussed in more detail. The targets were:

1. Achieve access to safe drinking water, sanitation and hygiene for all.
2. Improve (by a locally relevant percentage) sustainable use of water resources and development of water resources.
3. Initiate and reinforce equitable, participatory and accountable water governance with stronger management actions and administrative systems.
4. Reduce discharge of untreated wastewater and agricultural nutrient pollution by locally relevant percentages, while increasing wastewater reuse by a supportable percentage.
5. Reduce illness and death from waterborne disease by a supportable percentage and limit economic loss from natural and human-influenced water-related disasters by locally supportable percentages.

Safe Drinking Water, Sanitation and Hygiene

The water, sanitation and hygiene access target had four components: (1) elimination of open defecation; (2) achieving universal access to basic drinking water, sanitation and hygiene for households, schools and health facilities; (3) to reduce by 50% the proportion of the population without access at home to safely managed drinking water and sanitations services; and (4) to progressively eliminate inequities in access (UN-Water 2014, 15). The target added key elements to the original water goal. The first was the addition of hygiene to the original water and sanitation elements; the second was the expansion of sectors covered from just households to also cover other sectors such as schools and medical institutions. A third key addition was the goal of total elimination of open defecation, which has been shown to be a major source of the spread of disease and environmental damage.

Sustainable Water Resources

This target allows individual nations to develop, enhance and protect the water resources that are right for their specific conditions, but also calls for cooperation between nations that are served by cross-border sources. Progress toward achieving the self-set target metrics entails progress on these three components: bringing freshwater withdrawals in line with the sustainability of available water resources; restoring and protecting ecosystems to ensure the ability to provide water-related services; and increasing productivity by all sectors (primarily agriculture, industry and energy, and municipalities).

Excessive water withdrawals are currently increasing faster than the ability of many surface water and groundwater to meet demands. Globally, agriculture accounts for 70% of all water withdrawals. However, current inefficient irrigation methods results in only approximately a third of the water withdrawn for agriculture being used by the crops for which it was intended. Industry (including energy production) accounts for 20% of global water withdrawals. Much of that water is returned to rivers and streams either highly polluted or, when turned to steam, is lost from evaporation during cooling. Much of the water withdrawn for all sectors is often lost—typically after being expensively treated and purified—due to leaks in obsolete infrastructure, misuse, or inefficient agriculture or industrial applications.

Strengthening Water Governance

Water governance has been defined as: "The range of political social, economic and administrative systems that are in place to develop and manage water resources, and the delivery of water services to different levels of society" (UN-Water 2014, 18). The target for water governance is clear: "All countries [will] strengthen equitable, participatory and accountable water governance." For this to happen, countries are to be guided by targets in these four areas:

- Implement integrated water management policies and activities, with participatory decision making at the local, basin and national levels;
- Deliver clean, safe drinking water, sanitation and hygiene services in an affordable, accountable and financially and environmentally sustainable manner;
- Ensure that appropriate and equitable regulatory (and enforcement) systems exist for managing water resource standards, maintaining adequate purification and delivery infrastructure, and services to improve public authorities and water operator performance;
- Strengthen knowledge transfer and skills development training programs.

Strengthening water governance requires a broad commitment to fair and equitable governing. To inculcate the philosophy, public service will require investments in education, knowledge and skills development and transfer, all of which are relevant to equality for women and children. For nations in which different standards apply, the following guidance is included for achieving this target: "The collection, analysis, and use of sex-disaggregated data, gender statistics and other social data are necessary to check that meeting the targets remove inequalities" (UN-Water 2014, 19).

Importantly, the proposed post-2015 goals did not presume to restrict all countries to meeting a specific set of metrics for sub-targets. Rather, as proposed, the target was couched in terms that made it possible for each country to set its own sub-targets, and to adopt sustainability pathways based on its own strengths and limitations.

Managing Wastewater, Pollution and Water Quality

The planners recognized that the water quality target element of the social and economic development objective could not be reached unless wastewater and other pollutants were no longer permitted to be discharged untreated into water supply sources. Managing wastewater provides social, health, economic and environmental benefits that far exceed the costs.

Human activity is recognized at the main cause of water pollution and the spread of waterborne disease. Moreover, human-related water pollution makes many already limited water resources unusable without expensive treatment. Changes in many traditional practices are required if these goals are to be achieved. An important element in these changes is acceptance of the idea of managing treated wastewater not just as a pollutant, but also as a valuable resource, particularly in areas of high and extreme water stress and in many urban centers.

To emphasize the importance of this target, planners adopted this wording regarding targets: "Reduce wastewater pollution and improve water quality by reducing untreated domestic and industrial wastewater by [a locally sustainable percentage]; increasing wastewater reuse safely by [a sustainable percentage]; and reducing [agriculture] nutrient by [a sustainable percentage] to maximize water resources availability and improve water quality" (UN-Water 2014, 20). To meet their targets, action will be needed on achieving self-established change percentages for the following three tasks: (1) reduce untreated domestic and industrial wastewater; (2) increase use of safe recycled safe water; and (3) reduce water pollution by nutrients.

Mitigating Water-Related Disasters

From 2000–2006, globally occurring water-related disasters numbered 2163; more than 4000 disasters have occurred since 1970, according to the Emergency Disasters Database. Those events resulted in the deaths of more than 290,000 people, affecting in some way more than 1.5 billion people, and inflicting damages of more than US$422 billion (Adikari and Yoshitami

2009). The UN-Water organization considers water-related disasters to the most economically and socially destructive of all natural hazards. Floods and droughts are expected to increase due to global warming. Increased hydro-hazard impacts and costs are attributable to such factors as increased event frequency and magnitude, unplanned urbanization, degradation of ecosystem services, vulnerable livelihoods, and inaccurate public perception of risk. The challenge facing planners, government leaders and water professionals is how to identify appropriate measures for preserving sustainable water supplies in continuously changing internal and external environments.

The trends in natural disasters since 1980 show they are occurring with greater severity in most regions of the world. Floods, droughts and windstorms (hurricanes and cyclones) account for nearly 90% of the most damaging events. The costs of water-related disasters have grown significantly since 1980. The expected increase in air and water temperatures due to climate change and urbanization is likely to see continuation of extreme weather events. Since 1980, the threat of economic loss from storms and resulting floods have increased by more than 160%, while the losses due to hurricanes and tropical cyclones have increase by 265% in developed countries (NOAA 2017).

Among all observed natural disasters related to human action and water, floods are the most regularly occurring disasters. During the first ten months of 2017, 15 weather and climate disaster events occurred in the USA with losses of more than US$1 billion each (Fig. 3.1). These events included one drought event, two flooding events, one freeze event, seven severe storm events, three tropical cyclone events, and one wildfire event. These 15 events resulted in the deaths of 282 people and had significant economic effects on the areas impacted. The costs for the 2017 hurricanes Harvey, Irma and Maria were still being assessed at the time of writing this book.

Figure 3.1 includes the top 10 countries with most all types of disasters from 2005 to 2014. The totals include all water, climate and weather disasters, including hurricanes, cyclones, tsunamis, rainstorms, floods and earthquakes. China led the list with 286 disasters during the ten-year period, followed by the USA with 212 disasters. The most destructive disasters occurred in the USA, where disasters cost a total of US$443 billion. With 62 disasters, Japan was ninth in terms of the number of events, but third in terms of the cost of the damage, exceeded only by China and the USA.

The proposed post-2015 target for water-related disaster damage mitigation is to reduce the resulting mortality and economic cost by a sustainable percentage. The sub-targets, again to be established by individual countries to match their individual resources are:

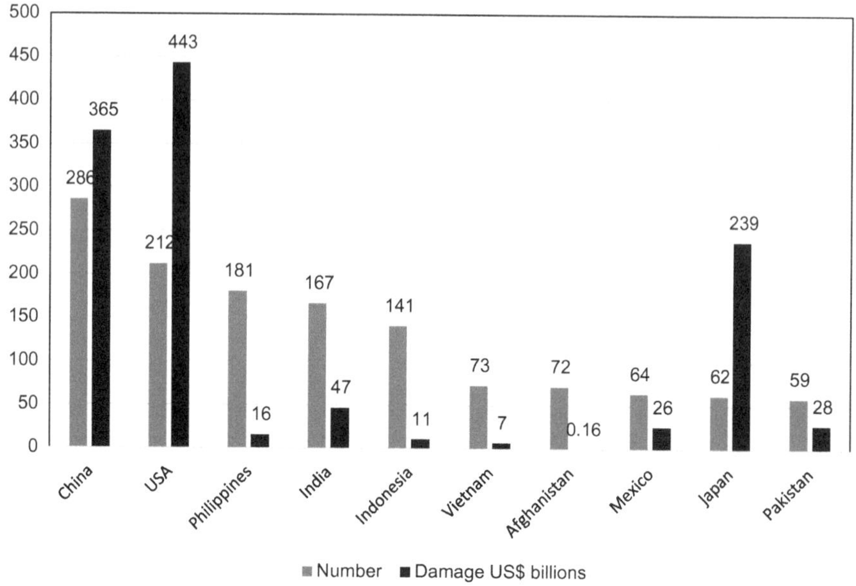

Fig. 3.1 Top ten countries with disasters and the damage amounts, 2005–2014 (*Source* UNISDR 2015)

1. Collect and distribute increased knowledge and understanding for communities at risk of water-related disasters; of particular importance are disasters that are likely to occur as a result of climate change.
2. Adopt integrated water-related disaster risk management planning that includes an appropriate mix of structural and nonstructural mitigations in order reduce mortality and economic losses.
3. Adopt and initiate early warning systems for communities most at risk of water-related disasters.
4. Apply a comprehensive preparedness approach to water-related disaster management that meets all the needs of at-risk communities.

Summary

The barriers to achieve sustainable water supplies are more than regional management problems; they affect the livelihoods and health of millions of people around the globe. Water providers are following a variety of paths in their efforts to overcome these sustainability problems. Four of these paths were discussed in this chapter: adoption of water management policies that go beyond local concern to look at the problem from all aspects of the preservation, supply, distribution and use of finite supplies of water; implementation of new management and governance models; an openness to collaborative innovation and best practices developed elsewhere; and the recognition that the sustainability problems will never go away and, therefore, the continued development of new sustainability goals.

Global and regional water management organizations warn us that implementing the activities proposed for achieving water sustainability goals will be difficult and costly. However, the commitment to investing in sufficient infrastructure for acquiring, treating, delivering and properly reusing treated wastewater will not go away. Rather, external forces beyond the control of governments and water suppliers make this commitment to sustainability critical. Achieving stable and sufficient water supplies in both developed and developing regions requires developing a cadre of skilled water and wastewater operators and professional planners, engineers and administrators. These men and women must be given the authority to implement changes to traditional agriculture methods and inefficient and polluting industrial water applications.

The powerful forces of climate change, population growth and urbanization demand development of new and improved water storage facilities. Societies that depend on access to sufficient and stable supplies of clean, safe water must make it possible for their water suppliers to have the capital for implementing needed infrastructure. Climate change, population growth and urbanization require a global effort to deal with problems of growing water scarcity, the pollution of existing surface and groundwater supplies, excessive withdrawals of finite water supplies, and inefficient irrigation methods.

References

Abramson, Mark A., Jonathan D. Breul, and John Kamensky. 2006. *Six Trends Transforming Government*. Washington, DC: IBM Center for the Business of Government.

Adikari, Yogamath, and Junichi Yoshitami. 2009. Global trends in water related disasters: An insight for policymakers. United Nations Educational, Scientific and Cultural Organization (UNESCO). Accessed December 19, 2017 from http://unesdoc.unesco.org/images/0018/001817/181793e.pdf.

Borins, Sandford. 2014. *The Persistance of Innovation in Government*. Washington, DC: Brookings Institution.

Cohen, Steven. 2011. *Sustainability Management*. New York: Columbia University Press.

Král, Pavel, and Věra Králová. 2016. Approaches to changing organizational structure: The effects of drivers and communication. *Journal of Business Research* 69 (2016): 5169–5174.

Lall, Upmanu, Tanya Heikkila, Casey Brown, and Tobias Siegried. 2008. Water in the 21st century: Defining the elements of global crises and potential solutions. *Journal of International Affairs* 61 (2): 1–17.

NOAA (National Oceanic and Atmospheric Administration). 2017. Billion-dollar water and climate disasters: National Centers for Environmental Information. Accessed December 19, 2017 from www.ncdc.noaa.gov/billions/events/US/1980-2017.

Sroufe, Robert. 2017. Integration and organizational change towards sustainability. *Journal of Cleaner Production* 162 (217): 315–329.

UNISDR (United Nations Office for Disaster Risk Reduction). 2015. The economic and human impact of disasters in the last 10 years. Accessed December 19, 2017 from https://www.unisdr.org/we/inform/disaster-statistics.

UN-Water. 2014. A post-2015 global goal for water: Synthesis of key findings and recommendations from UN-Water. Accessed December 16, 2017 from www.un.org/waterforlifedecade/pdf/27_01_2014_UN-Water_paper_on_a_post2015_global_goal_for_water.pdf.

World Bank. 2017. Water resource management. Accessed March 29, 2018 from http://www.worldbank.org/en/topic/waterresourcesmanagement.

Part II

Barriers on the Path to Sustainability

4

The Barrier of Climate Change

The complex system we know as climate is shaped by, and interacts with, both internal and external geo-atmospheric phenomena. Internal components of the system include the atmosphere; the oceans; sea ice; the land and its characteristics, such as glaciers, forests, farmlands, ecosystems, snow and ice cover, surface water and groundwater. External factors include the sun's energy, the relationship between the sun and earth and the earth's changing orbit about the sun, characteristics of the earth such as the distribution of land and water, topographical characteristics of the earth such as mountains and deserts, characteristics and composition of the earth's atmosphere and oceans, and other similar characteristics (Karl and Trenberth 2005).

A major barrier to developing and implementing a common strategy for achieving sustainability that helps make climate change so difficult to deal with is the variability of the impacts it has on the world's water supplies; for example, land areas that lie farther from the equator are likely to experience generally warmer weather in all seasons. Warmer weather results in less snowfall, more rain and rising ocean levels. On the other hand, areas near the equator such as North Africa, the Middle East and Central America are expected to receive less than normal rainfall. As a result, they will experience great water scarcity.

Overall, the USA is projected to receive more rainfall than normal. The arid Southwest will see more and more prolonged droughts. The water-rich areas in the Northeast will experience more extreme weather events with more flooding. The Northwest will see less snow and more rain. Globally, many of these changes are already underway and adding to water management problems. For

D. E. McNabb, *Global Pathways to Water Sustainability*,
https://doi.org/10.1007/978-3-030-04085-7_4

example, in the USA regions of the American Southwest that depend upon Colorado River water are already seeing water withdrawal restrictions due to earlier snow melting in the Rocky Mountains. This is having an impact on the agricultural and municipal water sector needs. Wetter conditions are also expected in northern Europe, with dryer conditions occurring closer to the Mediterranean, and warmer temperatures throughout Europe. The eastern and central portions of Asia are experiencing periodic declines in precipitation and prolonged periods of drought. Globally, land nations are being hit by more and more severe extreme weather events, while also having to cope with rising sea levels. Simulations with various levels of global warming and socioeconomic conditions indicate continuation of a general increase in water scarcity in much of the world, particularly when evaluated with population and urbanization trends. Meanwhile, severe water scarcity is expected in China, India, Bangladesh, Mexico, the Middle East and North Africa. The effects of these changes are global and are particularly worrisome to the water suppliers in the world's urban centers. As researchers Jaramillo and Nazemi (2018, 10) noted in their study on measuring the impact of change on city infrastructure, "Climate change already is, and will continue to be, one of the major environmental threats to cities around the world."

Factors Contributing to Climate Change

The earth's climate has been warming over the past two centuries, largely a result of the industrial revolution activities of humans. Smoke from burning hydrocarbons such as coal and oil, and exhaust emissions from motor vehicles release CO_2 and other gases into the atmosphere. These human-caused contributions to climate change act as 'anthropogenic forcers' that add to concentrations of greenhouse gases from naturally occuring events such as volcanic eruptions. Among their sources are burning fossil fuels, sulphate aerosols from industrial byproducts, and increases in atmospheric water vapor from human-caused changes to land surface conditions. Together, human activities and natural causes result in the collection of greenhouse gases in the atmosphere. Greenhouse gasses act as a thermal blanket over the earth, causing the earth's temperature to rise. Climate is always changing, although changes typically occur over the course of many centuries. These changes have resulted in both cold and warm periods. The Ice Age that ended some 11,000 years ago is an example. And, as climate changes, so does weather.

Weather is the characteristics of the atmosphere at the time it is observed and measured. It may be hot or cold, wet or dry, calm or stormy, clear or cloudy, calm or windy. Weather is what is going on outside at any location at any time. The combination of all the earth's weather averaged over long periods of time is climate. Climate change is measured from century to century, while weather changes occur in relatively stable patterns such as season to season and region to region. Satellites make it possible to predict changes in weather on a daily basis and often hourly.

Climate's Impact on Water Supplies

Changes in weather have short-term impacts on water. Rain and snowstorms, floods and droughts are weather events. Climate, on the other hand, has long-term impacts over the earth's finite supply of fresh water. The melting of the world's glaciers and the continuing expansion of the Sahara Desert are climate-related phenomena. In many areas of the globe, the influence of climate change is apparent in the form of longer droughts, while other areas are experiencing more extreme weather events, such as more severe rainstorms, hurricanes and more rapid snowmelts. Adding to the impact of climate change is the growing population's increasing demand for water needed for agriculture, heating or cooling, drinking and cleaning (Jain 2011). As water demand continues to grow, many regions are encountering shrinking water supplies while in other areas shortages are less of a problem than increases in runoff, flooding, or sea-level rise.

Water managers must simultaneously meet the needs of growing communities, sensitive ecosystems, farmers and ranchers, energy producers, and manufacturers—all while climate change alters the hydrologic cycle around them. Climate change is already affecting the quality and the supply of water in many parts of the world. Changes in the water supply occurring in the USA were described in this 2016 report by the EPA:

> Over the past 50 years, the amount of rain falling during very heavy precipitation events has increased for most of the United States. This trend has been greatest in the Northeast, Midwest, and upper Great Plains, where the amount of rain falling during the most intense 1% of storms has increased more than 30 percent. Warming winter temperatures cause more precipitation to fall as rain rather than snow. Furthermore, rising temperatures cause snow to begin melting earlier in the year. This alters the timing of streamflow in rivers that have their sources in mountainous areas … Many areas of the United States, especially the West, currently face water shortages. The amount of water available in these areas is already limited, and demand will continue to rise as population grows. Many areas in the West have experienced less rain over the past 50 years, as well as increases in the severity and length of droughts; this has been especially of concern in the Southwest. (EPA 2016)

Water Scarcity

Despite the already apparent damaging effects of climate change on all sections and socio-economic sectors around the globe, the connection between climate change and the growing scarcity of water has not received much attention outside of universities and a few government agencies. This may be because the changes in the global freshwater supply are only now coming to the public's attention. More people are concerned over the damages from extreme storm events and floods from heavy downpours than are affected by water scarcity. The changes occurring in the patterns of changes in the world hydroclimate and the probability of extreme weather events are occurring slowly. Moreover, a consensus on the potential consequences of climate change on the global water supply has not yet been reached (Miralles-Wilhelm 2017).

Water quality suffers in many areas due to increases in rainfall and the resulting storm runoff and flooding. At times during 2016 and 2017, Hawaii, the Pacific Northwest, the Northeast and Midwest of the USA experienced heavy rainfall events that caused problems for human health, as wastewater treatment and water treatment systems were often overwhelmed by the sudden heavy volumes of water. Urban storm runoff from heavy downpours can increase the amount of pollutants into rivers and lakes. Runoffs carry sediment, nutrients, pollutants, trash, animal waste, and other materials into water supplies, making them unusable, unsafe, or in need of water treatment.

Damage in Coastal Areas

Many of the world's largest urban areas are located in areas subject to damage from coastal flooding. In the USA, nearly 25 million people live in coastal areas. Freshwater and wastewater facilities in coastal areas are particularly threatened by sea-level rise. U.S. areas susceptible to water pollution are shown in Fig. 4.1. The oceans have risen 8 inches over the last century, and the rise is occurring at an increasing rate. Current estimates forecast global sea-level rises of from 1 to 3 feet by the end of this century (EPA 2017). In many coastal areas where the land is subsiding due to oil production and aquifer withdrawals and other actions, the increase is even greater. As the sea rises, saltwater is also encroaching on contiguous freshwater aquifers. This is forcing coastal cities and towns to either seek other sources of fresh water or

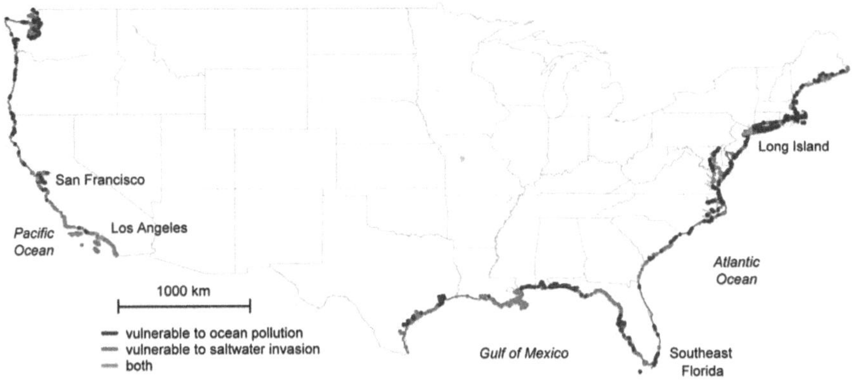

Fig. 4.1 U.S. coastal areas vulnerable to water pollution (*Source* NASA 2016)

increase costly desalination of coastal groundwater from aquifers that have long served as sources of drinking water supplies.

The impact of climate change on infrastructure in coastal areas is adding to the problems already facing many areas. Infrastructure for water and wastewater services is highly susceptible to damage from heavy rainstorms, hurricanes and storm surges. Shoreline erosion, coastal flooding and drinking water pollution are also among these common problems. The global warming occurring with climate change is also having an impact on surrounding land areas; for example, warmer temperatures in mountain areas is causing earlier runoff from melting snow. This is changing the ecosystems of the rivers and streams leading to the coasts; for example, areas of the Gulf of Mexico and Chesapeake Bay have already seen dead zones from inflows of land-based pollution such as agricultural fertilizers contributing to algae blooms. When algae decompose, it results in depletion of the oxygen in the water, resulting in damage to fish and other aquatic species. And, as more freshwater is removed from rivers for human use, saltwater is moving farther upstream and intruding into freshwater aquifers. Droughts cause coastal water resources to become more saline as freshwater supplies from rivers are reduced.

Water infrastructure in coastal cities, including sewer systems and wastewater treatment facilities, face risks from rising sea levels and the damaging impacts of ocean surges.

Resulting water scarcity levels for each scenario were measured for geographic regions and individual nations and reported as levels on a water scarcity index (demand divided by runoff plus inflow). Climate change impact estimate summaries for seven major regions were reported as follows (Miralles-Wilhelm et al. 2017):

- North America: With no expected major changes in overall runoff, the regional trend is for a small decrease in water availability in Canada and the USA, with some increases in water availability in parts of the USA by 2050.
- Latin America: A continuing trend of declining runoff, with two of the five scenarios suggesting extreme shortfalls in Brazil, Colombia, Ecuador, Peru and Bolivia.
- Europe and Central Asia: A consistent precipitation decrease across all scenarios for declining runoff, with sharp drops in the Russian federation after 2050.
- East Asia: Runoff is expected to be stable and high in some areas, but runoff decreases are expected in China, Bangladesh, Cambodia, Myanmar, Thailand and Vietnam.
- India: Results were mixed across the five scenarios. Overall, runoff increases are expected by 2050, with some estimates beginning in 2025.
- Middle East and North Africa: Continuous large drops expected in runoff, with Iran a possible exception with a somewhat lower decline.
- Africa: Great variation in the south, but overall a continuous decline in runoff across the continent, with larger declines for some areas in the last half of the century.

Predicting Water Scarcity

Recognizing that nothing is certain in predicting the impact on global water availability by these computed climate change scenarios, the team was able to reach agreement on three key trends in the impact on the water supply. First, a trend in increased water scarcity is clearly expected to occur, but by varying degrees in all regions of the earth; climate change is not the sole contributor to these increases, however. Second, the models suggest that water scarcity is demand-driven, rather than by such influences and climate, population growth, urbanization, increases in irrigated agriculture, and water withdrawals for energy production contribute more to water availability. And third, some water scarcity will occur globally, with significant increases occurring between 2025 and 2050; areas expected to suffer the greatest levels of scarcity include China, India, Mexico, Eastern Siberia, the Middle East and North Africa.

Water Scarcity Impacts to Sectors

In a 2017 World Bank Group report, Fernando Miralles-Wilhelm and a team of six researchers from the University of Maryland and Pacific Northwest National Laboratory described the results of a study of a variety of physical impacts of climate change on global fresh water resources. In brief, their series of multiple scenarios was designed to find answers to three questions relating to the potential impact of continued climate change on the world's water supply. The first question was: what physical impacts on global water scarcity and surface water supplies will result from climate change? The second question was: how will different levels of future development affect water runoff and scarcity? And the third question was: what impact will different levels of efforts to moderate greenhouse gas emission and carbon sequestration efforts have on water scarcity, the effects of stormwater runoff on surface water sources, and annual amounts of water in rivers and streams, lakes from rainfall and snow melting.

The analysis looked at three levels of possible climate history (wet, dry and normal) with the open-source Global Change Assessment Model (GCAM), with five levels of possible socioeconomic and technological development, and the impact of water scarcity on three water-dependent sectors: agriculture, energy and municipalities. The conclusion was that the climate models suggest significant decreases in surface water resources in some areas and increasing supplies in others—all with the caveat that there is large uncertainty in the results for many areas. Overall, water supply is expected to decrease in many mid-latitude and dry subtropical regions. High agreement was also reached on projected rainfall for many areas farther from the poles, increases for areas farther from the poles, as well as in many humid mid-latitude regions. In addition to water scarcity, the quality of many sources of surface water is likely to suffer as warmer temperatures and agriculture pollutants result in algae-producing toxins.

Agriculture Water Demand

Global water scarcity is expected to be an even more critical problem in the future, albeit at different levels in many parts of the world. However, the report emphasized that climate change is generally not the only reason for much of that scarcity. Increases in population and urbanization may have greater influence on water scarcity. Withdrawals for irrigated agriculture are expected to increase as much as 40% in Europe, North America and parts of Asia. On the other hand, some scenarios indicate decreased withdrawals for agriculture in parts of India, Pakistan and southeastern China due to expected increased precipitation in those regions. Overall, increases in water withdrawals for irrigated agriculture have also been predicted to increase anywhere from 7 to 21% by 2018.

Agriculture is also vulnerable to changes in precipitation patterns. Rainfall agriculture in exceptionally susceptible changes in the amount and unpredictability of precipitation. The effects of these expected changes on different regions and crops are both positive and negative. Rice farmers in Southern Asia may need less surface water for rice lagoons due to increase in rainfall. Dryland farmers of wheat and other grains in Central Asia, North America and South America that depend on rain and/or snow during the spring growing season will suffer with the projected decreases in precipitation.

A potentially larger problem is the expected increases in water pollution from agriculture fertilizers, urban runoff and soil erosion taking place as a result of increased flooding from extreme weather events.

Water for Energy Production

Water in large and consistent quantities is necessary for hydropower production and thermal power plants must have reliable supplies of water for steam turbines and for cooling. These and other energy-related activities make the energy sector the second largest user of water in developed nations. Climate change affects the water supply for power production in many ways. Changing seasonal streamflow patterns from earlier melting of feeder snowpacks, together with the warming of surface water, requires more water storage; longer periods in the cooling process will result in less water being returned to aquifers, rivers and lakes, and greater water losses due to evaporation. Reductions in river water availability for cooling thermal power production facilities in warmer and dryer regions in the USA and Europe will result in fewer days when withdrawals are permitted, shutting down power generation. Moreover, warmer surface water used for power generation cooling has been shown to lower power production efficiency in some European locations.

Power plants located along rivers will be more likely to be affected by streamflow variability and reduced streamflow. Hydropower generation in the thousands of United States Bureau of Reclamation dams in the western half of the USA and other parts of the world are expected to see reduced precipitation from the hydrographic cycle changes caused by climate change. Streamflows in areas where winter snowfall has served as a freshwater reservoir, releasing water during spring and summer melting, will see earlier melting and alerted flow patterns.

Municipal Water Needs

Municipalities, the third sector covered by Miralles-Wilhelm et al. in their (2017) analysis of the impact of climate change on water supplies, are affected in two main ways by what happens to the water supply. First, humans must have water to survive. It is used for drinking, cooking and cleaning. It is also need for the production of food, clothing and many industrial products manufactured for the consumer market. Municipalities are responsible for acquiring, storing, purifying and distributing clean, safe water to homes, hospitals, schools, factories and other businesses. In many municipalities, they are also responsible for collecting treating and discharging clean, safe wastewater. Climate change is affecting all these activities.

Most municipalities depend on surface water sources for their main supply. As these surface water supplies dry up, municipalities must depend more on groundwater to meet growing demand. However, groundwater supplies are not eternal; aquifers must be recharged. Too often, the water for recharging is polluted water from streets and other paved locations, as well as from nearby farms and livestock facilities.

The water from rivers, streams and underground aquifers is often stored in aboveground reservoirs and lakes. Water quality often suffers in these open storage facilities. More precipitation is falling as rain than as snow, while higher mean temperatures expected with climate change is melting the winter snowpacks resulting in less natural water storage earlier in the year and higher rates of evaporation from lakes and reservoirs, as well as from groundwater aquifers. Box 4.1 presents a proposal for adaptive approaches for mitigating the impact of climate change on water management at the national and possibly the regional level.

Climate Change and Municipal Water Systems

In the regions of the earth where rainfall is decreasing, the amount of water stored in natural storage facilities is dropping at an alarming rate. Table 4.1 includes a brief descriptions of how climate change and other factors are

Table 4.1 Major cities running out of drinking water (estimates only)

Cities	2018 population	Selected discussion
Tokyo City Metro	 13,930,000 38,310,000	Main source is surface water, but majority of rain falls in 4 months; city has insufficient storage capacity to meet dry season demand of very large population; Japan experiencing more severe storm damage
Miami City Metro	 2,496,435 6,158,824	Main source is groundwater, but excessive withdrawals and history of draining Everglade have resulted in intrusion of saline water into aquifers, more extreme storms with resulting flood damage
London City Metro	 8,187,892 18,000,000	Continued population growth means demand exceeds supply of traditional surface water supply. Climate changes is resulting in more extreme storms and urban runoff to enter surface water sources
Cairo City Metro	 12,000,000 20,440,000	Almost all municipal water taken from Nile River, population growth, industrialization and dumping of raw sewage into surface sources has resulted in excessive pollution of all raw water resources
São Paulo City Metro	 12,120,000 21,730,000	Climate-change-influenced drop in precipitation, resulting in prolonged drought; rapid population growth and urban migration have resulted in demand exceeding supply
Beijing City Metro	 49,600,000 23,000,000	City located in arid north, limited rainfall and polluted surface water sources has resulted in overexploitation of groundwater resource; population growth adds further cuts in traditional supplies
Bangalore City Metro	 8,900,000 11,900,000	City losing more than half its drinking water in leaks from antiquated plumbing systems; new housing construction outpacing ability of city to provide infrastructure. Surface water polluted
Mexico City City Metro	 8,219,000 21,500,000	Population and migration to urban area outpacing ability to meet demand; from 30 to 40% of treated water is lost to leaks and spills. Long distant sources of supply add losses from leaky pipes

Source EcoWatch (2015) and WWAP (2015)

contributing to the inability of water suppliers to keep up with growing demand in eight of the world's largest cities. The municipalities in these areas are also facing higher demand for water as temperatures rise and more drought conditions appear. Higher water temperatures also result in more algae blooms, thus raising the costs for water purification treatments. The barriers to achieving sustainable water supplies make their tasks staggering. It is not surprising that there have been hundreds of studies published on the impact of climate change on municipal water systems.

Box 4.1 Proposals for preparing water management for the impact of climate change

Water utility and resource managers must plan now for the effects of climate change. High uncertainty of the impacts involved require that strategic planning be extended over several decades. A variety of possible solutions with favorable outcomes regardless of climate change should be prepared. These should include solutions that may be implemented quickly and adapted to variations of actual events. This allows policy to evolved with changing conditions and programs that build on, rather than supplant, earlier investments. "Adaptive measures that may prove particularly effective include:

- rainwater harvesting,
- conservation tillage,
- maintaining vegetation cover,
- planting trees in sloping fields,
- mini-terracing for soil and moisture conservation,
- improved pasture management,
- water recycling and reuse,
- desalination,
- more efficient soil and irrigation water management.

Restoring and protecting freshwater habitats, and managing natural floodplains, are additional adaptive measures that are not usually part of conventional management practice."
Source Miralles-Wilhelm et al. (2017, 253)

Summary

Research leaves little doubt that the earth's current climate change and observed warming is having a major impact on the global freshwater supply. To determine the scope of that impact, researchers around the globe have conducted studies and run computer simulations of condition changes to the end of this century. Historical studies show that over the last century, the earth has undergone a slow but steady increase in temperature along with a roughly 8-inch rise in sea levels. That trend has taken a dramatic increase in the pace of the increase in the earth's weather's surface temperature, including that of the world's oceans. Simulations studied around the globe have resulted in suggested that much of the world will see severe pressure on their water resources by 2050, and extreme water scarcity by 2100.

The hundreds of studies and simulations examined by the Miralles-Wilhelm et al. team noted that, while climate change is having a significant impact on world's water supplies, it is not the only, and probably not the primary, cause for the current and predicted water scarcity. Rather, they suggested that increases in water scarcity are more likely to be led by growth in water demands from agriculture, energy and municipality sectors. Also, while efforts to reduce greenhouse gas emissions will indeed bring about reductions in freshwater withdrawals in some regions, other regions are likely to see increases in withdrawals.

An adaptive, flexible approach to strategic planning for dealing with the impacts of climate change over periods as long as 30 years has been recommended by the IPCC. They include developing several alternative scenarios in which potential solutions are necessary for individual and related problems that are likely to occur at different times over several decades in the future.

Climate change is here and must be dealt with. Efforts to mitigate the effects of climate change on the world's water supply begin with managing future risks associated with extreme water scarcity and adopting strategies that emphasize resiliency and sustainability. These strategies must be designed for the time and place in which they will occur; no single strategic approach is possible.

References

EcoWatch. 2015. Eight major cities funning out of water. Accessed July 14, 2018 from www.ecowatch.com/8-major-cities-running-out-of-water-1882061458.html.

EPA (Environmental Protection Agency). 2016. Climate impacts on water resources. US Environmental Protection Agency. Accessed November 24, 2017 from https://19january2017snapshot.epa.gov/climate-impacts/climate-impacts-water-resources_.html#main-content.

EPA (Environmental Protection Agency). 2017. Climate impacts on coastal areas. Environmental Protection Agency. Accessed November 28, 2017 from https://19january2017snapshot.epa.gov/climate-impacts/climate-impacts-coastal-areas_.html.

Jain, Sharad, K. 2011. Population rise and growing water scarcity in India—Revised estimates and required initiatives. *Current Science* 101 (3): 271–276. National Commission for Integrated Water Resources Development (NCIWRD), India Ministry of Water Resources.

Jaramillo, Pablo, and Ali Nazemi. 2018. Assessing urban water security under changing climate: Challenges and ways forward. *Sustainable Cities and Society* (Forthcoming): 1–12.

Karl, Thomas R., and Kevin E. Trenberth. 2005. What is climate change? In *Climate Change and Biodiversity*, ed. Thomas E. Lovejoy and Lee Hannah, 15–28. New Haven: Yale University Press.

Miralles-Wilhelm, Fernando, Leon Clarke, Mohamad Hejazi, Sonny Kim, Kelly Gustafson, Raul Muñoz-Castillo, and Neal Graham. 2017. *Physical Impacts of Climate Change on Water Resources*. Washington, DC: Global Water Practice, World Bank Group. Accessed October 16, 2017 from file:///C:/Users/David/Downloads/112726-WP-PUBLIC-Physical-impacts-of-climate-change-on-water-resources.pdf.

NASA (National Aeronautics and Space Administration). 2016. Study maps hidden water pollution in the U.S. National Aeronautics and Space Administration. Accessed November 13, 2017 from www.nasa.gov/feature/jpl/study-maps-hidden-water-pollution-in-us-coastal-areas.

WWAP (United Nations World Water Assessment Programme). 2015. *The United Nations World Water Development Report 2015: Water for a Sustainable World*. Paris: UNESCO. Available at http://unesdoc.unesco.org/images/0023/002318/231823E.pdf.

5

The Population Growth Barrier

Population growth is one the most significant barriers to achieving water sustainability. It is not that there is not enough water to meet the needs of the billions more people now inhabiting the earth. But rather that most—98%—of that water is unfit for human consumption; it is salt water. Or it falls in either the wrong places, at the wrong time, or in the wrong quantities. And despite a significant decline in the rate of population growth, every year for the rest of this century more people will demand a share of the limited water resource that is available. Another 83 million people are added to the world's population every year. The United Nations reports that in 2017 there were 7.6 billion people in the world, and more than half of them resided in an urban area. Figure 5.1 displays world population growth since 1750.

Population Growth Rate

The world population growth is rapidly overcoming the ability of many regions of world to meet food and water needs. The world population was 7.3 billion in 2015 and 7.6 billion in 2017. The world population will increase by approximately one billion people by 2030, bringing the total up to 8.5 billion. With more people reaching child-bearing age and high fertility rates in Asia and Africa, the world population is expected to reach 9.7 billion in 2050 and 11.2 billion by 2100. More people result in increasing overall demands for all water sectors.

D. E. McNabb, *Global Pathways to Water Sustainability*,
https://doi.org/10.1007/978-3-030-04085-7_5

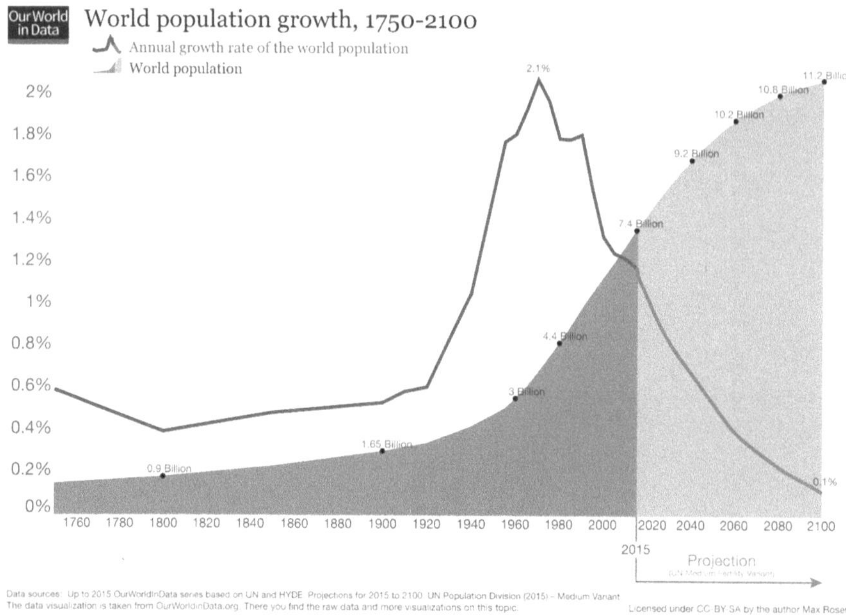

Fig. 5.1 World population growth and global growth rate, 1750–2100 (*Source* Roser and Ortiz-Ospina [2017]. From UN and the Netherland Environmental Assessment Agency's HYDE database)

Much of the growth will occur in regions and countries least capable of providing the investments need for improving water access for the growing population. In 2011, more than 2 billion people in these regions were already living in areas suffering from severe water stress. With agriculture in these areas accounting for more than 70% of freshwater from all sources, the numbers of people living in regions with less than 1000 cubic meters of available freshwater (extreme water stress) per person will only increase.

Countries with declining populations face a different set of problems, led by the declining numbers of a population over which to spread the cost of repairing and replacing water and sanitation infrastructure. Syria led the list of 20 countries with the highest rate of population loss in 2017 with a decline of 9.73%. Oman, with a population 4.6 million in 2017, was one of the fastest growing countries on the globe, with a one-year population increase of 4.78%. At least six African countries had one-year growth rates greater than 3% from 2016 to 2017: Angola, Burundi, Malawi, Uganda, Niger and Burkina Faso (Table 5.1). Eleven of the remaining nations with declining populations were in Eastern Europe, led by Moldavia with a 1.02% decline and Bulgaria with a 0.83% loss. The three Baltic countries all had declining populations: Estonia, 0.68%; Latvia, 0.62%; and Lithuania, 0.29%.

Table 5.1 Selected countries with the high population growth rates in 2017

Country	% 2016–2017 growth	2017 population	Urban population (%)
Oman	4.78	4,636,262	73
Angola	3.50	29,784,193	35
Burundi	3.30	10,864,245	13
Malawi	3.30	18,622,104	16
Uganda	3.20	42,222,947	16
Niger	3.20	21,477,348	18
Chad	3.10	14,899,994	21
BR Congo	3.31	81,339,988	39
Burkina Faso	3.00	19,193,382	29
Mali	3.00	18,541,980	37
Zambia	2.90	17,094,130	39
Ethiopia	2.90	104,918,090	19
Tanzania	2.80	57,310,019	30
Benin	2.79	11,175,692	44
Madagascar	2.72	25,570,895	35
Iraq	2.60	38,274,618	67
Guinea	2.60	12,717,146	37
Egypt	2.50	97,553,151	38

Source Statista (2018) and Worldmeters.info (2018)

The UN's 2010 Water for Life report on the water supply problems of attaining water sustainability under the adverse conditions associated with urbanization included the following warning:

> Two main challenges related to water are affecting the sustainability of human urban settlements: the lack of access to safe water and sanitation and increasing water-related disasters such as floods and droughts. These problems have enormous consequences on human health and well-being, safety, the environment, economic growth and development. The lack of adequate water and sanitation facilities leads to health issues such as diarrhea, malaria and cholera outbreaks. Though water supply and sanitation coverage increased between 1990 and 2008, the growth of the world's urban populations jeopardizes those results. While between 1990 and 2008, 1052 million urban dwellers gained access to improved drinking water and 813 million to improved sanitation, the urban population in that period grew by 1089 million people. (UNDESA 2014)

The annual global growth rate peaked at 2.1% in the 1960s and has declined ever since. By the end of this century, the global growth rate is expected to be less than one tenth of 1% per year. The world's population is expected to grow to 8.6 billion by 2030, 9.8 billion in 2050, and 11.2 billion in 2100. Providing a sustainable supply and the infrastructure needed for distributing adequate supplies of clean, fresh water, as well as the facilities for collection, treatment and discharge of wastewater into already polluted ecosystems will tax the capabilities and resources of city governments to meet the demand city dwellers will require.

Nations with the Largest Populations

Ten nations are expected to account collectively for more than half of the world's projected population increase over the period 2017–2050: India, Nigeria, the Democratic Republic of Congo, Pakistan, Ethiopia, the United Republic of Tanzania, the USA, Uganda, Indonesia and Egypt (ordered by their expected contribution to global growth). The top ten fastest-growing countries accounted for three-quarters of the total population estimates in 2017. The largest three nations in both periods are China, India and the USA.

Growth in Water Demand

In 2018, half of the world's population lives in cities. This percentage is expected to continue to increase at about the same rate at least until 2050. The water use of that growing urban population will occur at a rate faster than the population. Demand for water will grow from population growth, growth in industrial activity to support city life, and for distributing the energy needed for cooling and for producing more energy. Cities are particularly adversely impacted by climate change; the average annual temperature in cities is often from two to six degrees higher during daylight hours than nearby rural areas, and up to five degrees warmer during the night. More water will thus be need for cooling, washing, drinking, and in the thermoelectric plants that produce electricity.

The Supply–Demand Gap

The gap between the growing population demand for water and the shrinking supply of freshwater will require greater water-use efficiency. The highest population growth rate will occur in three countries that are among the ten nations with the largest populations and, coincidently, the greatest need for water: China (1), India (2), and Nigeria (7). What these countries are doing to mitigate the impact of population growth on their water supplies range from greater use of recycled water, changes in agriculture water supplies, and an on-going search for alternate supplies, including increased use of desalinated seawater and brackish surface water.

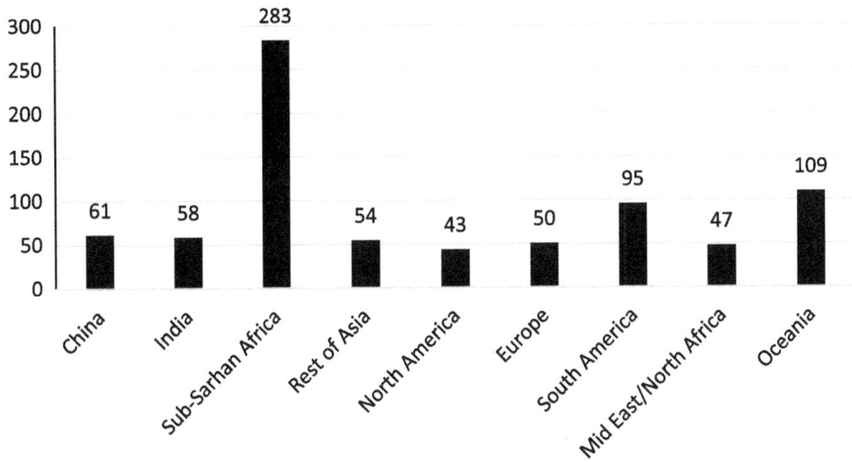

Fig. 5.2 Estimated increases (%) in annual water demand, 2005–2030 (*Source* 2010 Water resources global water demand)

The combination of population growth and the increases in demand for irrigation water by the agriculture sector to feed the growing population, is expected to increase global demand for water by approximately 50% above the total demand in 2050. The distribution of the percentage increases in demand are shown in Fig. 5.2. Demand increases in sub-Saharan Africa are nearly three times as great as demand increases in the next two regions listed: 109% in Oceania and 95% in Latin America. The smallest expected increase—43%—is in North America.

Most Water-Stressed Regions

This section looks at population growth and water demand in areas where the greatest increases are projected, beginning with the combined Middle East and North Africa region. Data sources range from the United Nations, the World Bank, World Population Review, and additional U.S. government and public domain sources. The most water-stressed regions and nations are identified using World Resources Institute (WRI) water stress index (WRI 2015). The WRI index is a measure of competition for and depletion of surface water in 167 countries by 2020, 2030 and 2040. In 2014, 33 countries were in states of high water stress. Chile, Estonia, Namibia and Botswana could face an especially significant increase in water stress by 2040.

Water and Growth in the Middle East and North Africa

The Middle East and North Africa (MENA), already facing high water stress, is the region expected to be most severely impacted by a water shortage by 2040. Population data for the Middle East and North Africa varies according to the date of publication and the regional inclusions of the nations included. Published MENA statistics range from 19 to 22 nations, spreading from Morocco on the West to Iran in the East. Other maps include predominantly Muslim countries in Southwest Asia in this region.

The 2017 UN population statistics for the region includes data for 19 countries. The MENA region holds from 5 to 6% of the world's population, but only 1% of the world's renewable water resources (Pedraza and Heinrich 2016). It also contains 60% of the world's oil reserves and 45% of the world's natural gas reserves, although not all countries in the region benefit from this carbon-based resource.

In addition to not having a secure and sustainable water supply, the MENA region also suffers from inefficient water use and mismanagement, recurring drought, surface water pollution, and what are described as 'inappropriate' legal political and economic frameworks incapable of managing the region's water problems. Some 85% of the region's available surface and groundwater is now taken up by the inefficient agricultural sector, which depends on groundwater almost exclusively, and is rapidly depleting the region's aquifers. Water researchers also consider the entire region to be extremely vulnerable to the adverse effects of climate change. In addition to all of this, the region continues to be subject to extensive civil strife.

Fourteen of the world's nations facing or about to face severe water stress are in the MENA region; nine are considered extremely highly stressed with maximum scores of 5.0: Bahrain, Kuwait, Palestine, Qatar, United Arab Emirates, Israel, Saudi Arabia, Oman and Lebanon. From 1950 to 2000, population grew rapidly in most of the MENA region, averaging from 2% to nearly 4% annually. However, that growth rate is expected to slow to an average of 1% per year from 2020 to 2030, after which time it is expected to be closer to 0.5%. The region depends heavily on groundwater and desalinated sea water.

Water and Population Growth in Sub-Saharan Africa

Sub-Saharan Africa (including central Africa) is the fastest growing population region on the globe. The sub-Saharan African continental area includes 51 nations, from Western Sahara in the Northwest to South Africa in the Southeast. Also included are the Indian Ocean: the Seychelles, Mauritius and Comoros islands and the large off-shore island of Madagascar. The population in 2017 was 1.02 billion, which was 13.64% of the total world population. The major countries in this region and their 2018 estimated populations are Nigeria, 193.5 million; Ethiopia, 106.3 million; DR Congo, 83.6 million; Tanzania, 58.3 million; and Uganda, 43.6 million. By 2050, Nigeria is projected to be larger than the USA by about 30 million people.

Sub-Saharan Africa includes some of the most water-stressed nations on the globe. Overwhelmingly dependent upon rainwater for irrigation, the region lacks the infrastructure to collect water from where it is available and distribute to where it is needed. At the same time, the region's population growth and high urbanization rates make it even harder to provide the needed freshwater and sanitation services. Also contributing to the high degree of water stress in some states are weak governments, corruption, mismanagement of resources, inefficient irrigation practices, poor long-term investment strategies, and a lack of environmental research and a commitment to environmental protection (CFR 2017).

Regional Water Stress

Areas particularly hard hit by lack of water supplies are the northern regions that extend across the continent and which have long had to live with encroachment of the Sahara Desert. Additional stress is due to historic conflicts between settled farming groups, nomad, and livestock grazing groups, all of which compete for restricted water resources. Lack of access to clean, safe water and sanitation facilities continues to result in widespread exposure to waterborne diseases such as cholera throughout much of the region.

With independence and subsequent border restructuring in the twentieth century, existing water courses were designated borders between groups. In many cases, this has contributed to competition for water supplies. Inability to reach agreement on water access and storage have resulted in failure to invest in the necessary storage and other infrastructure development.

Although construction of large dams has been possible in much of the region, capital has not been available to carry out large water management actions. For example, of the 980 large dams that exist in sub-Saharan Africa, close to 589 of these are in South Africa. Tanzania, with roughly the same area and water management potential as South Africa, has only two large dams.

Water and Population Growth in Latin America

The United Nations groups these 33 nations in their statistical reports on the Latin America and Caribbean countries (UN 2017). The impact of population growth on water supplies and wastewater services are similar in some parts of North America, Central America and South America. By 2030, population growth in these countries will add close to 100 million people to their 2018 population of 645.6 million. Most of this growth will occur in cities. The large Latin American nations are urbanizing at a rapid pace. The population of Colombia was estimated to be 75 million in 2018, more than 62 million in Argentina; both Peru and Venezuela was estimated to be more than 42 million each; and Chile's population will be greater than 20 million. Most of the smaller countries are already highly urbanized.

While much of the western hemisphere has plenty of water, population growth and climate change are creating supply problems for some. In North America, Mexico and the Southwestern region of the USA are experiencing water scarcity. In Central America, El Salvador is having the greatest overall difficulty meeting municipalities' demand for water, although citizens in poorer regions of some states are facing similar difficulties.

Water Stress in Mexico

The population of Mexico, the world's tenth largest nation, was more than 130 million in January of 2018. Table 5.2 contains population statistics for Mexico from 2010 to 2018. At present rates of growth, Demographers predict Mexico will have a population of from 148 to 150 million by 2030, and 164 million in 2050.

In 2018, the population of the metropolitan area of Mexico City was 21.2 million, making it one of the Western Hemisphere's largest urban areas. Metropolitan Mexico City consists of the city proper and some 70 other districts or municipalities. Conaqua, Mexico's national water commission, distributes water to the metro area. Conagua was established in 1989 to manage

Table 5.2 Population growth estimates for Mexico, 2020–2050

Year	Population estimates	Yearly change (%)	Urban (%)	Urban population estimates
2020	133,870,027	1.24	79.4	104,297,564
2025	141,131,503	1.06	80.1	112,981,354
2030	147,540,127	0.89	80.7	119,016,210
2035	153,060,606	0.74	81.2	124,245,356
2040	157,689,665	0.60	81.6	128,609,814
2045	161,436,237	0.47	81.9	132,138,956
2050	164,279,302	0.35	82.1	134,827,995

Source From material in Worldmeters.info.org (2018) and World Population Review (2018)

and protect the country's water resources and to ensure their sustainable use. In 2008, the organization announced that 88% of the water it delivered was chlorinated and safe for human consumption. However, subsequent research indicated that only 60% of the treated water was suitable for human use; the 40% not delivered was due to leaks in the system (Spring 2011).

> Many poor communities (in the Mexico City metropolitan area) not only lack treated water access service, they lack access to any supply of water suitable for human consumption. Exacerbating the problem are the environmental effects of water policies that have put the whole metropolitan area at risk. These policies have impoverished the ecosystem and are threatening the precarious equilibrium that determines the rate at which the city sinks into the lakebed below. Historical neglect, political venality and ignorance have contributed to this crisis, permitting the systematic deforestation of the surrounding mountains and the construction of communities in the ravines and alongside the lakebeds through which water descends during the rainy season. The city discharges virtually all of its wastewater through a costly drainage system into the rivers of a nearby state, where forage and food crops are grown for the Mexico City market. Thus, the city's water management problems not only threaten the viability of Mexico City, but also jeopardize the well-being of outlying communities and environments. (Barkin 2014)

Water Stress in Central America

Water problems in the seven countries of Central America, Belize, Costa Rica, El Salvador, Guatemala, Honduras, Nicaragua and Panama, are being impacted by a combination of population growth, deforestation, urbanization and extreme weather events from climate change-influenced changes in precipitation patterns. Despite the heavy rainfall the isthmus receives every year, population growth and climate change are contributing to water shortages in sections of the region. El Salvador, with a population of 6.1 million, is facing a general water shortage (Ayala 2010). El Salvador population growth rates were 1.75% in 2016 and 2.16% in 2017, with nearly 70% residing in urban areas.

If these relatively high growth rates continue in El Salvador, the demand for water due to population growth alone will increase by 300% by 2050, and by more than 1600% by 2010. When local effects of climate change are added, an additional 20% could be added to the demand for water, while the total availability of renewable water could drop as much as 63% from 2010 levels. El Salvador is the most de-forested of the Central American countries. This results in rapid runoff, reduced aquifer recharge, and low-quality water (Vidal 2017).

Other countries with population growth and water problems are Honduras, with an estimated population of 9.3 million in 2018, a growth rate of 1.67% and 50.1% urbanized; and Nicaragua, with a population of 6.3 million, a growth rate of 1.08% and 61% urbanized. The central water problem for these two nations is the lack of access to freshwater by large elements of their rural populations.

Water Stress in South America

South America is one of the world's most water-blessed regions. However, for many reasons, populations in many parts of the continent have long had to deal with freshwater shortages. Water supplies in Ecuador, Bolivia, Chile and Peru are all dealing with problems of allocating water between mining operations, agriculture and municipalities. Adding to the water access problems from this unequal distribution of water supplies are pressures being put on the region's water resources by population growth, climate change, urbanization, unregulated water uses by mining firms and other industries, pollution of surface and groundwater resources, reduction in river flows from construction of many high dams, and commercial and inefficient agriculture (Barlow and Clarke 2004).

South American leaders have long been aware of these and other water scarcity problems. John Vidal discussed many of them in a 2017 article published in *The Guardian*: "In South America, human-induced salination of groundwater is causing desertification in significant parts of Peru, Bolivia and northwestern Argentina ... In most large cities, over 50 percent of the water supply is lost through infrastructure leakage. Some cities lose almost 90 percent through leaky pipes ... The destruction of water sources, combined with inequitable access, has left most Latin Americans 'water poor.' Millions live without access to clean water at all. While the region's available resources could provide each person with close to 110,500 cubic feet of water every year, the average resident has access to only 1010 cubic feet per year. This compares to North America's annual average of 4160 cubic feet and Europe's 2255.6 cubic feet" (Vidal 2017).

Chile is the only Latin American country in the World Resource Institute's list of the top 33 countries expected to suffer from water stress by 2040. It was one of the first South American nations to privatize its water systems. That water stress danger is considered to be a result of a combination of climate change, shifting precipitation patterns, and industrial water withdrawals, rather than population growth or urbanization. Local farm groups protest that water use in Ecuador, Peru and Bolivia traditionally used for agriculture is now being re-directed to industries and mining firms (Vidal 2017).

Water and Population Growth in Asia

Five of the world's most populous countries in 2017 are in Asia: China, India, Indonesia, Pakistan and Bangladesh. Large portions of the population in these countries are expected to see significant growth in the total numbers of people living under conditions of high to extreme water stress. Other non-Middle East Asian countries with some degree of water scarcity include Singapore, Kyrgyzstan, Macedonia, Kazakhstan and Turkmenistan (WRI 2015).

China, with an estimated population in 2018 of 1.42 billion, was the world's largest country (Table 5.3). However, it may soon lose that distinction to India, largely due to China's one-child policy. That policy and economic development are said to have resulted in a growth rate close to less than 0.5%. Over the same period, India, with a population of 1.347 billion in 2017 has a growth rate of 1.33%. This will result in India having a larger population than China in 2022 (previous estimates said this would not happen until 2025).

The rate of population growth in China has been declining since 1985. Annual growth was relatively stable at near to or slightly above 2% throughout most of the 1960s and 1970s. The peak rates occurred in 1968, 1969 and 1970, when annual growth percentages were 2.80, 2.84 and 2.71%, respectively. As currently indicated, beginning around 2035, the population of China is no longer expected to be increasing. Rather, declines are expected of from −0.11% in 2035 to −0.43% in 2050.

Table 5.3 Population growth in China, with forecasts to 2050

Year	Population	Growth rate (%)
1975	896,856,603	2.05
1980	970,996,305	1.43
1985	1,043,975,491	1.60
1990	1,144,866,891	1.80
1995	1,221,954,152	1.03
2000	1,260,344,112	0.58
2005	1,302,086,831	0.54
2010	1,337,387,900	0.54
2015	1,372,342,307	0.50
2020	1,424,548,266	0.39
2025	1,438,835,627	0.20
2030	1,441,181,813	0.03
2035	1,433,608,888	−0.11
2040	1,417,472,814	−0.22
2045	1,394,456,025	−0.33
2050	1,364,456,223	−0.43

Source China Statistics, World Bank, Worldmeters.info

The computer model used in the MIT study projected surface water demands for agriculture, industry and municipalities, as influenced by changes in three factors: climate change alone, population growth alone, and a combination of the two. The final analysis suggested that population and economic growth will have greater adverse effects than climate change shifts on water stress. Population growth will result in many more people currently living under moderate water stress will be living under extreme water stress in the near future. As expected, the combined effects will result in the greatest increase in the numbers of population living under water stress. The only model-predicted decrease in the numbers of people not moving to greater water stress occurred only as a reduction in total population rather than the expected increase (Table 5.3).

Possible Relief

The MIT study concluded with a positive note, although it suggested that the projected results, while certainly of great concern for leaders in Asia, "Do not necessarily imply an insurmountable future for this region. Through climate mitigation, and perhaps more importantly, proper planning and financing of adaptive and protective measures for these anticipated shortages ... these future water systems can be augmented to better ensure their resiliency and sustainability. Addressing these options for the future, however, will require substantial research and additional experimentation with integrative tools [as the models used in the analysis]" (Fant et al. 2016, 31).

Summary

Population growth, together with urbanization and the effects of climate change are resulting in water scarcity and other water-related problems in many parts of the globe. Particularly hard hit is the MENA region, sub-Saharan Africa, China, India, small islands of Oceania, and the Southwestern USA. The estimated world's population of 7.63 billion in 2018 is projected to grow to 8.6 billion in 2030, to 9.8 billion in 2050, and to more than 11 billion by the end of this century. Political leaders are already having difficulty achieving a sustainable balance of the water needed to maintain economic growth, water for agricultural irrigation, and the needs of the millions of people crowding into cities, town and mega-urban concentrations. Ten nations will account for more than half of the world's population in 2050; the five largest in 2017 were China, India, the USA, Indonesia and Brazil. The nations expected to have the greatest difficulty in achievieng water sustainability in the future were China, India and Nigeria.

Led by China, many countries are implementing a variety of policies to mitigate the impact of population growth on their water supplies. The amount of water used for agriculture in China has decreased significantly. These range from greater use of recycled water, desalination of salt water and brackish water, shifts from high water use crops and more efficient irrigation methods, and an on-going search for alternate supplies, including increased use of ground water. Over-dependence upon groundwater for all sectors without sufficient recharging is depleting many aquifers. In addition to finding sustainable water supplies, developing the infrastructure necessary to provide freshwater, sanitation facilities, and the food needed by the world's fast-growing population make achieving sustainability difficult, but not impossible.

References

Ayala, Edgardo. 2010. El Salvador: Most water-stressed country in Central America. International Press Service (IPS). Accessed January 25, 2018 from www.ipsnews.net/2010/12/el-salvador-most-water-stressed-country-in-central-america/.

Barkin, David. 2014. Mexico city's water crisis. North American Congress on Latin America (NACLA). Accessed January 29, 2018 from https://nacla.org/article/mexico-citys-water-crisis.

Barlow, Maude, and Tony Clarke. 2004. The struggle for Latin America's water. *Global Policy Forum*. Accessed January 26, 2018 from www.globalpolicy.org/social-and-economic-policy/global-public-goods-1-101/46052-the-struggle-for-latin-americas-water.html.

CFR (Council on Foreign Relations). 2017. Water stress in sub-Saharan Africa. Accessed January 12, 2018, from www.cfr.org/backgrounder/water-stress-sub-saharan-africa.

Fant, Charles, C. Adam Schlosser, Xiang Gao, Kenneth Strzepek, and John Reilly. 2016. Projections of water stress based on an ensemble of socioeconomic growth and climate change scenarios: A case study in Asia. MIT Joint Program on the Science and Policy of Global Change. Accessed January 13, 2018 from https://globalchange.mit.edu/sites/default/files/MITJPSPGC_Reprint_16-6.pdf.

Pedraza, Lisdey Espinoza, and Markus Heinrich. 2016. Water scarcity: Cooperation or conflict in the Middle East and North Africa? *Foreign Policy Journal* (September 2). Accessed January 25, 2018 from https://www.foreignpolicyjournal.com/2016/09/02/water-scarcity-cooperation-or-conflict-in-the-middle-east-and-north-africa/.

Roser, Max, and Esteban Ortiz-Ospina. 2017. World population growth. Published online at OurWorldInData.org. Accessed December 3, 2017 from https://ourworldindata.org/world-population-growth/.

Spring, Ursula Oswald. 2011. *Water Resources in Mexico: Scarcity, Degradation, Stress, Conflicts, Management, and Policy*. New York: Springer.

Statista. 2018. The twenty countries with the largest population growth rates in 2017 (compared to the previous year). Accessed January 7, 2018 from www.statista.com/statistics/264687/countries-with-the-highest-population-growth-rate/.

UN (United Nations). 2017. World population prospects. Accessed October 20, 2017 from https://esa.un.org/unpd/wpp/Publications/Files/WPP2017_KeyFindings.pdf.

UNDESA (United Nations Department of Economic and Social Affairs). 2014. Water and cities. Accessed January 8, 20178 from www.un.org/waterforlifedecade/water_cities.shtml.

Vidal, John. 2017. As water scarcity deepens across Latin America, political instability grows. *The Guardian*, March 1. Accessed January 26, 2018 from www.theguardian.com/global-development-professionals-network/2017/mar/01/water-scarcity-latin-america-political-instability.

Worldmeters.info. 2018. Countries in the world by population. Accessed January 7, 2018 from http://www.worldometers.info/world-population/population-by-country/.

WPB. 2017. Sub-Saharan population. *World Population Review*. Accessed January 12, 2018 from http://worldpopulationreview.com/continents/sub-saharan-africa-population/.

WRI (World Resources Index). 2015. Environmental democracy index. Accessed December 6, 2018 from www.wri.org/publication/environmental-index.

6

The Urbanization Barrier

Global urbanization is another of the most severe barriers to achieving water sustainability; it has been described as "one of the most important demographic trends of our time" (Senn and Spuhler 2018). In 2008, the number of people living in urban areas surpassed those in rural areas for the first time in history. The percentage is continuing to climb. The urban population in 2015 accounted for 54% of the total global population, up from 30% in 1950, and is expected to be 60% of world population by 2030. This makes urbanization one of the most pressing problems for the world's water managers (WHO 2016; Bai et al. 2017).

The last chapter examined the problems that much of the world faces in the efforts to supply enough water to meet the needs of a fast-growing population for which the existing methods and sources of supply will not be sufficient. This chapter seeks to explain how urban migration is exacerbating present and future water problems. It begins with urbanization and water problems in India, followed by China, Pakistan, Iran, Indonesia and in states of the south and southwestern USA.

Urbanization Concentration

The majority of population growth taking place in this century will occur in urban areas, many of which are located in less developed regions. Urbanization in these regions add to the developing world's water and sanitation services problems. Water suppliers are generally unable to provide

D. E. McNabb, *Global Pathways to Water Sustainability*,
https://doi.org/10.1007/978-3-030-04085-7_6

the resources for supplying needed freshwater even where sources of supply are or might be available. Nor are they able to keep pace the growing need to supply their populations with the necessary wastewater collection and treatment facilities. One result is heightened pollution from discharging untreated sewage into existing water courses. Untreated sewage is a major source of waterborne disease in these regions.

In developed urban areas, the high cost of repairing and replacing aged water and wastewater infrastructure can result in money-saving management decisions, which in turn can result in providing highly polluted water to residents, the unintended high cost of which were illustrated in the water source change that took place in Flint, Michigan, USA. Much of the world's developed urban areas must deal with water and wastewater infrastructure that is more than 50 years old—even more than 100 years old in some locations.

Urban water supply and wastewater service providers in many developing countries are already strained under current demands and face formidable obstacles to achieving water sustainability (WSP 2018). This lack of service is often simply a product of the financial inability of providing the infrastructure needed to keep up with the pace of this growing demand. Often, unsanctioned slums spring up on the fringes of urban centers, adding to the difficulties. In many water-stressed regions informal providers and small-scale independent providers (SSIPs) are popping up to meet the demands of the population for water supply, especially in the slums. In most cases, no effective regulation of the informal providers exists. The World Bank's Water and Sanitation Program reports that sewage collection and treatment present an even greater challenge; where public toilet facilities have been constructed, they are often not properly managed and soon fall into disrepair and disuse.

Urbanization Defined

It is hard to pin down what is a city and what is a town. Governments reporting the statistics apply widely different criteria for determining whether an area is urban. Urbanization describes the movement of people from rural environments to a centralized urban area. People move to urban areas to benefit from the economic activity missing in rural areas. The degree of a nation's urbanization is measured by a change in the percentage of a distinct population that resides in an urban area. Urban areas are cities and towns, and, in some regions, urban areas also include villages, many of which face three main water-related sustainability challenges:

1. Ensuring a stable, reliable and adequate supply of clean, safe water for their residents and other sectors.
2. Securing provision of sanitation facilities that include systems for treating wastewater before returning used water to supply sources.
3. Providing aid to residents suffering from physical harm, disease and loss of habitat from water-related extreme weather events such as hurricanes, tornados, floods and droughts. The urban poor suffer most from these challenges.

Urbanization has become one of the most significant trends of the twenty-first century, impacting global development, energy production and consumption, agriculture and water resources (McDonald et al. 2014). The world's largest urban area, Tokyo, Japan, must meet the water and sanitary needs of nearly 38 million residents. However other cities are not far behind, as indicated by the present and future data for the world's 30 largest cities in Table 6.1. In 2030, 15 of the world's largest cities will be in Asia; five each will be in China and India. Most of the sustainability problems resulting from the urbanization trend will occur in countries in regions of Africa and Southeast Asia. The two European cities on in 2017, Moscow (21) and Paris (25), were no longer among the world's 25 largest cities 2030 list.

Two of the 25 most densely populated urban areas in 2017 were in the USA: New York City (9), with an estimated 18.6 million residents, and Los Angeles (20), with an estimated population of 12.3 million. In the projections for the 25 largest urban areas in 2030, New York City remains on the list, moving to fourteenth place with an expected population of 19.9 million.

Box 6.1 The race to an urbanized world

Urban areas are expected to absorb all the world's population growth, at least as far in the future as 2050. Two demographic trends are shaping this mass movement of people: population growth and continued rural-to-urban migration.

According to the United Nations, "The vast majority of [these new town and city dwellers] will be living in crowded slums with inadequate or non-existent, water and sanitation services.

Safe drinking water systems and adequate sanitation that effectively disposes of human waste will be essential to ensure cities and towns grow sustainably. Extending these services to the millions of urbanites currently unserved will play a key role in underpinning the health and security of cities, protecting economies and ecosystems and minimizing the risk of pandemics."

Although significant global progress has occurred in bring fresh water and sanitation services to developing nations, many government water and wastewater infrastructure planners "have been unable to keep pace in many regions. Today, 700 million urbanites live without improved sanitation, contributing to poor health conditions and heavy pollution loads in wastewater, and 156 million live without improved water sources."

Source UN Water (2017)

Urbanization Problems

The litany of problems ensuing from this demographic trend is large and growing larger (Box 6.1). Senn and Spuhler compiled the following list accumulated from studies by the United Nations, the World Health Organization and other published sources in 2017. The list was subsequently published online by the private, non-profit organization, Sustainable Sanitation and Water Management (SSWM). Table 6.2 lists the issues in the original order of importance.

Table 6.1 The world's 25 largest urban areas in 2017 and projections to 2030

	City	2017		City	2030
1	Tokyo, Japan	37,800,000	1	Tokyo, Japan	37,200,000
2	Delhi, India	25,000,000	2	Delhi, India	36,100,000
3	Shanghai, China	23,000,000	3	Shanghai, China	30,800,000
4	Mexico City, Mexico	20,800,000	4	Mumbai, India	27,800,000
5	São Paul, Brazil	20,800,000	5	Beijing, China	27,700,000
6	Mumbai, India	20,700,000	6	Dhaka, Bangladesh	27,400,000
7	Osaka, Japan	20,100,000	7	Karachi, Pakistan	24,800,000
8	Beijing, China	19,500,000	8	Cairo, Egypt	24,500,000
9	New York, USA	18,600,000	9	Lagos, Nigeria	24,200,000
10	Cairo, Egypt	18,400,000	10	Mexico City, Mexico	23,900,000
11	Dhaka, Bangladesh	17,000,000	11	São Paul, Brazil	23,400,000
12	Karachi, Pakistan	16,100,000	12	Kinshasa, DER Congo	20,000,000
13	Buenos Aires, Argentina	15,000,000	13	Osaka, Japan	20,000,000
14	Kolkata, India	14,800,000	14	New York, USA	19,900,000
15	Istanbul, Turkey	14,000,000	15	Kolkata, India	19,100,000
16	Chongqing, China	12,900,000	16	Guangzhou, China	17,600,000
17	Rio de Janeiro, Brazil	12,800,000	17	Chongqing, China	17,400,000
18	Manilla, Philippines	12,800,000	18	Buenos Aires, Argentina	17,000,000
19	Lagos, Nigeria	12,600,000	19	Manilla, Philippines	16,800,000
20	Los Angeles, USA	12,300,000	20	Istanbul, Turkey	16,700,000
21	Moscow, Russia	12,100,000	21	Bangalore, India	14,800,000
22	Guangzhou, China	11,500,000	22	Tianjin, China	14,700,000
23	Kinshasa, DER Congo	11,100,000	23	Rio de Janeiro, Brazil	14,200,000
24	Tianjin, China	10,900,000	24	Chennai (Madras), India	13,900,000
25	Paris, France	10,800,000	25	Jakarta, Indonesia	13,800,000

Sources United Nations (2017) and Rosenburg (2017)

Table 6.2 Sustainability problems associated with rampant urbanization

Problem	Discussion
Disruption of the natural water cycle	The conventional water cycle involves evapotranspiration and natural aquifer recharging. Sealed soils and impervious services in urban areas results in a cycle that involves clean water transportation in and wastewater out of urban areas
Pollution of surface water sources	Urban centers are a major point source of pollution of surface waters. More than 90% of untreated urban sewage in the developing world is discharged directly into the same water sources from which water is withdrawn. In many low-income areas, untreated wastewater seeps into aquifers, polluting groundwater and spreading disease
Depletion of groundwater sources	In many urban areas around the globe, water is drawn from both surface water and groundwater sources. Shallow local wells in many areas dominate sources in many developing areas, resulting in withdrawals that exceed any natural or recycled water recharging
Broken soil nutrient cycles	In most urban areas, water once used for non-potable purposes is discharged without reuse. This results in potential soil nutrients being lost to surface water bodies, where concentrations result in algal blooms and oxygen depletion
Waste of resources	Water treated for human consumption is used for dishwashing and bathing. This graywater is commonly discharged with black water, with little water recycling. Fixed rates for fresh water provide little incentive for conservation
High water demand	When large numbers of people are concentrated in high-density urban areas, the need to provide fresh water and sanitary services is very high. Moreover, research shows that per capita water use is higher in urban areas than in rural areas
Cost-intensive infrastructure installation and maintenance	Urban growth calls for continuous expansion of water and wastewater infrastructure. The current systems of centralized networks are very costly. Older urban systems often suffer from freshwater leakage as high as 60%

Source From material in Senn and Spuhler (2018)

Megacities, Cities and Towns

The UN defines megacities as urban centers with 10 million or more residents. Few governments of these very large urban centers are able to provide new and repaired water and infrastructure fast enough to keep up with the largely unofficial growth. In megacities, the quantity and quality of water are often threatened by population growth, land use changes, inefficient farming practices, deforestation, erosion, pollution of surface water ecosystems, missing infrastructure planning, poor or missing water management, political conflicts, and permissive or corrupt government policies or practices (Spring 2011).

However, repairing, replacing or expanding water and wastewater services are not just a concern for the growing number of very large urban centers. Rapid urban growth is also taking place in the thousands of smaller urban and semi-urban communities (Tortajada 2008). It is in these smaller cities and towns where urban growth is often creating the greatest need for new water and wastewater infrastructure. However, few of the local governments have the resources to finance the improvements.

An example of what many of these communities' face is the experience of a small community of roughly 550 residences in a western U.S. state. The community had to borrow nearly US$1.5 million for a new 185-foot deep well and water filtration system. Overall, the community's water and wastewater systems include assets valued at more than US$4.5 million, but few banks are willing to take a water system as collateral.

This migration of more and more people to urban areas is a global phenomenon. The UN's population division reports that in 2016, nearly half of the world population lived in urban centers. Urbanization increases have occurred in all major geopolitical regions of the world over the last two-thirds of a century. In 2016, North America was the most urbanized of the nine major areas of the world with 82% of its population living in urban areas. This was closely followed by Latin America with 80%. The least urbanized regions of the world were sub-Saharan Africa at 38%, and South Asia at 33% urbanized.

Urban Management Challenges

Six of the challenges urban center governments must overcome to become and remain sustainable described by a multi-regional team of seven researchers in 2017 were: air pollution, ecosystems damage, land use, solid waste management, the effects of climate change, and water pollution and water-related infrastructure, including wastewater infrastructure and stormwater runoff (Bai et al. 2017). With more of the world population residing in urban centers, national development policies necessarily include goals for integrating plans for the infrastructure for urban wastewater and drinking water supply systems into sustainability objectives. Regardless of the location and ways in which urban water systems are supplied from surface water sources, supply planning must be carried out at the basin level at the minimum. This includes planning for stormwater collection and keeping urban stormwater discharges from polluting surface water resources. Research has shown that urban road and roof runoff increases the level of flood damage and contamination from pollutants, including heavy metals that are non-biodegradable (Mikovits et al. 2018).

Urban areas must often import freshwater from a distant surface water source. This imported water is often augmented by water locally withdrawn from rapidly depleting groundwater aquifers. In many areas of the world, the infrastructure to supply and treat clean freshwater and effectively treat wastewater has not kept pace with the rapid migration of people to urban areas. As a result, fringe regional slums often lack access to a sustainable supply of clean water or collection and treatment wastewater facilities. The supplies of freshwater that do exist are often polluted. Untreated sewage is discharged into nearby surface waters. Human and animal waste, together with increased stormwater runoff from impervious surfaces in urban areas, results in serious harm to local ecosystems and spread deadly diseases.

Much of the world's urbanization is occurring near coastlines, a result of which has been excessive withdrawals of available surface water and extreme pollution discharged into adjacent wetlands, river and seas. Some of the effects of this pollution are:

> In addition to chemicals, pollution by fecal and pathogenic bacteria and antibiotic resistance gene is also widespread in urban environments, particularly urban aquatic environments. Pollution of fecal and pathogenic bacterial mainly results from poor waste treatment facilities and overflows under flooded conditions in urban systems. Researchers suggest that fecal indicator bacteria are the top cause of river and stream impairments in the United States … A high portion of antibiotic resistance genes is often detected in sewage sludge. (Bai et al. 2017, 222)

Urbanization, along with agriculture to feed the many millions flocking to the world's cities, is among the socioeconomic factors with the greatest impact on global water supplies. India, the second most populous nation in 2015 was still only 32.7 percent urbanized, was the world's greatest user of

Table 6.3 Highly urbanized countries with the highest use of freshwater, 2017 estimates

Rank	Country	Est. 2017 population	Urbanized (%)	Annual freshwater withdrawals
1	India	1.34 billion	32.7	761.0 bcm
2	China	1.48 billion	55.6	554.1 bcm
3	USA	324 million	81.6	478.4 bcm
4	Pakistan	197 million	38.5	183.5 bcm
5	Indonesia	264 million	68.0	113.3 bcm
6	Iran	81 million	73.9	93.3 bcm
7	Vietnam	96 million	33.6	82.0 bcm
8	Philippines	105 million	44.4	81.6 bcm
9	Japan	127 million	93.5	81.5 bcm
10	Mexico	129 million	89.8	80.3 bcm

Source Data collected from (USA) Worldatlas.com and misc. UN and USA official sources
Note billion cubic meters (bcm)

freshwater with annual withdrawals of 761.0 billion cubic meters. China, with 55.6% of its 1.48 billion people residing in urban areas, is the world's most populous nation and the second greatest water user with withdrawals of 554.1 billion cubic meters of freshwater. The USA, which was 81.6% urbanized in 2017, was the third largest withdrawer of freshwater with 478.4 billion cubic meters. Pakistan, Indonesia and Iran ranked fourth, fifth and sixth, respectively, among the ten nations with the greatest total freshwater withdrawals (Table 6.3). Descriptions of the effects of urbanization on selected water-related problems now follow, beginning with problems in India, followed by case histories in China, Pakistan, Iran, Indonesia, and the USA.

Urbanization in India

India continues to largely an agricultural nation. In 2011, the World Bank estimated India's population of 1.247 billion will increase to 1.583 billion by 2030. In 2016, 33.14% (an estimated 410 million) of the total population of 1.324 billion resided in urban areas. India is projected to become the world's largest nation in 2022, when its population reaches 1.4 billion.

India's estimates of municipalities' water requirements of 7% in 2025 will increase to 9% in 2050. Agriculture demand over this 25-year period will continue to be the greatest drain of increasingly stressed supplies.

Urbanization in China

In 2018, less than 58% of China's population resided in urban areas. However, the nation had been on a path of urbanization growth since 1955. Increasing the share of the population residing in urban areas has long been an official policy of the Chinese government. Beginning from an urban population of just 13.6% in 1955, China embarked on what became a rapid industrialization that represents one of the world's largest migrations. The growth in city dwellers included increases in the numbers of registered city residents and a large floating population of unregistered residents mainly from nearby agricultural areas. City growth has driven many farmers from their traditional small farm plots. With no other survival alternative available, they migrate to the city to find work. Cities have not been able to build the needed housing, transportation systems, and water and sanitation infrastructure fast enough to provide for new immigrants. The result has often been a peripheral slum of temporary structures without power, water or sufficient sanitation services.

Population Growth Estimates

Estimates are that China's urban population of 282 million will increase to 580 million by 2030. In 2030, China's population expected to reach 1.5 billion; by then, 69.3% of the population will live in urban areas. The population in 2050 is expected to decline to 1.4 million, with 79% living in cities. The adverse effects of this rapid urbanization are extensive (Cui and Shi 2012). They range from temperature increases and air pollution from loss of open areas and heavy traffic to over-taxed water supplies and insufficient treatment of wastewater that further pollutes diminished surface water sources as well as depleted aquifers. Water stress is occuring in many parts of China. The north of China is experiencing the greatest water shortages. The research of Shen et al. (2005) illustrates the underlying reasons that the urban water problems came to be. Excessive groundwater withdrawals began in about 1950 in support of growing industrialization, intensive agriculture and urbanization in the Shijiazhuang region of China.

While noting that the exploitation of groundwater resources has resulted in social and economic benefits by providing low-cost, drought-reliable and high-quality water supplies for urban areas, the excessive draw-downs of supplies needed for industries and cities in semi-arid and semi-humid regions have brought about many problems in water resource management. Current water supplies and drainage management in these regions call for new solutions to overcome the water scarcity, deteriorating water quality, missing infrastructure, inappropriate wastewater treatment problems that have often followed. Water withdrawals, uses and issues have risen from urbanization in this part of North China since 1950.

A medium-sized city of 287,999 in 1950, the provincial capital city of Shijiazhuang grew to 1.6 million in 1999. The province had a population of more than 9.4 million in 2017. Most of the branches of the Hutuo River—once the area's main surface water supply—have dried up, with water visible only during major floods. Increasing volumes of treated wastewater is used for farm irrigation to supplement the declining surface water sources. Where this has occurred, subsequent research has found some nitrate pollution introduced in the soil. The four major findings of the Shen et al. study of the region and its water issues were:

1. Surface water supplies have either shrunk dramatically or dried up entirely, with the main river now showing only a seasonal flow.
2. With less surface water available, excessive pumping of groundwater with insufficient recharging has exhausted aquifers in some areas.
3. Pollution of diminished surface water supplies and some groundwater exists. Large quantities of untreated wastewater are drained into dry riverbeds, also contaminating groundwater.
4. Little water management activity was present, traceable to the traditional flood irrigation methods still predominant. Little water use efficiency was present.

Urbanization in Shanghai

The combination of economic growth, industrialization and population increase has resulted in rapid urbanization in most Chinese cities. The experience of China's largest city, Shanghai, is an example. The population of Shanghai grew from 5.03 million in 1950 to an estimated 25.1 million in 2017. The government imitated a no-more-growth policy for this and other cities. In most Chinese cities, official housing registration and permits are needed to live, work, receive medical care and educate children in public schools, but this rule is often circumvented by private entrepreneurs and a large floating population (temporary migrant workers) that goes uncounted (Cui and Shi 2012). Almost 40% of Shanghai's 25.1 million residents are believed to be unregistered migrants from rural areas. Throughout all of China, more than 230 million people belong to the so-called 'floating population', people without permission to reside permanently in urban areas.

Shanghai is located near the mouth of the Yangtze River, suggesting the city is not threatened by a shortage of surface water. However, upriver industrial pollution has resulted in Shanghai being added to the United Nations' list as one of the six cities predicted to experience severe drinking water problems in this century. Shanghai receives an adequate supply of rainfall every year, and is seldom affected by the droughts that are common in northern China. However, local water supply is not the problem. Of more concern is the uncontrolled pollution of surface water resources that makes them unfit for human consumption. Reprinting a report by *The Economist*, life science assistant editor Marc Llanilla (2013) wrote that "more than half of China's surface water is so polluted it cannot be treated to make it drinkable ... and one-quarter of it is so dangerous it can't even be used for industrial purposes." He added that most of the China's groundwater is not any safer for human consumption. In 2013, close to 40% of China's farmers depended on groundwater for irrigation, with more doing so every year. As a result, an estimated 90% is of all irrigation water from underground aquifers in China is polluted to some degree, and nearly 60% of the groundwater under withdrawn for the growing cities throughout China has been reported to be "severely polluted."

Urbanization in Pakistan

Pakistan is fast becoming one of the most urbanized countries in Asia. Factors driving the urbanization include population growth, an influx of refugees from Afghanistan and other Muslin countries. This faced-pace growth has resulted in most urban areas being under-resourced, particularly in terms of fresh water and sanitation infrastructure. In 2017, agriculture consumed 90% of the available water, most of which was withdrawn from already water-stressed aquifers. The country's urbanization grew from 39.9% in 2013 to 40.5% in 2017 and is expected to be 50.0% in 2030.

Pakistan has eight cities with populations greater than 1 million. The largest is the coastal city of Karachi, which in 2017 was the 12th largest city in the world. With a population of 16.1 million, this sprawling megacity suffers from periodic water riots due to limited water supplies and wastewater infrastructure. The urban area is continuing to grow, largely due to population growth. By 2030, Karachi may become the world's seventh largest city with a population of 24.8 million. Lahore, the second largest city in Pakistan, has an urban population of 11.13 million. Islamabad, the capital, has a population of 2.01 million.

A combination of rapid urbanization, population growth, shifts in socio-economic patterns and a disproportionate share of benefits directed to large commercial farm organizations have placed Pakistan in the water-stressed class of nations (Mustafa et al. 2013). This combination of internal and external forces is projected to leave the nation water scarce by 2035. The amount of water available for all sectors dropped from 5260 cubic meters per person in 1951 to approximately 1040 in 2010. In 2017, it was less than 1000 cubic meters per person.

Allocation of the country's water is likely to become a major point of controversy in the future. Currently, from 80 to 90% of all water withdrawals are absorbed by the agriculture sector. Major crops include high water-consumption wheat, cotton and rice. Industrial and municipal or domestic withdrawals are minuscule compared to agriculture. Insufficient infrastructure is resulting in some 90% of industrial and municipal wastewater often discharged untreated, to be allowed to percolate into the soil, further polluting aquifers.

Urbanization in Iran

Iran, among the world's most arid nations, had an estimated population of 81.6 million in 2018, and an annual growth rate of 1.07%. Nearly three quarters of the population (74.6%) live in urban areas. The population of the four largest cities in 2017 was: Tehran, 7.15 million; Mashhad 2.31 million; Isfahan, 1.55 million; and Karaj, 1.45 million. Table 5.3 lists Iran's projected national and urban population to 2050.

Population growth surged to 4.3% in 1985, followed by a steep decline to 1.5% in 1995, briefly rising again in the late 2000s to nearly 2.0% per year, but then declining every year since. Population increase in 2017 was 1.10% and is projected to be 1.05% in 2018.

Urbanization in Indonesia

In 2016, Indonesia became 54% urbanized, having gone from just 15% urbanized in 1961. Jakarta, the capital of Indonesia, a city of more than 30 million people, is dealing with extreme water-related problems from this unchecked urbanization. The most commonly encountered problems of inadequate infrastructure, including inadequate housing and the antiquated and insufficient water and sanitation infrastructure all contribute to the adverse effects of unregulated private wells. Rising sea levels from climate change is now adding to their water-related problems. Essentially an island that spreads along the northwest coast of Java, Jakarta lies between the Java Sea to the north and 13 rivers that flow through the city from the south.

Indonesia receives nearly 300 inches of rainfall annually; yet, only a small portion of the rain needed in the city is brought in from the southern mountains. It also lacks an effective sewage collection and treatment system. Climate change is adding to the city's problems in two ways: flooding from more and heavier rainstorms and the wind and ocean-surge flood damage from more extreme weather events. In addition to a combination of unchecked urban sprawl, polluted air and water, extremely poor sanitation, increasingly heavy monsoon rains, rising sea levels, no freshwater and other challenges, the metro area of 30 million people is sinking into the sea. With no piped drinking water available, large numbers of individuals, businesses, industries and developers continue to dig illegal wells, further depleting the aquifers that were the cushions that supported the land area. Sections of the city have fallen 14 feet in just the last few years. If changes are not implemented soon, much of Jakarta will be under water; close to 40% of the city is already below sea level, with only a leaking concrete wall protecting them from storm surges. Meanwhile, development continues in Jakarta, the sea keeps rising, and large portions of the people must still wash themselves and their clothing in water streams and canals that are nothing but open sewers (Kimmilman 2017).

Urbanization in the U.S

Wherever urban growth occurs, the demand for water increases at a faster rate than the population increases. Urban residents use more water for bathing, cooking, watering landscaping and other uses. A close relationship between water supplies, urbanization, climate change and sustainability exist in varying degrees in most of earth's populated landmasses, including the Southwestern and Southeastern USA (Yigzaw and Hossain 2016). Research has shown that the increased economic opportunity in the region has improved the level of well-being reported by many city dwellers. However, the increases in urbanization are accompanied by increases water use that are greater than the rate of population growth. City dwellers use more water for food preparation, washing and sanitation, and for the industries that supply them. The seven states that make up the U.S. Southwest are among the regions where water supplies most stressed by urbanization.

This region's water stress is largely the product of two phenomena: the migration to the cities of the Sunbelt states, and the climate change that is altering precipitation patterns. Other factors affecting the regions fragile water resource are excessive withdrawals of groundwater from limited aquifers for agriculture, and reduced precipitation and earlier snowpack melting from climate change.

California is the most urbanized state in the nation. Other Southwest states and their water-use ranks are Nevada, 3; Utah, 9; Arizona, 10; Colorado, 15; Texas, 16; New Mexico, 22; and Oklahoma, 36. Los Angles, California, with 17.8 million residents, is the largest urban area in the Southwest and the largest in the nation. Others are Houston, Texas with a population of 9.9 million, and Phoenix, Arizona with 4.5 million residents.

Major metro urban areas in the Southwest and their estimated populations are (data source estimates range from 2015–2017). Major Southwest-state metro areas and their population are:

- California: San Francisco, 3.3 million; San Diego, 2.6 million; Riverside/San Bernadino, 1.9 million; Sacramento, 1.7 million; and San Jose, 1.7 million;
- Texas: Dallas, 4.9 million; Houston, 6.9 million; and Ft. Worth, 2.5 million;
- Arizona: Phoenix, 5.5 million; and Tucson, 1.01 million;
- Colorado: Denver, 2.8 million;
- Nevada: Las Vegas, 1.9 million;
- Utah: Salt Lake City, 1.15 million;
- New Mexico: Albuquerque, 0.9 million.

Most of the region depends on the already over-allocated Colorado River. The Colorado River is a main source of water for millions of urban dwellers in Arizona, California, Colorado, Nevada, New Mexico, Utah, and Wyoming. Southern California from Los Angeles to San Diego is the most urbanized area of the nation and a major user of Colorado River water. California, Nevada, Arizona and New Mexico have suffered a recent decade or longer drought, while extreme weather events such as hurricanes and tornados are relatively common in the urban areas of Texas and Oklahoma. Recent examples include the August 25–29, 2017 Hurricane Harvey in 2017 that dropped more than 60 inches of rain on some urban areas of Texas, making this storm the highest single-storm recorded rainfall in the USA. Texas is hit with an average of 126 tornadoes each year, while an average of 52 tornados strike Oklahoma each year.

Summary

Migration of large numbers of the global population to urban area is one of the most severe barriers to achieving water sustainability; it is described as one of the most important voluntary demographic trends to have ever occurred. The number of people living in urban areas supassed those in rural area in 2008 for the first time in history. In 2015, the urban population accounted for 54% of the earth's total, up from 30% in 1950. It could be 60% of the world's population by 2030. Population growth is slowing and may result in stabilization or actual declines near the end of this century. However, no decline has been seen in the unofficial urban migration of agriculture workers displaced as a result of industrial development, construction of roads for the current growth in private ownership of automobiles, and other urban developments.

Among the many water-related problems brought on or contributed to by urbanization are: pollution of surface water supplies, depletion of groundwater resources, loss of agriculture land, higher per-capita water demand, and the critical need for water and wastewater infrastructure investments.

References

Bai, Xuemei, Timon McPhearson, Helen Cleugh, Harini Nagendra, Xin Tong, Tong Zhu, and Yong-Guan Zhu. 2017. Linking urbanization and the environment: Conceptual and empirical advances. *Annual Review of Environment and Resources* 42: 215–240.

Cui, Lin li, and Jun Shi. 2012. Urbanization and its environmental effects in Shanghai, China. *Urban Climate* 2 (2012): 1–15.

Kimmilman, Michael. 2017. Can Jakarts save itself from sinking in 10 years. *The Seattle Times,* December 27, pp. A1, 9.

Llanilla, Marc. 2013. China's top 6 environmental problems. *Live Science*, March 15. Accessed January 21, 2018 from www.livescience.com/27862-china-environmental-problems.htm.

McDonald, Robert J., Katherine Weber, Julie Padowski, Martina Flöke, Christof Schneider, Pamela A. Green, Thomas Gleeson, Stephanie Eckman, Bernard Lehner, Deborah Balk, Timothy Boucher, Günther Grill, and Mark

Montgomery. 2014. Water on an urban planet: Urbanization and the reach of urban water infrastructure. *Global Climate Change* 27: 96–105.

Mikovits, Christian, Wolfgang Rauch, and Manfred Kleidorfer. 2018. Importance of scenario analysis in urban development for urban water infrastructure planning and management. *Computers, Environment and Urban Systems* 68: 9–16.

Mustafa, Daanish, Majed Akhter, and Natilie Nasrallah. 2013. *Understanding Pakistan's water-security nexus.* United States Institute of Peace. Accessed January 18, 2018 from www.usip.org/sites/default/files/PW88_Understanding-Pakistan%27s-Water-Security-Nexus.pdf.

Rosenberg, Matt. 2017. *The Largest Cities in the World.* Accessed October 29, 2017 from www.thoughtco.com/a-list-of-the-largest-cities-in-the-world-1435767.

Senn, Dominique, and Dorthee Spuhler. 2018. *Water, Sanitation and Urbanization.* Accessed December 6, 2018 from https://sswm.info/node/7722.

Shen, Yanjun, Changyuan Tang, Jieying Xiao, Taikan Oki, and Shinjiro Kanae. 2005. Effects of urbanization on water resource development and its problems in Shijiazhuang, China. In *Proceedings of the S2 symposium of the Seventh IAHS Scientific Assembly at Foz do Iguaçu, Brazin*, April. Accessed December 27, 2017 from http://hydro.sjziam.ac.cn/resources/Effects%20of%20urbanization%20on%20water%20resources%20development%20and%20its%20problems%20in%20Shijiazhuang,%20China.pdf.

Spring, Ursula Oswald. 2011. *Water Resources in Mexico: Scarcity, Degradation, Stress, Conflicts, Management, and Policy.* New York: Springer.

Tortajada, Cecilia. 2008. Challenges and realities of water management of megacities: The case of Mexico City metropolitan area. *Journal of International Affairs* 61 (2): 147–166.

UN (United Nations). 2017. World population prospects. Accessed October 20, 2017 from https://esa.un.org/unpd/wpp/Publications/Files/WPP2017_KeyFindings.pdf.

UN Water. 2017. Water and urbanization. Accessed January 21, 2018 from www.unwater.org/water-facts/urbanization/.

WHO (World Health Organization). 2016. Urban population growth. World Health Organization. Accessed October 20, 2017 from www.who.int/gho/urban_health/situation_trends/urban_population_growth/en.html.

WSP (Water and Sanitation Program). 2018. Urban water and sanitation: The challenges. Water and Sanitation Program, World Bank. Accessed January 8, 2017 from www.wsp.org/topics/urban%20water%20supply%20and%20sanitation.

Yigzaw, Wondmagegn, and Faisal Hossain. 2016. Water sustainability of large cities in the United States from the perspectives of population increase, anthropogenic activities, and climate change. *Earth's Future* 4 (12): 603–617.

7

Agriculture and Inefficient Water Use

The idea of sustainable agriculture concept became a generally accepted focus of the U.S. Agriculture Department in the 1990s. The term 'sustainable agriculture' and its definition included in the 1990 Farm Bill is shown in Box 7.1. An early use of the term occurred in a *Journal of Soil and Water Conservation* article by Richard Duesterhous (1990, 4), citing an earlier use by University of Missouri agriculture professor Richard Ikert: "Sustainable agriculture describes farming systems that are capable of maintaining their productivity and usefulness to society indefinitely. Such systems … must be resource-conserving, socially supportive, commercially competitive, and environmentally sound."

Box 7.1 Sustainable agriculture in U.S. Food, Agriculture, Conservation, and Trade Act of 1990

Under that law, "the term sustainable agriculture means an integrated system of plant and animal production practices having a site-specific application that will, over the long term:

1. Satisfy human food and fiber needs;
2. Enhance environmental quality and the natural resource base upon which the agricultural economy depends;
3. Make the most efficient use of nonrenewable resources and on-farm resources and integrate, where appropriate, natural biological cycles and controls;
4. Sustain the economic viability of farm operations; and
5. Enhance the quality of life for farmers and society as a whole."

Source Findings, Purposes, and Definitions, U.S. Code, Title 7, Chapter 64

D. E. McNabb, *Global Pathways to Water Sustainability*,
https://doi.org/10.1007/978-3-030-04085-7_7

The Department of Agriculture's use of the term *sustainable agriculture* occurred again in 1994 in the following statement by U.S. Secretary of Agriculture Daniel Glickman (Secretary's Memorandum 9500-6: Sustainable Development September 13, 1996):

> USDA is committed to working toward the economic, environmental, and social sustainability of diverse food, fiber, agriculture, forest, and range systems. USDA will balance goals of improved production and profitability, stewardship of the natural resource base and ecological systems, and enhancement of the vitality of rural communities. USDA will integrate these goals into its policies and programs, particularly through interagency collaboration, partnerships and outreach. (USDA 2007)

Global Wasteful Agricultural Irrigation

Irrigated agriculture is essential in meeting much of the world's food needs. It is the world's largest water user, accounting for more than 70% (85% in the USA) of annual global water consumption. However, it is also wasted or polluted by farm and pasture applications. This irrigation water is freshwater that is not stored, delivered, scheduled or applied efficiently. Moreover, unrestricted aquifer or surface water withdrawals threaten the sustained availability of both groundwater and surface water. In some areas of the central USA, for example, the over-withdrawal of water from the Ogallala Aquifer (also known as the Great Plains Aquifer System) under the central U.S., excessive withdrawals for irrigated agriculture has resulted in the water table dropping more than 150 feet in many locations. The Ogallala Aquifer is a vast, low water table aquifer surrounded by sand, silt, clay and gravel located beneath the lands east of the Rocky Mountains to nearly the Mississippi River. One of the world's largest aquifers, it underlies an area of approximately 174,000 square miles from South Dakota to Texas (NRCS 2012).

Available irrigation water supplies in many parts of the world have similarly been reduced due to aquifer depletion, competition, regulation and or drought. Inefficient irrigation affects water quantity and quality of runoff from farms, pastures, school grounds and golf courses, parks and home lawns. Irrigation systems can waste water through over-watering and evaporation from open ditches, canals and reservoirs. Storm runoff from fields carries with it wrongly applied fertilizers and insecticides. Emerging urbanization problems further complicate agriculture water resource protection and water allocation.

Some of the problems directly or indirectly associated with inefficient agricultural water use include critical groundwater declines in agricultural production areas, saltwater intrusion into ground water supplies. Other problems include heightened competition for water among the many users of water, including thermoelectric power producers and drinking water suppliers. Reductions in stream flows also occur. The U.S. Agriculture Department has identified a variety of possible pathways as solution to problems from or associated with irrigated agriculture. They include the following:

- Conservation of the water used;
- Conversion to other crops that use less water;
- Conversion to other sources of water;
- Improvements in irrigation water use efficiencies;
- Off stream storage of water during periods of excess runoff;
- Water re-use and water recycling; and
- Ground water recharge.

Water and the Environment

By far the largest proportions of the world's available freshwater is used for agriculture (cropping activities, livestock and aquiculture) and for energy production (primarily for thermoelectric generation). Globally, estimates of as much of 70% of all the freshwater withdrawn from all sources is used for irrigation—more than twice that of the 20% used for industry and far more than the 8% used by municipalities. Agriculture is the major consumer of surface and groundwater in the USA, accounting for close to 80% of the water consumed in the USA (USDA 2017).

In arid regions such as much of the Middle East and the American Southwest, the percentage of water used by agriculture is greater than 90%. Most irrigation methods are notoriously inefficient; for many crops, only 10–20% of the water supplied is returned to the air as transpiration from the plants. The rest, polluted with chemical fertilizers and pesticides, is lost, either soaked into the soil or evaporated. The main causes of wasteful and unsustainable agriculture water use are leaky irrigation systems; wasteful field application methods and cultivation of thirsty crops not suited to the environment.

Through the 1990s, the world grew increasingly aware of agriculture's connection with the degrading of the natural environment and the need to include the accelerating pollution of water by extant agriculture methods in environmental protection planning. This connection was included in the Agriculture Department's report on the meaning of sustainable agriculture. Sections of that report are included in Box 7.2.

Box 7.2 Agriculture, the environment and water

Agriculture profoundly affects many ecological systems. Globally, negative effects of current practices include the following:

- Decline in soil productivity can be due to wind and water erosion of exposed topsoil; soil compaction; loss of soil organic matter, water holding capacity, and biological activity; and salinization of soils and irrigation water in irrigated farming areas. Desertification due to overgrazing is a growing problem, especially in parts of Africa.
- Agricultural practices have been found to contribute to non-point source water pollutants that include: sediments, salts, fertilizers (nitrates and phosphorus), pesticides, and manures. Pesticides from every chemical class have been detected in groundwater and are commonly found in groundwater beneath agricultural areas; they are widespread in the nation's surface waters. Eutrophication and "dead zones" due to nutrient runoff affect many rivers, lakes, and oceans. Reduced water quality impacts agricultural production, drinking water supplies, and fishery production.
- Water scarcity in many places is due to overuse of surface and ground water for irrigation with little concern for the natural cycle that maintains stable water availability.
- Agriculture's link to global climate change is just beginning to be appreciated. Destruction of tropical forests and other native vegetation for agricultural production has a role in elevated levels of carbon dioxide and other greenhouse gases. Recent studies have found that soils may be both sources of and sinks for greenhouse gases.

Source USDA (2007)

Agriculture Water Scarcity

By 2050, the world will need to feed more than 9 billion people: 2 billion more than the population in 2014 (World Bank 2014). This will mean the world must find ways to achieve a 50% increase in agricultural production. However, this will result in a 15% increase in already-strained water withdrawals only for agriculture. Globally, the agricultural sector is by far the largest global user of water; irrigated agriculture accounts for about 70% of total freshwater withdrawals worldwide. The major types of irrigation systems used throughout the world are (CDC 2016):

- Rain-fed farming: The natural application of water to the soil through direct rainfall.
- Surface irrigation: Furrow or border irrigation; basin irrigation (similar to flood irrigation as in rice paddies). Water is distributed over and across land by gravity or natural flow, no mechanical pump is involved.
- Flood irrigation: Water is turned into a field without any flow control such as furrows, boarders or corrugations. This is the least efficient, least uniform and least effective method of irrigation, with much water wasted.
- Localized irrigation: Water is distributed under low pressure, through a piped network and applied to each plant.
- Drip irrigation: Localized irrigation; drops of water are delivered at or near the root of plants. In this type of irrigation, evaporation and runoff are minimized. One of the most efficient users of water.
- Sprinkler irrigation: Water is distributed by overhead high-pressure sprinklers or guns from a central location in the field or from sprinklers on moving platforms. More efficient than surface irrigation; less efficient than drip irrigation.
- Center pivot irrigation: Water is distributed by a system of sprinklers that move on wheeled towers in a circular pattern. This system is common in flat areas of the USA.
- Lateral move irrigation: Water is distributed by sprinklers with water through a series of pipes, each with a wheel and a set of sprinklers, which are usually automatically rotated. The sprinklers move a certain distance across the field and then need to have the water hose reconnected for the next distance. This system tends to be less expensive but requires more labor than the others.
- Sub-irrigation: Water is distributed through a system of pumping stations, canals, gates, and ditches. This type of irrigation is most effective in areas with high water tables.
- Manual irrigation: Water is distributed across land through manual labor and watering cans. This system is very labor intensive.

The potential ways for developing new sources of water in these regions are few or nonexistent. Therefore, improving the efficiency of water use for agriculture is necessary to meet existing and future demand. Current agricultural methods also result in soil salinity and water consumption from evaporation, mainly resulting from inefficient surface irrigation. Inefficiencies also persist in municipal water supply, where estimated losses from urban water networks range from 20–50%. Inefficient flood or sprinkler irrigation systems are still in use in some water-stressed areas of the American West. The traditional crop flood irrigation system that is also still commonly used throughout the world. Cheap and easy to produce and operate, these systems are among the most common methods of row crop irrigation used in developing regions.

The flood crop method is one of the oldest irrigation systems still in use and most likely evolved with the earliest domestication of farm crops in the Fertile Crescent. Water is brought to the field in a canal, ditch or pipe and then released to simply flow among the growing crops. It may be collected at the end of the rows for reuse or collected as polluted wastewater. Although effective, it is one of the most inefficient methods of irrigation. Typically, no more than half of the water released is actually used by the crop. The rest is either lost to evaporation, runoff, infiltration of parts of the field without crops, or it is absorbed by weeds. To ensure that crops at the end of the field receive adequate water, some of this 'tailwater' is often collected and redistributed. Collecting that water and return pumping it for reuse can greatly increase water efficiency.

Although the divide between nations with low water availability and those with high water availability appears to follow the divide between developed and developing nations, the problems of water access, water scarcity and water quality are universal. Conditions such as climate change, population growth, urbanization, and water scarcity that are beyond the ability of many governments to resolve will become progressively more global. Because agriculture use dominates the list of uses for freshwater everywhere, improving agriculture water efficiency is a universal requirement (Moss 2008). Globally, significant academic and commercial effort has been devoted to finding ways to improve agricultural water-use efficiency. Much of that work has focused on improving irrigation methods in drylands and regions of severe water scarcity. Some of the major reasons for agricultural water use inefficiency in developing regions are well documented, and include the following (Mahto 2013; Bhalage et al. 2015):

- Missing or poor maintenance of canals and other distribution systems resulting in weed and other vegetation growth, buildup of silt, and other losses;
- Distortion of canal section from silting or collapse on sloped fields, resulting too little or too much water in channels;
- Water loss from missing lining in canal reaches crossing permeable soil strata;
- Leakages in channel gates and shutters;

- Missing flow regulation gates, resulting in uneven water distribution;
- Over-irrigation due to missing control structures and facilities for the volumetric supply of irrigation water to farmers;
- Poor irrigated field management practices;
- Failure to shift from crops requiring excessive water; and
- Lack of awareness among farmers on correct irrigation practices and cropping patterns.

Most wasted irrigation water occurs simply because too much water is applied to the field. Inefficiency occurs because of excess runoff from the field or wasted deep percolation below the root zone. More water is lost from leaks in piping systems and evaporation from open channels transporting water from the source to a field. The U.S. Department of Agriculture identified the following preventable activities that continue to increase the inefficiency of agriculture water use in water-stressed agriculture areas of the USA (USDA 2012):

- Not stopping the flow of water to the field when the amount of water needed has been delivered;
- Not knowing the exact quantity of water (head) flowing from the water source to the field;
- Not knowing the soil moisture level at the time of irrigation;
- Not knowing the water-holding capacity of the soils being irrigated;
- Not applying water based on the intake characteristics of the soil;
- Not knowing the daily water use requirements of the crops being grown;
- Not using the proper set sizes for the available head of water;
- Non-uniform slopes that cause poor water distribution across the field; and
- Excessive slopes, which creates high runoff losses.

Rainfall-Only Agriculture

Rainfall is the only significant source of water for crops in many of the world's dryland agricultural regions. Agriculture researchers have developed many ways to improve the efficiency of dryland farming where surface or groundwater is scarce, and rainfall is the sole water source or for crops. A variety of methods that have been found to be successful in increasing yields in dryland grain-farming in Australia and Syria were described by Neil Turner in a 2004 paper on rainfall-use efficiency. Listed were the adoption of procedures such as minimum tillage, appropriate fertilizer use, improved weed/disease/insect control, timely planting, and a range of rotation options. In conjunction with new varieties of crops, these activities have the "potential to increase the yields and rainfall-use efficiency of dryland crops ... [Research has shown] that most of the agronomic options for improving rainfall-use efficiency in rainfed agricultural systems decrease water losses by soil evaporation, runoff, throughflow, deep drainage, and competing weeds, thereby making more water available for increased water use by the crop" (Turner 2004, 2413). Turner also noted that other researchers have suggested that more efficient water-use practices were responsible for at least two-thirds of the crop increases and new wheat strains were responsible for one-third of the of the increases that have occurred in dryland wheat yields in Australia since 1980. Wheat, rice and other grains are considered to be crops that are large water users.

Water efficiency is defined as the amount of biomass produced per unit evapotranspiration (plant transpiration plus soil evaporation). Harvest index is the ratio of harvested yield to total above-ground biomass. The relationship between water (rainfall), transpiration and crop yields is: "As rainfall falling at a particular site can be transpired by the crop, transpired by weeds, lost by soil evaporation, deep drainage, runoff or throughflow (subsurface flow), or stored in the soil for subsequent use by a crop, the yield and rainfall use efficiency in dryland cropping systems can be improved by decreasing losses of water from the soil and weeks, and maximizing the water use (transpiration) by the crop itself" (Turner 2004, 2415).

Agriculture Over-Use of Groundwater

Groundwater resource depletion describes the falling water tables in much of the world as a result of sustained groundwater pumping. Groundwater is the sole source of water supply for much of the world, including the USA. Groundwater supplies the water for half of the population of the USA and nearly all the rural population, where self-supplied wells are the main water source. Supplies of well water for crop irrigation results in more than 50 million gallons withdrawn every day from already overdrawn aquifers. Water for agriculture that is withdrawn faster than nature can resupply it is taking irreplaceable water from wells in most of the high plains of the USA, the Middle East, Asia, South America and Asia. As world population and urbanization continues to grow in these regions, water withdrawals are increasing at an even faster rate.

Fig. 7.1 Overhead irrigation system watering a potato crop in the arid U.S. West (*Source* Washington State Department of Ecology photo)

For many years, groundwater resource depletion has been a concern in the California Central Valley, the American Southwest, and the High Plains from North Dakota to Texas. However, increases in groundwater withdrawals have overstressed aquifers in many areas of the nation, not just in arid regions. The long drought in California has resulted in many new wells being dug and the agriculture water consumption rate shifted from 40% groundwater to 60% groundwater in the state. California Central Valley farm well drillers that used to strike good water at 500 feet deep must now drill down 1000 feet or more, at a cost of more than $300,000 for a single well (Dimick 2014). Box 7.3 describes the outcome of the all-too common disregard for over-withdrawals from aquifers. Figure 7.1 is a state Department of Ecology photograph of irrigation methods typically adopted for irrigating the types of crops described in Box 7.3.

Box 7.3 Landowners fined for allegedly pumping 500 million gallons from Odessa Aquifer

Landowners near Moses Lake, Washington have been fined for illegally pumping more than 500 million gallons of groundwater from the severely stressed Odessa aquifer. In June 2017, the Washington Department of Ecology issued cease and desist orders requiring the landowners and their lessees to stop pumping groundwater. Ignoring the orders, the landowners reportedly continued pumping water illegally for three-and-a-half more months to irrigate 530 acres of crops.

The Odessa aquifer has been rapidly declining since 1980. Groundwater has dropped more than 200 feet, forcing local farmers and homeowners to drill wells deeper to reach the diminishing water supply. Recognizing the severity of the problem, the stage legislature passed a law in 2004 prohibiting using water from the dwindling Odessa aquifer for irrigation when water from the Columbia River is available through the irrigation district. The State Department of Ecology issued the following fines:

- Landowners and their lessees were fined US$103,000 for illegally irrigating 65 acres of alfalfa.
- Another landowner was fined US$206,000 for illegally irrigating 130 acres of timothy hay.
- Two other landowners and their lessee were fined US$309,000 for illegally irrigating 335 acres of alfalfa and potatoes.

The landowners allegedly disregarded warnings and orders from the Washington Department of Ecology, continued to illegally irrigate through the 2017 growing season, and took their high-value crops to market. The estimated value of crops grown on the illegally irrigated lands is more than US$1 million.

"These landowners willfully ignored the law and tapped into a vulnerable aquifer without a legal right to do so," said the Department of Ecology's water resources program manager. "This isn't fair to other irrigators who follow the law or to local communities and rural landowners who depend on this groundwater for their drinking water."

More than US$200 million has been invested by local landowners and public agencies in recent years to ease the pressure on the declining aquifer by developing sustainable surface water supplies. Penalties can be appealed to the Pollution Control Hearings Board within 30 days.

Source WSDOE (2017)

Globally, groundwater depletion occurs at scales ranging from a single well to aquifer systems underlying several states or nations. The extent of the resulting effects depend on several factors including pumpage and natural discharge rates, physical properties of the aquifer, and natural and human-induced recharge rates. Groundwater depletion is now a major contributor to water stress in many parts of the world. Some of the negative effects of this groundwater depletion are:

- Drying up of wells;
- Reduction of water in streams and lakes;
- Deterioration of both surface and groundwater quality;
- Increased pumping costs;
- Land subsidence;
- Concentration of harmful impurities in remaining resources; and
- Fresh groundwater supplies becoming contaminated from saltwater intrusion.

Sinking Surface Soils

Excessive groundwater withdrawals also result in land subsidence. According to scientists at NASA's Earth Observatory, excessive pumping of groundwater at thousands of agriculture and municipality wells has caused land to subside as much as 28 feet in sections of California's San Joaquin Valley. This subsidence is exacerbated during droughts, when farmers rely heavily on groundwater to sustain one of the most productive agricultural regions in the USA.

Land subsiding is also occurring in other parts of the world, but not always because of agriculture water withdrawals. China, for example, is faced with land subsiding from both agriculture water aquifer depletion and in large cities, as a product of population growth and urbanization. For example, with a population in 2016 of 24.16 million, Shanghai, is sinking—and has been for decades. Since 1921, this city has sunk more than 6 feet. Across China, land subsidence is a problem in more than 50 cities, and close to 49,000 square miles of land have dropped at least eight inches over the same period.

Agricultural Water Pollution

Water pollution is mostly caused by human activity: human communities, agriculture and industries are the major sources. The 2017 annual report of the United Nations' World Water Development Program (WWAP) noted that fully 80% of all wastewater from municipalities is discharged untreated into water bodies. Global agriculture is the largest contributor to water pollution, adding human and animal waste and fertilizer nutrients, pesticides, salts, sediments, pathogens, metals and drug residues to surface and groundwater (Table 7.1). Nitrogen from fertilizers is now the prevalent chemical contaminant in aquifers. Industries dump millions of tons of heavy metals, solvents, toxic sludge and other wastes into water bodies (FAO 2017b; WWAP 2017). As the world population continues to increase, increased crop production must occur, thereby increasing the pressure on water supplies and more water pollution from agriculture. The impact of water pollution on human health is extensive.

According to the UN'S Food and Agriculture Organization (FAO 2016), the main agriculture water quality problems worldwide are salinization and nutrient and pesticide pollution. Salinization is commonly cited as the most widespread groundwater quality problem and as having the greatest environmental and economic impacts. Eutrophication of lakes and rivers is a result of high nutrient pollution (mainly nitrogen and phosphorus). High amounts of agriculture-associated nutrients results in excessive algae growth that depletes dissolved oxygen and kills aquatic species.

Table 7.1 Categories of major water pollutants from agriculture and impact by source

Pollution category		Impact by source		
		Crops	Livestock	Aquaculture
Nutrients	Primarily nitrogen and phosphorus from chemical and organic fertilizers; animal excreta	High	High	Low
Pesticides	Herbicides, insecticides, fungicides and bactericides, including organic phosphorous carbonates, pyrethroids and others	High	Low	Low
Salts	Ions of sodium, chloride, potassium magnesium, sulphate, calcium, bicarbonate in water or dissolved in soils	High	Low	Low
Sediment	Soils and organic matter dissolved in water	High	High	Low
Organic matter	Oxygen-demanding substances such as organic materials and plant matter, livestock excreta	Low	High	Medium
Pathogens	Bacteria and pathogen indicators such as *Escherichia coli*, total coliforms, fecal coliforms and enterococci	Low	High	Low
Metals	Metals such as lead, copper, mercury, arsenic and manganese	Low	Low	Low
Emerging pollutants	Pharmaceutical residues, hormones and feed additives	Low	High	Medium

Source From data in FAO (2017b)

Agriculture Water Pollution

Nutrients from farm fertilizer and livestock raising are considered to be the major water quality problem for surface water in the USA. In an effort to protect all surface waters of the country, President Barak Obama approved legislation that would keep most of the country's freshwater sources from pollution by agriculture, industry and development activities. The Environmental Protection Agency and U.S. Corps of Engineers then developed an adjustment to the Clean Water Act of 1972: The Clean Water Rule of 2015. The rule gives the agencies authority to regulate pollution in wetlands, some ponds, smaller and non-permanent streams and rivers that flow into larger bodies of fresh and salt water.

Supporters of the rule cited a variety of benefits that the rule provides, led by the point that the rule would put more than 60% of all bodies of freshwater in the continental U.S. under pollution protection (Devine 2017). An EPA spokesperson explained misconceptions about the rule: "The Clean Water Act made it illegal to pollute or destroy a covered water without a permit. If you're not doing either of those things, you don't need a permit. The rule does not regulate most ditches. It excludes farm and stock ponds, and grassed waterways and does not regulate groundwater, shallow subsurface flows, or tile drains. It does not make changes to current policies on irrigation or water transfers or apply to erosion in a field. The Clean Water Rule does not regulate land use or affect private property rights" (Kopocia 2015). The rule did not include any changes to permit rules regarding uses of pesticides or fertilizers.

Supporters of the rule highlighted that "more than 117 million Americans get their freshwater from public systems that draw their water from seasonal or rain-dependent streams" that were to be covered by the rule. They added that runoff from farms and livestock operations are making drinking water less safe (Devine 2017). Others point to the toxic algae blooms that are occurring in many U.S. waters from nitrogen and phosphorus from agriculture fertilizers as problems that must be stopped.

Organized Opposition

The rule was opposed by the agriculture industry, oil and gas developers, pesticide and fertilizer makers, and golf course owners, all of whom claimed the regulation infringed on property owners' rights. In 2015, the American Farm Bureau Federation (AFBF) led a lawsuit against the rule, arguing that it puts the burden on farmers to get a permit for using fertilizers near ditches and streams. The case has been in the courts ever since, and the rule has yet to be implemented. Implementation was further delayed in June 2017 when the new EPA administrator, Scott Pruitt, announced that President Trump was going to sign an Executive Order starting a process of repealing the Rule and replacing it with a body of rules that weakened—essentially eliminating—the Clean Water Rule. Opponents of the rule argued that the plan's protection of all waters that in any way lead to the navigable water of the nation was unconstitutional.

The U.S. agriculture industry strongly opposed the extension of control over all surface waters. The AFBF actively opposes the rule, arguing that the rule goes too far in extending the definition of the Constitution clause giving the Federal government control over all navigable waters. It also contended that the rule "defines terms such as 'tributary' and 'adjacent' in ways that make it impossible for farmers and ranchers to know whether the specific ditches, ephemeral drains or low areas on their land will be deemed 'waters of the United States' … The rule gives the agencies sweeping new authority to regulate land use, which they may exercise at will, or at the whim of a citizen plaintiff" (AFBF 2017).

Rise in Food Security

The United Nations (FAO 2017a) estimates that the global demand for food will rise by 70–100% by 2050. To meet this need, estimates are that food production in developing countries will need to almost double. Those vital food increases will not be possible without improved water use management. Many governments and international organizations have long been cognizant of the agriculture water problem. As early as 1990, an many as 20 UN organizations were involved in water issues, most of which were related to agriculture (Falkenmark 1990). As a result, they have made significant strides in helping developing areas in their efforts to approach food sustainability. For example, this aid resulted in the U.S. and other G8 countries launching the Feed the Future program in 2010. The program supports agriculture development in target countries and regions, including Ghana, Kenya, East Africa, Bangladesh, Haiti, Guatemala and Central America. Since 2009, USDA's international food aid programs has benefited about 34 million individuals globally with assistance valued at nearly US$1.6 billion. However, much remains to be done. An estimated 870 million people lack sufficient food supplies, and a billion or more suffer periodic shortages.

In 2015, global agriculture with have to produce 60% more food than it did in 2015, and by 2050, the need for more food will be 100% more than it was in 2015. Producing that much food will not be possible without greater water withdrawals and vastly improved water use efficiency. As noted earlier, irrigated agriculture in most of the developing nations is highly inefficient; upwards of 60% of the irrigation water withdrawn is wasted or lost.

Steven Cohen described the necessity of implementing an integrated water-agriculture management plan if world leaders hope to achieve a sustainable food supply for this and future generations:

> The sustainability of food supply is a fundamental element of the field of sustainability management. Food is not an option, and nutrition and health are closely interconnected. Like water and air, food is a necessity for human survival. Because of the increased urbanization of our settlements, we live farther and farther from the sources of food ...We need to learn how to feed ourselves without destroying the land and water resources required for agriculture. This is one of the critical sustainability challenges of the twenty-first century. (Cohen 2011, 100)

Achieving a Water–Food Balance

Population growth and other factors assures us that the world will need more food and more water over the next three-quarters of a century. Much of the world has been adopting new and innovative ways to balance the increasing demand with limited sources of supply. Successes in the search for new supplies of freshwater have provided some consolation. However, solutions such as very deep wells and dam construction remain beyond the ability of poor nations to finance. Moreover, these new sources are becoming harder and more expensive to find. As a result, there has been a widespread turn to other strategies, such as desalination where possible, and recycling wastewater for agricultural and uses. Acceptance of recycled water for domestic uses, while growing, has been slower. Attempts at adopting other strategies such as forced reallocation of irrigation water to domestic use and the shift from traditional water-intensive cereals face political barriers to widespread adoption but may be necessary in the future. Shifts from inefficient irrigation methods to drip methods have resulted in impressive reductions in water use while greatly improving crop yields have been implemented in many areas of high water scarcity. One of the least successful strategies has been voluntary water conservation policies.

Summary

Among the chief barriers to achieving the quantity and quality of the food needed to feed the increasing population is that gaining more food requires more surface and groundwater for irrigation. Most of the population growth is occurring in areas where water scarcity already exists. Groundwater sources are already being over-withdrawn. Many surface water sources in high productivity areas of grain supplies cannot compete with new demand for water from the growing urbanized areas. And climate change promises to bring devastating changes in the precipitation patterns of much of the globe. Clearly, it is highly likely that if farming methods continue the way they are practiced now, the food and water produced will not be enough to meet all demands.

In many areas of the world, water scarcity will result in severe food scarcity. The UN estimates that in 2025, 60% more food than was produced in 2015 will be needed to meet the needs of the world's growing population. In 2050, it will be necessary to produce 100% more food than was produced in 2015. Producing that much food will not be possible without greater water withdrawals and vastly improved water use efficiency. An increasing number of developing nations with restricted access to sustainable water supplies are adopting new approaches to achieving a necessary water–food balance.

References

AFBF (American Farm Bureau Federation). 2017. Clean Water Act, WOTUS. 2017. Accessed November 21, 2017 from https://www.fb.org/issues/regulatory-reform/clean-water-act/.

Bhalage, Pradeep, B.B. Jadia, and S.T. Sangale. 2015. Case studies of innovative irrigation management techniques. *Aquatic Procedia* 4: 1197–1202.

CDC (Center for Disease Control). 2016. Types of agriculture and water use. Accessed January 6, 2017 from https://www.cdc.gov/healthywater/other/agricultural/types.html.

Cohen, Steven. 2011. *Sustainability Management*. New York: Columbia University Press.

Devine, Jon. 2017. Trump's attack on clean water: What you need to know. Natural Resources Defense Council (NRDC). Accessed November 21, 2017 from https://www.nrdc.org/experts/trumps-attack-clean-water-what-you-need-know.

Dimick, Dennis. 2014. If you think the water crisis can't get any worse, wait until the aquifers are drained. Accessed November 13, 2017 from https://news.nationalgeographic.com/news/2014/08/140819-groundwater-california-drought-aquifers-hidden-crisis/.

Duesterhous, Richard. 1990. Sustainability's Promise. *Journal of Soil and Water Conservation* 45 (1): 4.

Falkenmark, Malin. 1990. Water issues confronting humanity. *Journal of Peace Research* 27 (2): 177–190.

FAO (Food and Agriculture Organization of the United Nations). 2016. Nigeria: Geography climate and population water report. Food and Agriculture Organization of the UN, Aquastat. Accessed December 9, 2017 from http://www.fao.org/nr/water/aquastat/countries_regions/NGA/.

FAO (Food and Agriculture Organization of the United Nations). 2017a. Agriculture and water quality interactions: A global review. Food and Agriculture

Organization of the UN, Aquastat. Accessed November 21, 2017 from http://www.fao.org/fileadmin/templates/solaw/files/thematic_reports/TR_08.pdf.

FAO (Food and Agriculture Organization of the United Nations). 2017b. Water pollution from agriculture: A global review (executive summary). Food and Agriculture Organization of the UN. Accessed November 20, 2017 from http://www.fao.org/3/a-i7754e.pdf.

Kopocia, Ken. 2015. The facts about the Clean Water Rule and agriculture. Environmental Protection Agency blog. Accessed November 21, 2017 from https://blog.epa.gov/blog/2015/06/clean-water-rule-and-agriculture/.

Mahto, Shankar. 2013. Present status of water use efficiency on irrigation projects in India and action taken for its improvement including role of role of farmers. Training Program on Increasing Water Use Efficiency (WUE) in irrigation sector, NWA, Pune, January 21–February 1, 9–19.

Moss, Brian. 2008. Water pollution by agriculture. *Philosophical Transactions: Biological Sciences* 363 (1491): 659–666.

NRCS (Natural Resources Conservation Services). 2012. Inefficient use of irrigation water. U.S. Department of Agriculture. Accessed December 4, 2012 from https://www.nrcs.usda.gov/Internet/FSE_DOCUMENTS/stelprdb1187280.pdf.

Turner, Neil C. 2004. Agronomic options for improving rainfall-use efficiency of crops in dryland farming systems. *Journal of Experimental Botany* 55 (407): 2413–2425.

USDA (United States Department of Agriculture). 2007. Sustainable agriculture: Definition and terms. Accessed November 10, 2017 from https://www.nal.usda.gov/afsic/sustainable-agriculture-definitions-and-terms.

USDA (United States Department of Agriculture). 2012. Improving irrigation water use efficiency. Natural Resources Conservation Service, U.S. Department of Agriculture. Accessed November 13, 2017 from https://www.nrcs.usda.gov/wps/portal/nrcs/detail/az/soils/?cid=nrcs144p2_065174.

USDA (United States Department of Agriculture). 2017. Irrigation and water use: Overview. Economic Research Service, U.S. Department of Agriculture. Accessed April 7, 2018 from www.ers.usda.gov/topics/farm-practices-management/irrigation-water-use.aspx.

World Bank. 2014. Water resources management: Sector results profile. Accessed November 13, 2017 from http://www.worldbank.org/en/results/2013/04/15/water-resources-management-results-profile.

WSDOE (Washington State Department of Ecology). 2017. Landowners fined for illegally pumping 500 million gallons from Odessa Aquifer. Washington State Department of Ecology. Accessed December 14, 2017 from https://www.ecology.wa.gov/About-us/Get-to-know-us/News/2017/Dec-19-Landowners-fined-for-illegally-pumping-500.

WWAP (United Nations World Water Assessment Programme). 2017. *The United Nations World Water Development Report 2017. Wastewater: The Untapped Resource.* Paris: UNESCO. Available at http://unesdoc.unesco.org/images/0024/002471/247153e.pdf.

8

Water-Thirsty Energy Production

The connection between water and energy, both of which are critical resources for development everywhere, also constitutes an inescapable barrier that must be overcome before global water sustainability is achieved. What have been called "inextricable linkages" bind water and energy together as drivers of economic growth (Schuster-Wallace et al. 2015). Each is dependent upon the other for achieving sustainability in both fuel and food production—that connection is the "water-energy nexus" (Sprang et al. 2014; Carter 2013).

Energy is used for the purification of the raw water used in power production. It is also needed to power the distribution of water to fields for irrigating food and bioenergy crops and the movement of treated water in large, reliable volumes to farms, factories, villages, towns and cities. More energy is then needed for treating raw sewage and transporting sanitary effluent to discharge points. Understanding the water–energy connection begins with a review of the role of water in the production of the fuels used to produce energy; the production of energy is a close second to agriculture in terms of the most extensive users of water. Understanding the nexus then requires an understanding of the many ways water is used during the production of power.

The USA produces and uses more energy than any other nation. In 2016, the primary source of that energy is petroleum (37%), followed by Natural gas (29%) and coal (15%). Only 10% of all energy is derived from renewable sources. Figure 8.1 shows the distribution of all energy sources, with a detailed breakdown of the sources considered to be renewable.

D. E. McNabb, *Global Pathways to Water Sustainability*,
https://doi.org/10.1007/978-3-030-04085-7_8

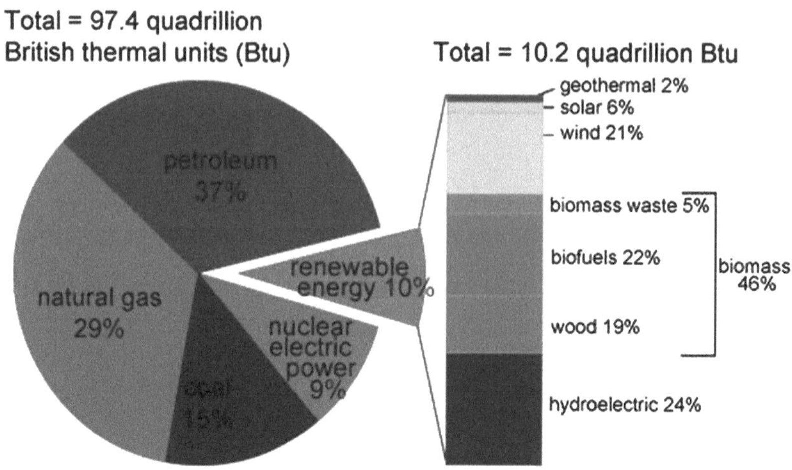

Fig. 8.1 U.S. energy consumption by source

Water's Role in Fuel Production

Research has shown that water and energy resources are closely interrelated (Wu et al. 2009; Sprang et al. 2014). Power is generated in many different ways, but the two main types of fuels used for producing energy are fossil fuels such as coal, petroleum and natural gas, and bio-mass that is processed into biofuels. The fuels needed for producing and transporting primary energy resources are highly dependent upon water, and water needs energy to be pumped from surface and groundwater resources, for treatment to potable stage, for distribution to final users in municipalities, industries and agriculture, and for the collection and treatment of wastewater. This relationship exists in both the provision of the fuels used to produce power and the final distribution of power. Fuels used for generating electricity in the USA from 2015–2017 are shown in Table 8.1.

The production of all major types of fuels used for transportation and electric power production, involves the use of water. The production of primary energy fuels such as coal, oil, natural gas, and many biomass renewables requires large amounts of water. Many of these processes also produce large amounts of wastewater. Water is then used in great volumes during the generation of the electricity that is used to power the pumps that withdraw water from surface and groundwater sources. More water is also consumed (lost by evaporation and therefore not returned to a surface or groundwater source) when it is cooled after its use in steam-powered electricity generating plants.

Table 8.1 Consumption of combustible fuels for electricity generation, 2015–2017

Fuels	2015	2016	2017
Coal (1000 short tons)	739,594	677,371	663,479
Petroleum[a] (1000 barrels)	45,101	39,418	35,330
Petroleum coke (1000 short tons)	4044	4253	3349
Petroleum total	*49,145*	*43,461*	*38,679*
Natural gas (billion cubic feet)	10,017	10,170	9441
Biomass[b] (trillion Btu)	720	665	673
All others[c] (trillion Btu)	310	273	262

Source EIA (2018)
[a]Includes distillate and residual oil and other liquids
[b]Includes wood and waste product fuels
[c]Includes other gases and other fuels

Water and Fossil Fuel Production

Large amounts of water are used in the production, refining and transportation of all types of fossil fuels, which are then used for producing electricity. Most fossil fuels are produced either through mining or drilling (UCS 2016). Coal has long been mined underground but is increasingly extracted by surface mining. Surface mining is only effective for deposits near the earth's surface, and is often highly damaging to local environments, including surface streams and other water resources.

Oil and gas are extracted by conventional and unconventional drilling methods; surface mining is the method for production from oil sands and some oil shale. Unconventional methods such as hydraulic fracturing methods have resulted in access to many otherwise fully exploited or otherwise unaccessable underground subsurface formations. When conventional methods are used, existing unground water is may also be brought to the surface. Called 'produced water,' it is often polluted with dissolved solids, heavy metals, hydrocarbons, and radioactive soils that make it unfit as potable water and harmful to the environment if simply discharged into existing surface water courses. Chemical-laden water is also inserted under pressure to help in the release of the trapped oil or gas. That water is polluted when brought back to the surface or lost. However, drillers are increasingly required to thoroughly treat the degraded water produced in nontraditional oil production rather than simply discharge it into the environment.

Water use in oil and natural gas production varies by drilling or recovery method. The following information was provided by the Texas Water Development Board (TWDB) and released by the Texas Railroad Commission in 2018. It is included here as an example of how water is used in oil and gas production elsewhere as well. Globally, water used for the exploration, development and extraction processes of oil, natural gas, aggregates (sand and gravel), and other materials. In Texas, these uses are reported to take up only 1% of all water withdrawals in the state. Currently, the largest amount of water is used in enhanced recovery. 'Enhanced recovery' is the term for the method of injecting fluids or gases to force more oil to the surface. Most of this water is saline or brackish water.

The second largest volume of water is used during the drilling and completion of oil and gas wells. "Water used for drilling [includes] drilling fluid preparation and make-up water for completion fluids, including cementing, in well stimulation, as rig wash water, as coolant for internal combustion engines, and for workers' on-site sanitary purposes. Fresh water is [also] used in oil and gas well stimulation. Stimulation methods include acidizing and/or hydraulic fracturing … One method of treatment to increase permeability is hydraulic fracturing treatment or 'fracing' (sic)" (RCT 2018) (Box 8.1).

Box 8.1 What is hydraulic fracturing?

"Hydraulic fracturing consists of pumping large volumes of water and other fluids into a subsoil formation. The water used has generally been treated with a chemical friction reducer, surfactant [a surfactant is a chemical product that forms a film over water molecules to increase the contact of the liquid to materials such as stone and sand; another term for the compound is a 'wetting agent'], and clay stabilizer to create a gel that is used to transport sand into the formation. The gelled fluid is pumped under pressure to create and extend a fracture or crack into the formation. The sand, known as proppant [a solid material such as sand or a coated sand forced into the fracture to keep it open], is carried in the gel and is deposited into the fracture to 'prop' or hold it open. The fracture treatments are designed to increase fracture length and minimize fracture height. The fractures result in increased surface area within the reservoir, which expands the productive area of the formation, and results in increases in the desorption of the oil and gas from the shale and increases in the mobility of the oil and gas. The result is lower completion costs and faster recovery of a larger volume of gas-in-place. The volumes injected during hydraulic fracturing treatment can range from 70,000 barrels in a vertical well to more than 90,000 barrels in a horizontal well. Hydraulic fracturing, where necessary, generally takes place immediately after drilling and periodically during the life of a well."

Source Railroad Commission of Texas (2018)

Nontraditional methods of oil production are common in many arid areas of the world. The central basin of California is one of those areas, and one that periodically suffers from prolonged droughts. Hydraulic fracturing has revitalized declining production in the oil fields surrounding Bakersfield. MembranePro Services, an oilfield service firm, announced they were doubling the production capacity of treated fractured-produced water at their Wasco, California treatment facility. The production capacity in mid-2018 was 210,000 gallons of this treated produced water intended for reuse by oil well producers, industry and agriculture in Kern County and other areas of the Central Valley. The facility was commissioned in February 2017. Production was scheduled to reach 420,000 gallons per day by the end of the year, further reducing the need to use severely limited freshwater supplies (Industrial Waterworld 2018).

Water and Renewable Fuels

Water is used in great quantities during the production of renewable energy fuels. Renewables for power production include bioenergy, geothermal, concentrating solar power (CSP), solar photovoltaics, wind power, and captured water that is then used by water-driven turbines. However, little water is lost by hydropower because all of the water is returned unadulterated to its source.

The growing of plant feedstocks for biofuels requires large amounts of water for irrigation and the final production of usable fuels. Major biofuel feedstocks include corn, sugarcane, switchgrass, sugar beets and agricultural residues. Other means of power production include nuclear fuels, wind, hydropower, and other unconventional and renewable resources and methods.

Globally, the production of ethanol from corn is the most water-thirsty of all fuels used for meeting society's demand for energy. While the water used in feedstock production varies from region to region, average water requirements for producing three major biomass feedstocks for each gallon of ethanol are 1500 gallons of water for corn, 1300 gallons of water for sugar beets, and 900 for the *jatropha* family of flowering shrubs.

Water and Biofuel Production

Large volumes of water are also used for growing biofuel feedstocks and for the turning of those feedstocks into the alcohol and biodiesel fuels that are used mostly for vehicular fuels. Biofuel feedstocks have evolved from the first generation of fuels that were developed from the food supply chain, such as corn and soybeans that produce ethanol and biodiesel. These first-generation plants are still the dominant feedstocks for biofuels. Second-generation biomass feedstocks that are beginning to contribute significantly to the total biofuel supply include cellulose from crop residues that are not part of the food chain, and non-food feedstocks such as switchgrass, some shrubs, and municipal biomass wastes. The future of biofuels may lie with the development of third-generation fuels from algae. Importantly, these can be produced in the laboratory and do not require land that can otherwise be devoted to food production.

Water and Electric Power Production

Power production includes hydropower, thermopower, nuclear generation, wind and photoelectric other renewable sources. Thermal power plants, the largest source of water demand, made up 70% of the total power production in 2014. The 29-nation Organization for Economic Cooperation and Development (OECD), led all regions in energy-related water withdrawals with 215 billion cubic meters of water (bcm; Table 8.2). This was followed by the Americas, with 165 bcm, of which the USA with 141 bcm was the major contributor. Measurements of water use are the water withdrawn from its source for any use; measurements of water consumption are the portion of water withdrawn that is not returned to the original water source after use.

Table 8.2 Water withdrawals and consumption by sector, billion cubic meters (bcm), 2014

Sector	Withdrawal (bcm)	Share of total energy water withdrawals (%)	Consumption (bcm)	Share of total water consumption (%)
Power	350	88	17	16
Fossil fuels	230	58	13	28
Nuclear	112	28	4	8
Renewables[a]	9	2	1	1
Primary energy production	47	12	30	64
Coal	11	3	10	22
Oil	8	2	6	13
Conventional	7	2	6	12
Unconventional	1	0	1	1
Natural gas	2	0	2	3
Conventional	1	0	1	2
Unconventional	1	0	1	1
Biofuels[b]	26	7	12	25
Totals	398	100	48	100

Sources EIA (2016b)

[a]Includes bioenergy, geothermal, concentrated solar (CSP), solar photovoltaics (PV), and wind

[b]Irrigated crops grown for biofuels

Water and Thermal Power Generation

Globally, most electricity is generated in thermal power plants. The most common fuels used by thermal power plants include all types of fossil fuels and nuclear energy. Together, all thermal power methods make up some 70% of the total global power-generating capacity. Almost every electricity generation process uses large amounts of water for the extraction, processing and distribution of fuels for generating electrical energy. Steam or heated and compressed air turbines in thermoelectric generation plants use either coal, oil, gas, geothermal or nuclear power as energy sources. Some of the heated water used in steam-driven power turbines is lost during cooling due to condensation. Water and energy are also critical components in the extraction, production and transportation of all fossil fuels and the growing and processing of feedstocks for biofuels. This means that sustainable energy would not be available without a sustainable supply of water. In the industrial sector, water for use in global thermoelectric generation makes up close to half of all the freshwater withdrawn (Wu and Chen 2017).

In 2014, almost 10% of all water withdrawals and 3% of all water consumption was for use by the energy sector, which includes primary energy production and electric power generation. At an estimated 350 bcm, 2014 withdrawals for power generation was the largest amount of all water withdrawn for the energy sector. Water withdrawals for the production of fossil fuels was second with annual withdrawals of 230 bcm, and withdrawals for nuclear power were 112 bcm.

Thermal electric generation requires water for steam-driven turbines and for cooling. In 2014, total withdrawals for thermoelectric power accounted for 88% of all water withdrawn for energy production, and for 36% of all energy water consumed. Other sectors that use water for production and related purposes include the production of fossil fuels, the nuclear power industry and renewable energy production.

Although the industry is fast converting where possible to natural gas, coal and nuclear power remain key fuels for power generators in many regions of the world, and particularly in China and the USA, where coal-fired thermoelectric generation is still the dominant user of coal. According to the U.S. Union of Concerned Scientists (UCS 2016), coal-fired power plants generate close to half of the electricity produced in the USA. They also have adverse effects on water quantity and quality. Water is used to extract, wash and sometimes transport the coal as a slurry by pipeline; to cool the steam used to turn the generator turbines; and to store coals as in storage ponds in an effort to control pollution from the plant. All of these actions can have major effects on water quality.

Table 8.3 Major producers of electricity from fossil fuels, 2014 data

Coal		Oil		Natural gas	
Country	TWh	Country	TWh	Country	TWh
China	4115	Saudi Arabia	152	USA	1161
USA	1713	Japan	116	Russian Federation	533
India	967	Iran	59	Japan	421
Japan	349	Iraq	50	Iran	196
Germany	285	Kuwait	43	Mexico	172
South Africa	232	Pakistan	42	South Africa	160
Korea	232	USA	40	Egypt	135
Russian Federation	158	Brazil	35	Korea	130
Australia	152	Mexico	33	Turkey	121
Poland	132	Indonesia	26	Thailand	119
Rest of the World	1372	Rest of the World	427	Rest of the World	2037
World Total	9707	World Total	1023	World Total	5155

Source EIA (2016a)
Note TWh = Terawatt hour (one trillion watts)

Water that is withdrawn from a surface or ground source and not returned to that same source for any reason is categorized as 'consumed water.' The loss is largely due to evaporation during cooling cycles. Primary energy production accounted for 30 bcm for 64% of all water consumption. The greatest amount of water consumed in the annual generation of electric power—17 bcm—is for coal production. The water withdrawals for energy production, electric power generation, and the sector's totals are shown in Table 8.2.

Outside the OECD, Asia accounts for half of water withdrawals in 2014 and 60% of consumption. By 2040, Asia will account for more than 55% of global water withdrawals for energy and almost 70% of its consumption. India, where the use of coal for power generation is increasing rapidly, is on the path to overtaking China as the largest source of energy-related water demand. The USA was the OECD's largest producer of electric power in 2014, accounting for 40% of all power generated in the organization nations, and for almost two-thirds of both water withdrawals and consumption in the OECD energy sector.

Thermoelectric power-generating facilities are becoming much more efficient, both in terms of the amounts of water they use and the energy they use to produce the power. Estimated 2010 U.S. thermoelectric withdrawals were 20% less than United States Geological Survey (USGS) estimates for 2005. On average, in 2010, 19 U.S. gallons of water were used to produce 1 kilowatt-hour (kWh) of electricity, compared with almost 23 gallons per kWh in 2005. Projects for achieving greater efficiency, and thus greater sustainability, are required for both water and energy production (Wang and Wang 2017).

Eastern U.S. states accounted for 86% of the total thermoelectric-power withdrawals in the USA and 75% of the related net power generation. The western states of California, Oregon and Washington account for the largest portion of hydroelectric power generation (Table 8.3).

Water and Thermopower Production

Research in the first decade of this century for a book on U.S. energy policy in the late years of the twentieth and early twenty first centuries led the authors to conclude that water use during the unconventional production energy such as fracking was only becoming an issue (Geri and McNabb 2011). The environmental focus then was on reducing air pollution by cleansing the coal-burning production of electrical energy. The topic of much energy-related research also focused on ways to improve the efficiency and reliability of the power transmission grid—both a security and an infrastructure problem.

There was little suggestion of reducing water use in coal or oil production at that time. Production of oil and gas in the USA and other oil and gas-producing nations was on a permanent slide to global scarcity. Nor was there much mention of a new method that would soon revolutionize oil and gas production around the globe. That method came to be known as hydraulic fracturing, or simply fracking. The term 'hydraulic' is a clear indication that water plays a major role in breaking open fissures in very deep geologic formations to release an unbelievably huge new supply of oil and gas. This process has made the USA a net exporter of oil once again.

The first fracking well was drilled in 1981 in Texas (Wright 2018). As other attempts followed, they too were unprofitable. It was not until 1998 that research came up with a new fracking formula would produce a profit. The process swept across the relatively stagnant Texas and Oklahoma oil fields to soon be taken up around the world. A USGS map showing levels of water used for fracking in the USA is shown in Fig. 8.2.

Every fracking well drill uses as much as 200,000 gallons of water to drill the well and start the fracturing, and ultimately requires a total of 3 million gallons of water. What is not absorbed is withdrawn as highly polluted wastewater. A 2015 Duke University research study reported that between 2005 and 2014, the wells then in operation generated an estimated 210 billion gallons of wastewater in order for the oil or gas to be extracted.

A typical 3-million-gallon fracking well could produce close to 15,000 gallons of chemical pollutants. Many of the chemicals used to improve the flow of oil or gas are proprietary and their composition is not published, complicating cleaning (Duke Staff 2015; Magill 2015). In addition, while underground, the water may be polluted with brine, heavy metals, radionuclides and organic materials that exceed drinking water standards and make treatment difficult and expensive (Soeder and Kappel 2009).

Fracking has many impacts on water supplies. Very large quantities of water are withdrawn from local resources for drilling, pumping, cleaning and transporting, which cuts into fresh water availability for agriculture, municipalities and other uses. The water taken from streams, rivers and other surface water sources, reduces habitat for the wildlife that depend on the water and natural foliage. The negative impact on water supplies is far greater in drought-prone regions such as the American Southwest. Many reports of the contamination of drinking water by wastewater and fluids that can flow back to the surface after injection have occurred (Gallegos et al. 2015).

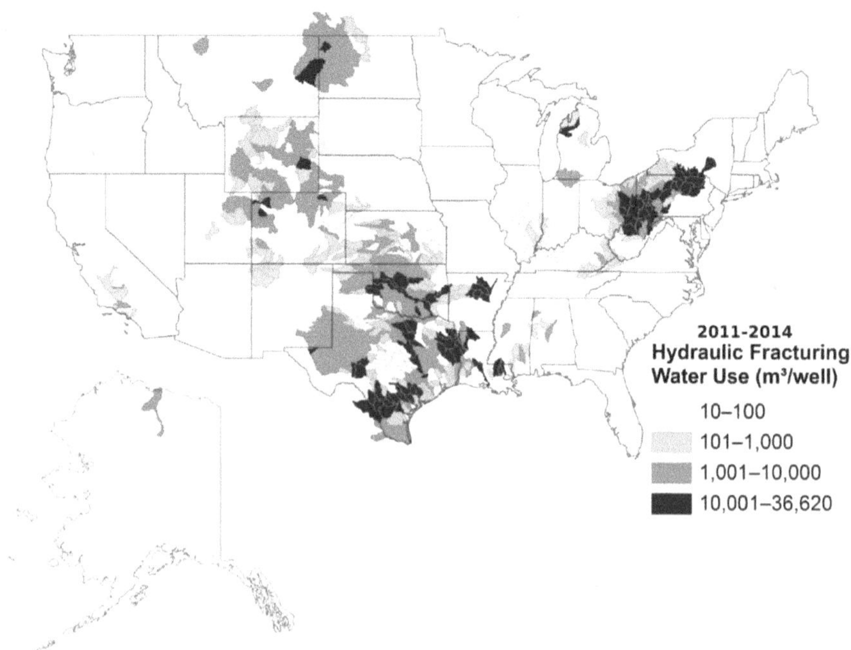

Fig. 8.2 U.S. regions with high water use for fracturing, 2011–2014 (*Source* USGS 2015)

Water and Nuclear-Generated Power

Of the close to 16,000 trillion kilowatts of electricity produced in the world every day, nuclear fission is the most water-intensive method of the chief thermoelectric generation methods. This includes both surface water and groundwater. According to the Union of Concerned Scientists, in 2008, nuclear power plants withdrew eight times as much freshwater as natural gas plants per unit of energy produced, and up to 11% more than the average coal plant (UCS 2011).

The USA was the major producer of nuclear-generated electricity, producing 831 trillion terawatts of electricity. In 2014, this was 32.8% of the world production of 2535 trillion TWh. The U.S. had 104 nuclear reactors producing electricity that year; 35 were boiling water reactors and 69 were pressurized water systems. Overall, a large nuclear power plant can withdraw from 800 million to 1 billion gallons of water a day. Water is for the steam that turns the generating turbines and for cooling the processes. Water for steam is usually cooled and re-used; water for cooling is either used once or recirculated.

Cooling Nuclear Power Water

In 2016, 60 of the 104 nuclear reactors in the USA used once-through cooling processes, 35 used wet cooling tower systems, and nine used dual cooling systems. A similar distribution was estimated for most of Europe. Nuclear plants in the UK (only using sea water), Sweden, Finland, Canada (using Great Lakes water), South Africa, Japan, Korea and China use once-through systems. Globally, 45% of nuclear plants use sea water and once-through cooling systems, 15% return their water to lakes, 14% to rivers, and 26% use cooling towers (WNA 2017).

Once-through systems return partially cooled water to the source from which it was taken. However, that returned water is usually as much as 30 degrees warmer than the source water. The resulting heat pollution can cause significant damage to fish and other aquatic life.

Energy and Water Shortages

Climate change has brought about global concern over water shortages in some areas, with the increasing number and severity of extreme weather events and changing precipitation patterns in others. Some of the phenomena exacerbating those concerns are the current and future consequences of receding glaciers, melting polar ice, droughts and diminishing mountain snow packs. At the same time, population increase and continuing global urbanization are creating greater water and wastewater stresses. This occurring at the same time that the demand for electrical energy in many parts of the world is growing.

That the increases in greenhouse gases are significant contributors to global climate change, and that the burning of fossil fuels for generating electricity and transportation is the major influence on climate change are both generally accepted premises. Furthermore, there is a growing demand for reducing greenhouse gases produced by burning fossil fuels and for placing greater reliance on renewable energy. However, the production of many types of renewable energy typically requires large amounts of water. Accordingly, the long-term sustainability of fresh water resources has become a concern of political leaders in much of the world. That concern has led to a shift the use of coal to natural gas and a slow shift to production power by renewable energy fuels.

Environmental protection and water conservation initiatives limit water availability for power generation. A variety of programs for alleviating the results of these trends have been implemented, such as requiring green development practices, the provision for controlling storm runoff by channeling storm water into open areas for recharging aquifers. There is also growing interest emerging in the co-locating of new and renewable energy projects with water reuse, reclamation and desalinization facilities. In addition, there is growing interest in new water conservation technologies, such as those developed in Israel and other countries.

Future Water and Energy Demand

The International Energy Agency (IEA) bases its future forecasts on two scenarios: The *New Policies Scenario*, which is based on ways governments see their energy sectors developing under current conditions, and the *450 Scenario*, which extends current policies with plans for limiting the average global temperature increase by 2100 to 3.6 °F (2 °C)—a low carbon concept. Table 8.4 shows the fuels used by the USA and the world for generating electricity.

By 2040, Significant declines in energy-related water withdrawals are expected to occur in the OECD, Americas, Europe and Eurasia. Increases in energy-related water withdrawals are projected to occur in all other areas, with India nearly doubling its water withdrawals from 35 bcm in 2014 to 68 bcm in 2040, and China increasing its withdrawals from 45 bcm to 55 bcm over the same period; withdrawals were more than double in the same period in Latin America.

Table 8.4 Fuels used in the USA and globally to generate electricity

Fuels	World (2014) (%)	United States (2016) (%)
Natural gas	22	33.8
Coal	38	30.4
Nuclear	11	19.7
Hydropower	17	6.5
Wind	–	5.6
Biomass	–	1.5
Solar	–	0.9
Oil	5	0.6
Other fuels	7	1.0
Totals	100	100

Source Data from EIA (2016a); world 2014 data the Shift Project 2015
Note Totals do not equal 100% due to rounding

By 2050, the demand for water will have increased by an estimated 30% compared with 2015. Demands for new water supplies will lead to greater demand for desalinated water and recycled wastewater. The electricity-intensive treatments required for this additional water will add to the growth in water demand that occurs with urbanization, and the extraction of more fossil fuels for thermoelectric generation of power. In addition, the long-term need for new and improved infrastructure to supply the water needed for greater energy production will require significant investments in more sustainable water and energy use.

Decentralized infrastructure as a way of self-generating electricity through the capture of methane generated during the wastewater treatment process is one approach for meeting this need. Additional application of renewable energy for power production such as wind power and hydropower will also be needed (Lee and Younos 2018).

Summary

Water plays many roles in energy production. It is an essential tool for conventional and nonconventional oil and gas production, it is necessary for steam-driven turbines that produce electricity, as well as for cooling in thermal and nuclear power generation plants. In all of these applications, the processes often severely pollute the water, which then may be discharged untreated into surface and groundwater supplies.

Water and energy are closely interconnected. For example, in most power plants, water is used for steam and for cooling the steam that drives power generating turbines. Water is also used during the mining of the coal, oil or gas that is used for generating the steam. Coal and water are often mixed in a slurry for pipeline transportation from mines to power plants. Water is used for refining oil and gas and for cleaning coal. Carbon-based fuels are also used for transporting water to where it is withdrawn to where it is need. Reducing the release of carbon into the atmosphere by converting it to biofuels requires very large amounts of water for growing and processing the biofuel crops.

The Union of Concerned Scientists in 2017 brought to light the problems associated with this interconnection when they cited that the links between energy and water ensure that the problems for one can creates problems for the other. Wherever using energy requires a large share of available water or where water resources are scarce or stressed, the connection can turn into a collision with dangerous implications for all sectors (UCS 2017).

References

Carter, Nicole. 2013. Energy-water nexus: The energy sector's water use. Congressional Research Service (CRS), Washington, DC. Accessed June 22, 2018 from https://fas.org/sgp/crs/misc/R43199.pdf.

Duke Staff. 2015. How much water does U.S. fracking really use? Duke University Campus Research. Accessed January 10, 2018 from https://today.duke.edu/2015/09/frackfoot.

EIA (U.S. Energy Information Administration). 2016a. What is U.S. electricity generation by energy source? Accessed February 8, 2018 from www.eia.gov/tools/faqs/faq.php?id=425&t=3.

EIA (Energy Information Administration). 2016b. World Energy Outlook 2016. Accessed December 7, 2018 from www.eia.gov/outlooks/ieo/pdg/0484)2016).pdf.

EIA (U.S. Energy Information Administration). 2018. Consumption of fuels for electricity generation, 2005–2017. Accessed June 22, 2018 from www.eia.gov/totalenergy/data/monthly/pdf/sec7_9.pdf.

Gallegos, Tanya J., Brian A. Varela, Seth S. Haines, and Mark A. Engles. 2015. Hydraulic fracturing water use variability in the United States and potential environmental implications. *Water Resources Journal* 51 (7): 5839–5845.

Geri, Laurance and David E. McNabb. 2011. *Energy Policy in the U.S.* Boca Raton, FL: CRC Press.

Industrial Waterworld. 2018. Produced water treatment plant to double capacity. *Water Briefs* 10 (3): 10.

Lee, Junesbok, and Tamin Younos. 2018. Sustainability strategies at the water-energy nexus: Renewable energy and decentralized infrastructure. *Journal of the American Water Works Association* 10 (2): 32–39.

Magill, Bobby. 2015. Study: Water use skyrockets as fracking expands. Climate Central. Accessed January 10, 2018 from www.climatecentral.org/news/fracking-water-use-skyrockets-19177.

RCT (Railroad Commission of Texas). 2018. Water use in association with oil and gas activities. Retrieved June 21, 2018 from www.rrc.state.tx.us/about-us/resource-center/faqs/oil-gas-faqs/faq-water-use-in-association-with-oil-and-gas-activities/

Schuster-Wallace, C.J., M. Qadir, Z. Adeel, F. Ranaud, and S.K. Dickin. 2015. Putting water and energy at the heart of sustainable development. United Nations University. Accessed July 19, 2018 from http://inweh.unu.edu/wp-content/uploads/2015/03/Putting-Water-and-Energy-at-the-Heart-of-Sustainable-Development-Web.pdf.

Soeder, Daniel J., and William M. Keppel. 2009. Water resources and natural gas production from the Marcellus shale. United States Geological Survey (USGS) Report. Accessed January 10, 2018 from https://pubs.usgs.gov/fs/2009/3032/pdf/FS2009-3032.pdf.

Sprang, Edward, W.R. Moomaw, K.S. Gallagher, P.H. Kirshen, and D.H. Marks. 2014. The water consumption of energy production: An international comparison. *Environmental Research Papers* 9 (10): 1–14. Accessed June 6, 2018 from iopscience.iop.org/article/10.1988/1748-9326/9/10/105002/pdf.

UCS (Union of Concerned Scientists). 2011. Nuclear power and water: Quick facts on nuclear power generation and water use. Cambridge, MA. Accessed November 1, 2017 from http://www.ucsusa.org/sites/default/files/legacy/assets/documents/nuclear_power/fact-sheet-water-use.pdf.

UCS (Union of Concerned Scientists). 2016. Energy and water use. Union of Concerned Scientists. Accessed June 19, 2018 from www.ucsusa.org/clean-energy/energy-water-use#.WylAIFVKiUk.

UCS (Union of Concerned Scientists). 2017. Energy and water use: Our energy choices have a direct effect on our water resources. Union of Concerned Scientists. Accessed December 6, 2017 from www.ucsusa.org/clean-energy/energy-water-use#.WigZ0FWnGUk.

USGS (United States Geological Survey). 2015. 2011–2014 hydraulic fracturing water use United States Geological Survey. Accessed January 11, 2018 from https://tse1.mm.bing.net/th?id=OIP.jWq9ymU_Kl8LqVB3hkDWhgHaFc&pid=Api&rs=1&c=1&qlt=95&w=141&h=104.

Wang, Sicong, and Shifeng Wang. 2017. Implications of improving energy efficiency for water Huiresources. *Energy* 140: 922–928.

WNA (World Nuclear Association). 2017. Cooling power plants. World Nuclear Association, London. Accessed November 3, 2017 from http://www.world-nuclear.org/information-library/current-and-future-generation/cooling-power-plants.aspx.

Wright, Lawrence. 2018. The glut economy. *The New Yorker* 93 (42): 42–53.

Wu, May, Michael Wang, and Salil Arora. 2009. Consumptive water use in the production of bioethanol and petroleum gasoline. Center for Transportation Research, Argonne National Laboratory (ANL). Accessed June 19, 2018 from www.great.es.anl.gov/publication-consumptive-water.

Wu, X.D., and C.Q. Chen. 2017. Energy and water nexus in power generation: The surprisingly high amount of industrial water use induced by solar power infrastructure in China. *Applied energy* 195: 125–136.

9

The Barrier of Failing Infrastructure

The American Water Works Association conducts a survey of the state of the water industry every year. Each year since 2013, water infrastructure renewal and replacement (R&R) has been the most pressing problem facing the many water systems surveyed (Murphy 2018). Closely related to the infrastructure issue was respondents' concern over where to secure the financing for capital improvements. Long-term water supply availability challenges were the fourth most pressing problem on the list of issues facing the water industry. Water supply and the infrastructure needed for the collection, transmission, treatment and delivery of water affected by the impact of climate change that have modified the world's hydrology patterns. Extreme weather events, hotter daily temperatures and shifting precipitation patterns are limiting many water systems' aging infrastructure to perform to their higher peak level needs.

Climate patterns considered during the design of new infrastructure can no long depend on reliable predictions of the future with the precision needed for these multi-million-dollar investments. Until the last half of the twentieth century, climate patterns appeared to change slowly, generally following historical records that enabled more or less predictable outcomes. Extreme water events were rare, and, although still damaging, were often less destructive than the extreme weather events occurring over the last several decades. Based on what they knew as stationary assumptions, meteorologists recognized that changes were occuring in long-term weather cycles and it was becoming increasingly difficult to predict when an extreme event, such as a "hundred-year storm."

D. E. McNabb, *Global Pathways to Water Sustainability*,
https://doi.org/10.1007/978-3-030-04085-7_9

Box 9.1 California high dam nearly collapses

California suffered through a prolonged drought from 2011 to 2017. The drought ended with heavy rain during the winter of 2016–2017. During the 1960s, the California Water Board had constructed the Oroville Dam on the Feather River, some 70 miles north of Sacramento, the state capitol. The heavy rain in January and February of 2017 caused the dam's water levels to rise rapidly, despite the drought-caused low level. The heavy rainfall filled the reservoir to dangerous levels.

"To relieve some of the pressure, the state released water down the main spillway at rates of up to 54,500 cubic feet per second (cfs). Then, on Tuesday, February 7, a crack in the spillway appeared and soon grew into a 250-ft crater. To prevent further damage, officials shut off water to the main spillway, but behind [the dam] the reservoir continued to fill. Officials were hesitant to rely on the dam's emergency spillway, because it had never been used and would wash earth and trees into the Feather River below. To test how much the dam managers could safely rely on the now-damaged main spillway, the state released relatively small amounts of water. This eroded the spillway's hole by another 50 ft. Because officials were unable to safely release enough of it, on Saturday, February 11, the reservoir's water rose to 901 ft—enough to flow over the emergency spillway, as it was designed to do".

"Unlike the main concrete spillway, the emergency spillway spills water over the lip of the dam wall. The water passes down an earthen hill and into the Feather River. On February 12, as water flowed down, it began to erode the hillside. Officials soon noticed that the erosion threatened to undercut the entire dam, which could cause a collapse and send a 30 ft wall of water into the valley and communities below. That same day, California officials ordered 188,00 people to evacuate the Feather River basin, which the reservoir now threatened to inundate. But the dam held, and the reservoir eventually dropped below 850ft. Two days after the evacuation order, the state downgraded the order to an evacuation warning and cautiously allowed residents to return to their homes. California borrowed $500 million from the federal government to pay for the Oroville Dam repairs, including $274 million that was previously approved by the President."

Source Water Deeply (2017)

The lingering concern of respondents over infrastructure issues in the 2018 report on the state of the water industry led the research designers to include a package of 11 infrastructure-related issues in the survey. Table 9.1 lists the items, their rank and rating on a 5-point scale, with 5 being the highest concern level.

The many operating environment challenges associated with the unprecedent changes occurring in the global climate are resulting in an increase in the intensity of storms, their duration and the frequency that weather extremes occur (Manocha and Babovic 2017; Hui et al. 2018). Water managers can no longer depend on traditional climate estimations. Planning for infrastructure

Table 9.1 Infrastructure concern scores of AWWA 2018 water system survey respondents

Rank	Challenging item	Weighted average[a]
1	Infrastructure reliability	4.4
2	Justifying R&R budgets to ratepayers	4.2
2	Access to infrastructure funding	4.2
3	Justifying R&R programs to commissions, boards, councils	4.1
3	System infrastructure emergency resilience	4.1
4	Establishing and following a policy for R&R	4.0
5	Prioritizing R&R needs	4.0
5	Coordinating R&R with other operations	3.9
5	Defining and maintaining appropriate levels of service	3.9
6	Meeting customer expectations	3.7
7	Infrastructure related regulatory constraints	3.6

Source From AWWA material in Murphy (2018)
[a]5-point scale; 1 = unimportant 5 = critically important

renewal or expansion must consider unpredictable climate-influenced changes in regional hydrology patterns. This is particularly true when planning for dams and reservoirs. More and longer droughts are increasing the need for more storage capacity, while severe flooding calls for releasing more water when heavy rains exceed existing storage capacity. Box 9.1 describes the unplanned-for infrastructure crisis that occurred when the nation's highest dam nearly collapsed in 2017.

The high cost of new water infrastructure does not just apply on the scale of replacing a potentially failing dam. It can also occur because of a need to shift water supplies from one region to another, such as the canals that move water from Northern California to Southern California, or that transport water from Central and South China to the more populous but water-scarce Northeastern China. It can also result from poor water management decisions. The harmful effects of a water system on the children of Flint, Michigan is a tragic example of such a failure. As a cost-saving measure, on Abril 25, 2014 the Flint water supply was shifted from purchasing water from the Detroit city system to withdrawing water from the heavily polluted and heavily corrosive Flint River. Not long after, Flint residents began to complain about foul-tasting and foul-smelling drinking water. Independent testing services reported that the reason for the bad water quality was because of dangerously high levels of lead in the water. After denying a problem existed for several months, Flint city and state official finally admitted a problem existed. It was then determined that the water was rapidly corroding the lead connection water pipes (Morris 2017). Nearly 9000 children were drinking water contaminated with toxic lead. Particularly dangerous to infants and toddlers, lead can impair learning ability, lower intelligence, and make children have trouble controlling their emotions. One year later, the Flint city council voted to shift back to Detroit water. However, the state-appointed financial emergency manager overruled the council, citing the US$4 million savings expected from using Flint River water, regardless of the lead.

Why Infrastructure Fails

There are more than 51,000 community water utilities in the USA. They range from the 500 or so very small operations to the urban center utilities that serve hundreds of thousands of customers (Table 9.2), and all must depend upon a mix of surface and buried in structure assets to do their job. They cannot fail to provide clean, safe water reliably and affordably to everyone that signs up to receive their services. It is no wonder, then, that utilities large and small worry about how to plan for, purchase, install and operate the miles of pipes and treatment facilities that they are responsible for.

Much of the world's water infrastructure is in bad shape. The USA is considered to be among the nations facing severe water infrastructure failure; the American Society of Civil Engineers gave the U.S. drinking water systems a grade of D in its 2017 Infrastructure Report (ASCE 2018). Most of the more than 1 million miles of iron, cement and plastic pipe that carries water to municipalities was installed in the early to middle years of the twentieth century. On average, these pipes have a lifespan of 75–100 years. Water suppliers are now facing the challenge of producing the billions of dollars needed to replace the miles of pipe that are literally falling apart or corroded beyond repair.

In total, U.S. water suppliers average an estimated 240,000 water main breaks annually, losing billions of gallons of treated water. Pipeline breaks and storm runoff and floods often allow polluted water to enter the stream of freshwater delivered to homes, schools and other public buildings, businesses and industries. The annual cost estimate for repairing freshwater and wastewater infrastructure and expanding delivery to the growing population is estimated to be US$1 trillion.

Achieving sustainability in the world population's access to a sustainable supply of clean, safe water is not going to be easy. Perhaps the biggest barrier is that the regions for which achieving this goal are precisely among the least able to pay for the infrastructure work that must be done before sustainability can come even close to fruition. Moreover, arriving at water

Table 9.2 Numbers of water supply organizations in the USA (% rounded)

Category	Very small 500/ less	Small 501–3300	Medium 3301–10,000	Large 10,001–100,000	Very large >100,001	Totals
Number of systems	28,595	13,727	4936	3851	426	51,535
Total systems (%)	55	27	10	7	0.8	100
Population served	4,738,080	46,688,745	28,758,366	109,769,366	137,250,793	300,205,288
Population total (%)	1.6	6.6	10	37	45.7	100

Source AWWA (2015) from EPA data

sustainability in these areas is also blocked by the additional barriers of high population growth, urbanization and water scarcity. Historically, municipal water utilities have not charged customers the full cost of providing water or wastewater services. This was possible because in the years after 1950, federal grants were available to fund water and wastewater infrastructure that had been deferred during the Second World War. Moreover, in the rush to provide needed housing, developers contributed substantially to support system expansion.

Funding for Infrastructure Repair and Replacement

To help water and wastewater utilities keep up with the infrastructure needed after the deferred maintenance necessary during World War II, the federal program to aid municipal wastewater treatment plant construction was authorized in 1972 with passage of the Clean Water Act (CWA). Established originally as a grant program in 1972, the grant program was replaced with a new system not long after. Prior to the 1987 amendments to the CWA, water and wastewater treatment assistance was provided in the form of grants made to municipalities (Copeland 2012). Cities had to find new ways of funding repairs to and replacement and expansion of their water infrastructure if they were ever to reach the cities' sustainability goals.

The World Economic Forum (WEF) defines infrastructure as consisting of two components: economic and social infrastructure. Economic infrastructure is what enables a modern society to function. It includes roads, railroads, ports and airports, energy for light and power, water and wastewater, solid waste, and telecommunications facilities. Social infrastructure consists of facilities for public services, including government buildings, hospitals, police and defense capabilities, housing and education and community facilities.

The global need for investments in water and wastewater infrastructure repair and new construction exceeds US$3.7 trillion annually; however, only something like $2.7 trillion is invested in infrastructure annually, resulting in a gap of $1 trillion between what is needed every year for water-related infrastructure and what is invested.

The USA, OECD and multinational development banks have extended billions for infrastructure investment to low and middle-income countries (Table 9.2). However, the consensus is that this aid and low interest loans have not been enough to close the gap created by the absence of a basis of quality infrastructure. One solution is for private capital to step into help meet the global infrastructure gap. Government agencies and development banks can help by identifying a number of bankable infrastructure projects (Moser and Nealer 2016). The USA participates directly in at least four organizations that fund infrastructure improvements: ISAID, USTDA, MCC and the IDB.

Water and wastewater projects are sometimes included in transportation-related infrastructure and sometimes in industrial development infrastructure. However, most U.S. utilities can no longer depend on a grant to fund their infrastructure repairs and replacements. They must now look to low-interest loans from state revolving infrastructure funds or pay for purchases from customer rate payments.

Globally, there has been a decline since 2012 in the number and extent of infrastructure projects initiated and in the size of the projects authorized. The World Banks's infrastructure spending peaked in 2010 at US$19.08 billion. The World Bank reported infrastructure spending of US$17.21 billion in 2014, US$14.63 billion in 2015, and US$15.42 billion in 2016; projected 2017 spending was US$13.92 billion. In the USA, the federal government's funding policy for water infrastructure has evolved through three different policy positions. Originally, funding for water systems was considered the responsibility of local cities and other incorporated communities, or left to individual commercial operators to self-fund.

Many private water systems still use the private ownership funding approach. Change began in the 1960s as more utilities began taking over private systems that could not keep up with the rapid population increases and urbanization. The Passage of CWA in 1972 resulted in billions of dollars in system construction federal grants. The federal grant policy again changed with passage of the Water Quality Act of 1987. Grants were changed to low-interest rate loans, funded by the federal government and operated by the states. This was the birth of the revolving loan fund model, the system that was still operating in 2018.

Infrastructure Repair, Replacement and Expansion

Water managers must plan for repairing, replacing and often expanding a wide variety of critical infrastructure. Much of that infrastructure is out of sight, buried deep underground. Water mains, the underground pipes that carry large quantities of water, often under pressure, from reservoirs or groundwater pumping stations and water treatment facilities, are the main component in water suppliers buried infrastructure. These pipes are made of many different materials, from the early wooden pipes, a small number of which are still in operation. The seven types of water mains in use today include asbestos cement, cast iron, cast iron-cement lined, ductile iron, reinforced concrete, pre-stressed concrete and polyvinyl chloride (PVC) plastic. Archaeologists have found that the Babylonians used clay sewer pipes in modern Iraq more than 6000 years ago. Ceramic and even lead water mains were common in ancient Roman cities, including Roman urban areas in England.

Water mains are almost always buried. Access to the break means cutting off service to surrounding areas. Digging through streets and other paved areas to get at a water leak can cause a serious disturbance to traffic, businesses and civic activities. As a result, when water mains break they cause great damage to a water utility's finances and its reputation. Much of the mains around the globe are old and subject to failure at any time. The average lifespan of seven different types and three different sizes of pipes used as water mains in the Western USA is shown in Table 9.3. Similar estimated lifespans exist for the states of New England and for the Midwest. Across all regions and types of pipe materials, cast irons pipes appear to have the longest install operating lives. Very small pipes (1-inch or smaller) appear to have the longest lifespans, with 135 years in the Midwest and 130 years for both the South and West USA. PVC pipes were listed as having the shortest lifespans compared with other materials.

Table 9.3 Estimated service lives by materials in Western USA (average years of service)

Pipe size			
Pipe material	Large	Medium and small	Very small
CI	115	105	130
CISL (LSL)	100	100	100
CICL (SSL)	75	75	75
DI (LSL)	110	110	110
DI (SSL)	60	60	60
AC (LSL)	105	105	105
AC (SSL)	75	75	65
PVC	70	70	70
Steel	95	95	95
Concrete and PCCP	75	75	75

Source From material in AWWA (2012)

CI, cast iron; CISL, cast iron cement lined; DI, ductile iron; AC, asbestos cement; PVC, polyvinyl chloride; PCCP, prestressed concrete cylinder pipe

LSL: Long service live from some combination of benign ground condition and improved laying practices, etc.

SSL: Short service life from some combination of harsh ground conditions and early laying practices, etc.

Water main sizes can run from 1-inch or less to more than one foot, with sizes depending on the intended use. For example, the following main size requirement are from the Lincoln, Nebraska Department of Public Works and Utilities, Water Main Design Standards (2000, Chapter 2.10): "The minimum main size is 6-inch (150 mm) diameter for residential areas, 8-inch (200 mm) diameter mains for commercial areas and 12-inch (300 mm) diameter for industrial areas shall be required unless it can be demonstrated that a smaller main will provide the proposed and future development with adequate pressure and fire-flow. Large industrial or commercial developments may require larger mains to assure adequate pressure.

The Lincoln water department has established a feeder network of larger mains on approximately one-half mile intervals. A 12-inch (300 mm) diameter or larger main shall be required in these locations." These standards comply with those of the American Water Works (AWWA) and other regional and state associations and governments.

Why Water Mains Break

Water mains break for many reasons. In many areas, a common cause of breaks in iron pipes is the result of external corrosion. Many soils are highly acidic and corrode the metallic pipe from the outside in. Extreme weather changes can then cause the ground to swell and contract, placing excessive pressure on the buried pipe, causing any weakened pipe to break. Water mains break when the water inside them expands when the weather gets close to freezing. This increased pressure then causes weakened pipes to rupture. Heavy road traffic is also a common cause of breaks in older buried pipes, much of which in urban areas has been underground carrying water for 60–100 years. It is no wonder that breaks in municipal water main are and will continue to be common.

A major problem for water suppliers is that too often the water leaking from those breaks does not reach the surface, resulting the loss of large amounts of treated water before the leak is found. However, modern technology has made locating leaks much easier. As water leaks from pipes it makes a noise. Small, one-person operated electronic equipment can identify a leak for its noise and repair can begin soon thereafter. In its 2012 study on water and wastewater infrastructure needs for the AWWA, the consultants came to the following conclusions:

1. The needs are large. Nationwide investments for buried infrastructure alone total more than US$1 trillion between now and 2040. Pipes at the end of their service life must be replaced or repaired and new pipes installed to meet the needs of a growing population.
2. Household water bills will go up. The money to pay for repairing and replacing old and installing new infrastructure must be found. Few if any grants or subsidies are available. Utilities must therefore borrow the funds and pay back the loans with increases in their rates. The AWWA report warned that in some cases, household water bills could triple what they were in 2012. However, some communities have adopted the requirement for developers to pay for the public service cost increase resulting from that growth in a pay-as-you go plan. Regardless of who pays and how, the total cost paid by the community will rise.
3. There are regional differences in cost increases. Generally, cost increases in the South and the West USA are likely to be greatest. This is because these are the regions with the greatest population increases, and the resultant need for additional infrastructure. Additionally, smaller communities will possible be forced to pay more for their water infrastructure bill than larger communities, owing to there being fewer people over which to spread the cost.
4. The costs keep coming. The national bill for water infrastructure replacement alone is expected to grow from approximately US$12 billion in 2010 to close to US$30 billion by 2040.
5. Postponing investment only makes the problem worse. Utilities can no longer defer replacement and repair. The longer infrastructure investments are postponed, the more failures can be expected in the already aged pipes and other assets.

Water mains are clearly not the only elements in a water or wastewater utility's infrastructure. To achieve sustainability, all that infrastructure has to identified, located, its vital needs determined, and repairs or replacements planned and initiated—all of which as to be paid for. Water managers are lucky; a pathway to accomplishing this task is tested, proved successful, and available. "This suggests that a crucial responsibility for utility managers now and in the future to develop the processes necessary to continually improve their understanding of the 'replacement dynamics' of their own water systems. Those dynamics should be reflected in an Asset management Plan (AMP) and, of course, in a long-term capital investment plan" (AWWA 2012, 130). A comprehensive asset management plan will not make paying for infrastructure repair and replacement any easier, but it will serve as a vital signpost on the path to water sustainability.

Repairing and Expanding Water Infrastructure

Water utilities in USA, like those in most of the other developed nations, are now facing the critical need to fund and implement a systematic schedule of infrastructure replacement. They must do this within a hostile environment that is fast being shaped by the effects of climate change and population growth. As a result, with every gain in their progress toward a competent and reliable water supply systems, they encounter a new barrier in the way of achieving sustainability. Still, work toward sustainability, they must. World sustainability planners point out that following a business-as-usual approach to water management will result in a global 40% deficit in freshwater by 2030 (Furlong et al. 2016). Water providers must integrate traditional water and wastewater infrastructure with a growing list of water supply technologies that include desalination, reuse at levels increasingly requiring treatment that meets and sometimes exceeds potable status, rainwater and stormwater management for supplementation of basic supplies and flood control, replacement of combined rainwater and sewage systems with systems that treat these water classes separately and eliminate sewer overflows. If that is not enough, water managers are increasingly required to contribute to the design and implementation of green architecture and urban design, ecosystem protection and the protection of endangered species in the peri-urban region directly adjacent to an urban area that is between urban and rural regions.

Recognition that the peri-urban areas of a city are constantly changing and moving was recognized as early as the 1970s. The OECD recognized the losses that occur to cities with disappearance of farmlands adjacent to cities in a discussion of the contribution of peri-urban area thus: "The impacts of economic growth and physical expansion of the urban area are not confined within urban boundaries; they reach into much wider areas surrounding urban centers, creating so-called 'rurban areas', 'urban fringe areas', or 'peri-urban areas'. While the peri-urban area retains the characteristics of the rural area, these are subject to major modifications: changes take place with respect to physical configuration, economic activities, social relationships and so forth" (quoted in UN Food and Agricultural organization report by Laquinta and Drescher 2000).

Peri-urban areas often withdraw their irrigation water from the same aquifers as the cities to which they are adjacent. In period of water scarcity, the result can be a conflict for water supply preferences between municipal, industrial and agricultural communities. The more that new arrivals settle in the peri-urban region, the greater the demand they hold for the existing water.

Sustainable development, the goal of all cities, is curtailed when water supplies are stressed due to the lack of water or the inability of the water and wastewater infrastructure to satisfy that need.

Water infrastructure includes the facilities for withdrawing raw water from a primary source, the pipes to move the water to a treatment facility, the facilities to store untreated reserves of water, the one or more treatment facilities for cleaning and making the water safe for human consumption, and the miles of underground pipes that bring the now treated water to where it is needed. Regardless of whether it is done by a water supplier or a separate entity, water infrastructure also includes the collection, treat and transportation of wastewater and sanitized sludge.

Urban growth requires expansion decisions on water infrastructure to center on what investments in water infrastructure should be made and where. The facilities to do all these tasks today must be constantly monitored and repaired or replaced when necessary. An urban water supplier has a responsibility to never run out of water or to distribute water that is unsafe. But, the existing infrastructure to do this is not able to meet all the needs of an expanded body of users. Thus, water utility planners must include planning for expansion and make infrastructure recommendations. Infrastructure recommendations are then analyzed and evaluated along with all other budgetary needs of the utility and the city. This further requires political, economic and social needs in an integrated approach to infrastructure planning. Water infrastructure planning has become "inherently subjective and affected by social and political dimensions, as well as prone to unavoidable conflicts" (Furlong et al. 2016, 2)

Clearly, the old engineering-based decision framework of the traditional way of planning infrastructure needs for urban water systems must be expanded to include all infrastructure needs, including the regions immediately surrounding the city. The need to provide service to new customers in peripheral regions of the city means significant new investments in water and wastewater infrastructure. Hence, population growth and urbanization expansion should be included in broader urban development objectives. Doing so before the expansion occurs will enable the expansion needs to be included in capital planning as well as in integrated strategic plans that consider water and urban expansion planning in combination.

Traditional meter readings are done by one or several technicians either walking or driving by a neighborhood. Individual readings are then tracked to determine water use for usage-based billing or to identify possible water leaks in the structure, New residential water meters uninstalled can range from US$150 for a regular meter to US$500 or more for a digital, 'smart' meter. One or two workers installing a new meter can take 1 or 2 hours, or as much as US$35,000 or more for installation of a new meter and trenching is required and pavement or sidewalks need to be replaced. For even small systems of a thousand customers. To replace 1000 meters at an average of US$450 per meter would mean a cost of US450,000, a very large budget item for most water suppliers.

U.S. Water Infrastructure Needs

The EPA has conducted a series of U.S. drinking water infrastructure needs and assessments. Its sixth report to congress on the condition of the Nation's water infrastructure was released in 2015. An estimated US$472.6 billion will be needed to maintain and improve the nation's drinking water infrastructure. Highlights of the EPA's projected costs of for needed water infrastructure in four main categories of water system repairs and replacements from 2015 to 2035 are:

- Water distribution and transmission: US$312.6 billion to replace or refurbish aging or deteriorating pipelines;
- Water treatment: US$83 billion to construct, expand or rehabilitate infrastructure to reduce contamination;
- Water storage: US$47.6 billion to construct, rehabilitate or cover water storage reservoirs; and
- Water source: US$21.8 billion to construct or rehabilitate intake structures, wells and spring collectors.

The EPA groups water infrastructure into four main categories with a fifth category for items that do not fall into any of the four main groups. Twenty-year projected infrastructure investment totals are made for three classes of water systems: large (serving more than 100,000 people), medium (serving from 3300 to 100,000 people), and small community systems (serving 3300 or fewer people). Three other classes are very small, not-for-profit non-community systems, American Indian and Alaska Native Village systems, and a category of costs of infrastructure associated with proposed and recently enacted water regulations. Table 9.4 lists total 20-year needs by system size and types of infrastructure.

Table 9.4 Total 20-year needs by system size and infrastructure categories, 2015–2035

	Infrastructure category (2015 billion US dollars)					
System type	Distribution and transition	Treatment	Storage	Source	Other	Total need
Large > 100,000	$120.1	$32.5	$13.4	$6.1	$2.3	$174.4
Medium > 3300	$144.9	$33.7	$20.0	$7.8	$4.3	$210.6
Small < 3300	$44.6	$11.0	$11.3	$6.7	$0.9	$74.4
Non-profit very small	$0.6	$1.0	$2.5	$1.0	$0.0	$5.1
Total states and US territories	$310.2	$78.2	$47.1	$21.6	$7.4	$464.6
American Indian and Alaska Native village systems	$2.4	$0.6	$0.5	$0.2	$0.1	$3.8
Infrastructure costs with regulatory needs	$0.0	$4.2	$0.0	$0.0	$0.0	$4.2
Total national need	$312.6	$83.0	$47.6	$21.8	$7.5	$472.6

Source EPA (2015)
Note Costs are in USD

The four categories are items associated with drinking water sources, transmission and distribution items, raw water treatment equipment, and untreated and treated water storage facilities. The main item in the other category are products that fit into system-wide supervisory control and data acquisition (SCADA) or emergency generators.

The under-esimated total of US$312.6 billion for the projected 20-year costs for transmission and distribution needs is 66% of the total projected needed amount; the total will be much greater. The category includes R&R of existing buried water pipes, installation of new pipes where it is necessary in order to eliminate stagnation, and installing or rehabilitating pumping stations. A large portion of the costs for this water-related infrastructure is for replacing vital aging and deteriorating transmission and distribution mains. The EPA estimates there is a total of 2.24 million miles of these mains for all system types in the USA. The category also includes replacement of meters for recording flow and water use, backflow protection, and valves needed for controlling flows and isolating sections during repairs of broken pipes.

Treatment needs to include new construction to house treatment facilities, for expansion to meet demands of a growing population, and rehabilitation of older treatment facilities. Treatments are necessary to make the water safe and palatable for consumption, and to improve the odor and the color of drinking water. The total 20-year investment needs for all classes of providers is estimated to be US$83.0 billion. The need varies with the quality and composition of the raw rater available to the system. Examples of treatments can be as simple as disinfecting the raw water by chlorination to filtration to remove sediments, treatments to eliminate lead, arsenic and many other naturally occurring and human-associated water contaminants. Facilities required for advanced treatment of wastewater for reuse vary in cost according to the intended final uses of the recycled water. However, water scarcity in arid regions of the country are increasingly common.

Investments in infrastructure at the source point include construction and rehabilitation of treated water storage facilities, drilling wells and other source collection facilities. The costs of dams or raw water reservoirs have not been included in the EPA's estimates of US$21.8 billion infrastructure needs for this category since 1995. The only exception is if the raw water basins are located at the treatment facility and storage is part of the treatment process. Storage requirements that are included in the EPA's 20-year infrastructure needs estimate of US$47.6 billion include building new finished water storage tanks and related equipment. Shares of the total 20-year infrastructure needs estimates by rehabilitation and new construction are displayed in Table 9.5.

Table 9.5 Shares of 20-year needs by new vs. existing infrastructure and by project category

Project category	Existing infrastructure[a] (%)	New infrastructure[b] (%)
Source	75	25
Treatment	91	9
Storage	90	10
Pipe	94	6
T&D other than pipe[c]	96	4
Other needs[d]	87	13
Overall	93	7

Source EPA (2015)
[a]Included rehabilitation, expansion, upgrade, regulatory required projects
[b]Drinking Water State Revolving Fund-eligible only
[c]Transmission and distribution (T&D) includes such infrastructure as pump stations, valves, meters, backflow prevention and service lines
[d]Includes SCADA systems, emergency generators not associated with other uses

Summary

Climate change, with population growth and urbanization, is changing the weather patterns the world has been familiar with for centuries. In doing so, it is altering the ability of the world's water suppliers to meet growing demand. The infrastructure needed to collect, store, treat and distribute water is stressed to the point of collapse in many areas. Much of that infrastructure is between 50 and 100 years old and is long past the time when it must be replaced. Meanwhile, large areas of the world are enduring growing water scarcity, while other areas are forced to develop ways of dealing with too much water.

Planning for achieving sustainability of clean, safe freshwater and the need for wastewater service requires rate increases to pay for the needed infrastructure repairs and replacements. Willingness to accept and pay for change is not a widespread characteristic among many ratepayers.

Estimates of the costs that will be incurred for repairing, replacing and expanding water infrastructure by 2035 vary. For example, in 2012, the AWWA-sponsored report on water infrastructure needs over the 25 years from 2010 to 2035 estimated that U.S water systems would require infrastructure investments totaling US$1 trillion to replace the existing water pipes that had reached its useful life and to fund new infrastructure needed to supply increased demand by all sectors brought about by population growth. That estimate did not consider the need to replace infrastructure damaged by climate change-associated hydrological conditions. Five years later, in the sixth of its series of reports to Congress on water infrastructure needs, the EPA estimate was less than half of the AWWA projection with a total national water infrastructure need from 2015 to 2035 at an estimated US$472.6 billion and included a longer period (25 years instead of 20). The reason for the discrepancy was that the EPA was restricted in its estimates from including needs resulting from population growth. Hence, the EPA total was based on 2015, when the U.S. population of was 321.37 million. It is projected to be 334.50 million in 2020, 359.40 million in 2030, 380.22 million in 2040, and 390.33 million in 2050.

References

ASCE (American Society of Civil Engineers). 2018. 2017 Infrastructure report card: Drinking water. American Society of Civil Engineers. Accessed May 21, 2018 from www.infrastructurereportcard.org/wp-content/uploads/2017/01/Drinking-Water-Final.pdf.

AWWA (American Water Works Association). 2012. Buried no longer: Confronting America's water infrastructure challenge. Accessed May 21, 2018 from www.awwa.org/portals/0/files/legreg/documents/buriednolonger.pdf.

Copeland, Claudia. 2012. Water infrastructure financing: History of EPA appropriations. Washington, DC: Congressional Research Service (CRS). Accessed August 2, 2018 from https://fas.org/sgp/crs/misc/96-647.pdf.

EPA (Environmental Protection Agency). 2015. Drinking water infrastructure needs survey and assessment: Sixth Report to Congress. Accessed August 10, 2018 from www.epa.gov/sites/production/files/2018-03/documents/sixth_drinking_water_infrastructure_needs_survey_and_assessment.pdf.

Furlong, Casey, Saman De Silva, Lachlan Guthrie, and Robert Considine. 2016. Developing a water infrastructure planning network for the complex modern planning environment. *Utilities Policy* 38 (2016): 1–10.

Hui, Rui, Jonathan Herman, Jay Lund and Kaveh Madani. 2018. Adaptive water infrastructure planning for nonstationary hydrology. *Advances in Water Resources*, In press (2018): 1–12. Accessed July 2, 2018 from https://doi.org/10.1016/j.advwatres.2018.05.009.

Laquinta, David L. and Axel S. Drescher. 2000. Defining the peri-urban: Rural-urban linkages and institutional connections. Research paper funded by the UN Food and Agricultural Organization (FAO). Accessed August 9, 2018 at www.fao.org/docrep/003/X8050T/x8050t02.htm.

Manocha, Nishtha, and Vladan Babovic. 2017. Development and valuation of adaptive pathways fore storm water management infrastructure. *Environmental Science & Policy* 77 (2017): 86–97.

Morris, John C. 2017. Planning for water infrastructure: Challenges and opportunities. *Public Works Management and Policy*, 22(1): 24–30.

Mosey, Helen and Eris Nealer. 2016. Barriers to bankable infrastructure: Incentivizing private investment to fill the global infrastructure gap. Washington, DC: Center for strategic and International Studies. Accessed November 14, 2017 from https://csis-prod.s3.amazonaws.com/s3fs-public/publication/160308_Moser_BarriersBankableInfrastructure_Web.pdf.

Murphy, Maripat. 2018. 2018 state of thewater industry: The challenge of building resilience. *Journal of the American Water Works Association* 110 (8): 61–71.

Water Deeply. 2017. Oroville dam incident explained: What happened, why and what's next. Accessed July 3, 2018 from www.newsdeeply.com/water/articles/2017/06/14/oroville-dam-incident-explained-what-happened-why-and-whats-next.

Part III

Pathways to Water Sustainability

10

Pathways to Water Supply Security

Water security is defined as the ability to access sufficient quantities of clean, safe water for municipal, industrial and agricultural water users to maintain adequate standards of food and goods production, proper sanitation, and sustainable health care. Water security is not just the absence of water scarcity; rather, as it is defined by the Sustainable Water Partnership (SWP), water security is "the adaptive capacity to safeguard the sustainable availability of, access to, and safe use of an adequate, reliable and resilient quantity and quality of water for health, livelihoods, ecosystems and productive economies" (SWP 2018).

Our world is indeed water rich, but only a little more than 1% of all that water is usable freshwater. And, much of this water is so polluted that it is almost impossible to get clean again. In many parts of even the most water-rich nations, water scarcity is the norm. As a result, nations try to do all they can to ensure the water supply they have now, and the water supply they will need tomorrow, is secure. Because, as the SWP points out, "Without water security there can be no national security … water is essential to the stability of every country on the planet" (SWP 2018).

This chapter looks at four examples of how areas with extreme water scarcity are working to achieve security and sustainability of water supplies. Case studies are described in this comparative study of alternative affordable methods of approaching the need for achieving sustainability in water scarce regions. A widening and possibly insurmountable gap between water supply and demand is projected to occur by 2030 in four locations: India, China, South Africa and São Paulo, Brazil. Total water demand already

D. E. McNabb, *Global Pathways to Water Sustainability*,
https://doi.org/10.1007/978-3-030-04085-7_10

exceeds current water requirements. A stable and secure balance between demand and supply is critical for these and other regions. Water demand is measured as the actual withdrawals from surface water, groundwater or nonconventional sources. The finite water supply for water-use sectors is the total of the renewable water resource, with a theoretical upper limit of all the surface water and groundwater that can be withdrawn for all uses in the region.

The Water Resources Group (WRG) is a private sector organization formed to study and report on the water supply problem and the estimated cost of achieving water sustainability, with funding by the World Bank. Their report, *Charting Our Water Future*, reviewed future gaps between water demand and water supplies in four countries: China, India, South Africa and Brazil, with an emphasis on the state of São Paulo. Since the 2009 release of the report, extended periods of drought, continuing population growth and urbanization have exacerbated the supply problems facing the four areas.

The WRG analysis included four estimates of: (1) the impact of economic and social growth on the nation's water resources; (2) costs and requirements of meeting future water demand; (3) comparisons of different growth scenarios; and (4) the implications for stakeholders. Recommendations included a mix of solutions to increase water supply and to improve water productivity. The group concluded that combinations of existing low-cost remediation approaches could eliminate up to 100% of the predicted 2030 demand–supply gap in the countries studied. However, nearly ten years after the release of that report, it can now be seen that the actions taken were not extensive enough to come even close to closing the regions' supply–demand gaps.

The sustainability pathways that countries adopt must be based on some combination of solutions to three underlying water supply and demand factors: (1) means of expanding supply; (2) increasing the efficiency and productivity of existing water uses; and (3) reducing demand by shifting the economy towards less water-intensive activities. Different combinations of these three options will determine which pathway nations will follow in achieving their sustainable water supplies. Each pathway is expected to have different costs and time frames for completing the processes that will bring about their distinct sustainability package.

Varying the country's economic outputs, the water intensity of its economy, and its reliance on supply-side infrastructure are all potential levers for achieving water security. The critical question to be answered is: what is the optimal mix for my country, its people and economic status? (McKinsey and Company 2009). In the beginning of arriving at the right answer to this question, these four classes of technical measures, broadly covering increases in supply and different types of efficiency and productivity characteristics, may be analyzed:

1. Agricultural productivity measures that could improve both the efficiency of water use in irrigation and crop yields on both rain-fed and irrigated lands.
2. Industrial efficiency measures.
3. Domestic and municipal efficiency measures.
4. Supply measures that increase accessible, reliable, environmentally sustainable supply.

The analysis followed cost-effective ways to close the 2030 supply–demand gap projected for the four countries using only two of the three approaches: increasing water supply and improving water productivity. The research assumed that the two solutions together could likely close 100% of the 2030 supply–demand gap.

Estimated annual capital costs to achieve this goal for the four countries was estimated to be US$19 billion in 2030. Expanded to cover the entire globe, the annual estimated costs were US$50 billion for integrated supply and productivity solutions for closing the projected water gap. This was thought to be one-quarter of the roughly US$200 billion that would be required for expansion of water supply solutions alone.

The cost curve developed for each of the four regions led the researchers to conclude that "different local costs and impacts of otherwise similar measures, [indicate] that the stand alone potential for technical measures to close a water supply–demand gap is unique for each country" (McKinsey and Company 2009, 73):

- India's cost-effective solution rests squarely on improving the water productivity of agriculture;
- China's solution focuses on better managing the rapidly growing demand from industrial and urban uses;
- Brazil's solution would depend heavily on industrial as well as municipal and agriculture efficiency;
- South Africa's least-cost solution would involve a combination of supply and industrial efficiency levers.

The Goal of Water Security in India

Although not on the 2008 list of the top most water-stressed countries, nor included among the nations on the projected 2040 water-stressed list, India's huge population justifies its inclusion in the McKinsey Group shortlist of regions facing a water crisis. India's population growth, climate change-based disruption in precipitation patterns, and uncontrolled urbanization together put the water supply issue high on the list of concerns of the Indian government. India is on the path to becoming the world's most populous country by 2022, making it highly dependent upon groundwater-irrigated agriculture, both for export and for feeding its growing population.

Pathways to a Sustainable Water Supply

India is concentrating on two pathways for achieving a sustainable water: (1) the more cost-effective path of greater agriculture groundwater irrigation efficiency; and (2) a very costly, large-scale surface water collection and distribution system known as the *National River Linking Project* (Luthra and Kundo 2013; Swain 2017).

India's water difficulties have been blamed on three factors. First, the country simply does not have enough water for its existing population, let alone for keeping up with the growing demand from population growth. Global water planners consider a country to be under water stress if it has access to less than 1700 cubic meters of water per person. India has less than 1000 cubic meters of water per person. Second, much of India's available water supply is of very low quality. This is largely a result of failure to keep up with needed investments in water and wastewater infrastructure, and agriculture and industrial pollution. The water in most of India's rivers is unsafe for human consumption, with little or no treatment facilities available. Water pollution is the result of agriculture fertilizers, animal wastes and insufficient treatment of human wastes and untreated industrial water discharges. The third contributing factor is the depletion of groundwater by the nation's agriculture sector.

The first two of these are social factors contributing to the country's water problem, and do not have direct water-related solutions. To solve its water problems, India is focusing on reducing water consumption and better distribution of its current supplies. Reducing water consumption is the first half of India's two-pronged approach to achieving water sustainability. Farmers are pumping out more water than can be recharged, with existing irrigation methods polluting water supply resources (Luthra and Kundu 2013). Of the two approaches, improving agriculture methods is considered to the be the most cost-effective and fastest path to improving water supplies.

Reducing Agriculture's Water Use

In 2014, irrigation accounted for some 84% of the total available water; industry took 12% and the domestic sectors consumed about 4% of the total available water. Nearly 55% of all agriculture water use was groundwater. Agriculture is the major contributor to the India's GDP. More than half of all workers in India are employed in the agriculture sector. In 2016, India exported close to US$39 billion in raw agricultural products and more than 4.4 million tons of milled rice. Rice and cotton, two of India's most valuable crops, are also two of the largest users of water.

For centuries, Indian farmers depended on rainwater for irrigation. The annual monsoon was dependable and sufficient for farmer's needs. Little planting survived the annual dry season. However, traditional agriculture began to change beginning with the advent of higher crop yields made possible by the green revolution during the late 1960s. New high-yield crop varieties such as dwarf wheat and rice, together with chemical fertilizers, chemical pesticides and controlled irrigation revolutionized agriculture around the globe.

Farmers with holdings of just several acres account for more than 40% of India's food grains, nearly all of which is grown in the dry season and irrigated with groundwater. To take advantage of the increase economic opportunity, new crops were planted during the dry season and irrigated with groundwater. Some districts even provided free or low-cost electricity for irrigation pumps. Subsidized power has contributed to the excessive withdrawals of groundwater. Water tables have dropped across India by as much as 25 to more than 50 feet since 2009.

Improving Agriculture's Efficiency

Improving agriculture efficiency includes shifting to new, less water-intensive crops such as more low-water use barley and less high-water use wheat and shifting from open channel row irrigation to controlled drip and sprinkler irrigation. During India's monsoon wet season, more than three feet of rain may fall. Little of that rainfall is harvested and stored; floods that spread agriculture and human waste pollution are common in many areas. Conversely, little precipitation occurs during the dry season, and without irrigation most crops cannot survive. Farmers typically base their planting on estimates of available water. If less rainwater or surface water is available, Indian farmers tend to increase the amount of land depending on groundwater for irrigation. There is little public demand for government construction of more and higher dams to store surplus monsoon water, and even less for increasing the cost of electricity to pump groundwater. Meanwhile, research has shown that increasing water efficiency does not reduce demand for ground water. Rather, farmers take advantage of the now extra water to increase planting in less productive soils, thus balancing out any conservation gains (Fishman et al. 2015).

Diverting India's Rivers

The many rivers of India play an important role in the lives of most of the people of India. They are the source of municipal water, transportation and electricity throughout the country. Because of these reasons and reasons pertaining to the Hindu religion, nearly all of India's largest cities are located by the rivers. Seven major rivers and their tributaries make up the country's river system. Table 10.1 lists the length of India's major rivers.

Surfacewater diversion is the second of India's two main pathways to water sustainability. In 1980, the Indian government announced a plan to connect the 14 major rivers from the Himalayas with the 16 rivers with lower in the peninsula. The Indian National Perspective Water Development Plan was designed to supplement low water in some rivers with excess water from rivers with too much water, thereby reducing flood damage while going far to overcoming the persistent water shortages in some areas.

Despite appearing to be a good plan, the linking was never completely implemented. Pointing to the 22-year delay in completing the project, in 2002, the Indian Supreme Court informed the government that it should go back to work and compete the project within a reasonable time period (Swain 2017). During the almost 40 years since it was announced, some progress has taken place. However, the plan is still nowhere near completion. At 2018 prices, the completed project has been estimated to cost somewhere between US$125 billion and $200 billion—money the Indian government has other pressing uses for.

Table 10.1 The ten largest river systems in India

River	Length in miles	Length in kilometers
Indus river system	1801	2900
Brahmaputra river system	1801	2900
Ganges river system	1568	2525
Godavan river system	910	1465
Krishna river system	869	1400
Namada river system	815	1313
Yamuna river system	752	1211
Mahanadi river system	528	851
Cauveri river system	497	800
Tapti river system	449	724

Source Wikipedia

China's Paths to Water Supply Security

China's water supply problems are the result of growing water consumption by its large and more financially stable population and the fact that while most people live in the agricultural and industrial north, most of its water is in its less-populated agricultural South. A bi-product of its rapid industrialization is the extensive pollution of its water and air environments. The country's lack of comprehensive regulation and control of non-point source pollution has resulted in China now facing what has been described as critical shortage of water for all usages (Tao and Xin 2014; Currell and Han 2017; Udimal et al. 2017). In 2014, the annual supply of freshwater was short by an average of 40 billion tons of water. As the nation's population continues to grow and urbanize, this supply–demand gap will continue to grow every year. Many aquifers can no longer meet water withdrawal demands. Analysts believe the water shortage largely exists because nearly half of the country's water sources are polluted. Both shallow and deep aquifers are contaminated. The major sources of pollution in both surface and groundwater is caused by unregulated sewage and municipal wastewater, industrial wastewater discharge, and agriculture fertilizers and pesticides. A 2011 study of 800 wells in nine provinces and autonomous regions, including Beijing and Guangzhou, found that nearly 79% of the water failed to meet government health standards. An estimated more than 200 million Chinese were forced to rely on unsafe raw water.

In 2014, the Chinese central government announced it was initiating a major war on water pollution. The goals of the program were to ensure that at least 70% of the surface water quality in seven of the country's major river basis would meet Class III standards (middle of the quality levels), and that no more than 15% of the groundwater would be rated Class V. Achieving the plan's goals was to be based on the following 10-steps (Currell and Han 2017):

1. Take central government control of, and reduce, pollutant discharges.
2. Promote transformation of the economic structure to lower the intensity of pollution.
3. Focus on the protection of water resources through water saving (conservation).
4. Strengthen support for water science and technology.
5. Allow market mechanisms to impact water and pollution levies.
6. Strict environmental law enforcement and supervision.
7. Strengthen the management of the overall hydrologic cycle.
8. Ensure security of water for ecological and environmental purposes.
9. Confirm and implement the responsibilities for all parties involved.
10. Strengthen public participation and social supervision.

Closing the Water Gap in China

Eliminating water pollution will make China's water safer, but protection alone will not close the supply–demand water gap. Alternative sources are needed as well. China is looking to increase augmentation of its water supplies with reused or reclaimed water (Udimal et al. 2017). China's reserves of water are dropping each year; between 2000 and 2009, water reserves declined 13%. The country is suffering prolonged droughts. Precipitation is also unevenly distributed. In normal years, southern provinces receive close to 80 inches a year, while the north and west only receive from 8 to 16 inches of the annual rain or snow. The country has taken two technological approaches to achieving water sustainability: improved coastal area desalination and the world's largest surfacewater diversion project to move water from the relatively water-rich south to its water scarce north.

China's Water Desalination

As of 2017, China had 131 seawater desalination plants operating with 66.6% of the desalinated water earmarked for industrial purposes and the remainder for municipal use.

Beginning in 2019, a new desalination facility on China's east coast will begin removing salt from 120,000 tons of seawater in Bohai Bay, producing 50,000 tons of potable water each day. This alternative supply of freshwater is transported to Beijing as part of China's commitment to relieve the North and Northeast's water scarcity problems. The other major alternative water supply is the fresh surface water carried hundreds of miles north in the mega-water diversion program.

China's Water Mega-Diversion Project

China has a mega-infrastructure project underway to more excess water from the south to the arid north and west. Begun in 2002, the US$62 billion South-to-North Water Diversion Project (SNWDP) is the largest such water diversion scheme ever attempted. When completed in 2050, it will link the Yangtze, Yellow, Huaihe and Haihe rivers, annually diverting 44.8 billion cubic meters of freshwater from the south to the north (Cho 2011).

The project is not without its critics. The south is becoming drier as climate change is melting Himalayan glaciers and lowering precipitation continue. Southern farmers and cities are afraid they will soon not have enough water to share with the north. China is also working with the energy sector to reduce its heavy use of water. It has become a global leader in the production of solar panel technology, investing in wind power and adding desalination facilities. China is replacing old coal-powered electricity generation facilities with nuclear power plants, along with substituting freshwater with saline and brackish water. However, coal mining and the use of coal in thermopower plants remain large users of China's freshwater.

Water Security Paths in South Africa

South Africa, a semi-arid country with low rainfall, limited underground aquifers, and a reliance on significant water transfers from neighboring nations, faces what the McKinsey and Company researchers saw as "difficult economic and social choices between the demands of agriculture, key industrial activities such as mining and power generation, and large and growing urban centers." However, the 2030 Water Resources Group, for which McKinsey provided analytical support, finds that sustainability "solutions are possible and need not be prohibitively expensive if they are addressed now" (Boccaletti et al. 2010).

Water scarcity in South Africa is the result of a combination of the characteristics of the atmospheric water cycle, economic and political factors. Thus, the combination of least-cost solutions for South Africa's water scarcity problems would be greater investments in water infrastructure to improve basic supply and improvements to water-use efficiency in the industrial and mining sectors. South Africa's water supplies are below the global average and are made more difficult to deal with due the effects of climate change and shifting precipitation patterns.

Box 10.1 describes South Africa's water supply crisis. Over the past two decades, all the southern portion of Africa in general has had an increase in the number of extremely warm days and nights and a decrease in the number of cold days and nights, which is probably the result of the decline in rainfall experienced over the same period.

Box 10.1 Cape Town faces severe water scarcity

April 16, 2018 was the day predicted for the water system in Cape Town, South Africa to run out of water to deliver its customers. According to a collection of information in the *Irish Times* and the *New York Times*, Cape Town is "an example of what can happen when lengthy drought and poor water supply management and conservation collide." In addition to what has been described as a once in 300-years drought, a third of the stored municipal water supplies are distributed to local agriculture and industry.

At the time of the writing, the city's six main reservoirs were dangerously low, and no remedy was in site. The city's 4 million residents were told they would have to collect their daily ration of 6.5 gallons (25 Liters) water in their own containers from one of 200 distribution locations. Piped water delivery was to be halted for up to six months while seasonal rain brought reservoir supplies are to be back up to a sustainable level. Water delivery was to be restricted to vital services such as hospitals and schools.

The city's long-time failure to find alternative sources of supply have also contributed the city's crisis. Meanwhile, plans to complete desalination plants to augment the climate change-influenced precipitation decline were reported to be proceeding too slowly to mitigate the existing water shortage.

Source Rogers (2018)

Water Scarcity in Brazil

By traditional water measurement methods, Brazil is a water-rich nation. This South American country has 12% of the world's freshwater reserves, accounting for 53% of all water resources in South America. However, Brazil shares access to much of its water with other South American countries. The country shares 83 boundary and transboundary rivers with other nations, in addition to sharing large-scale basins and groundwater aquifers. These transboundary river basins cover 60% of the Brazilian territory. The difficulty in achieving sustainable water supplies in Brazil are a combination of its size, deforestation, climate change, over exploitation by economic sectors, and insufficient management of the resource. Its size influences wide variation in available water resources, and climate change has brought about wide shifts in precipitation patterns. The Amazon basin in the north takes up roughly half of the country. The weather there is hot and humid, receiving more than 2000 mm (78.74 inches) of rain annually. On the other hand, the semi-arid northwest annually averages from less than 200 mm (7.87 inches) to 740 mm (29.13 inches) of rain. This area is subject to prolonged droughts.

Deforestation and the over exploitation of existing sources by agriculture, industry and fast-growing urbanization and climate change influences are affecting water supplies in all regions of Brazil. The report on Brazil's water problems by the World Bank explained that the sectors that contribute most to the economy—energy, agriculture and industry—are also the sectors most dependent on water, and the greatest contributors to the water scarcity problem. Maintaining dependable river flows are critical; for example, hydropower generates 62% of Brazil's energy, and high-water-demand biofuel crops for ethanol production are widely used in Brazil. Water is essential for agriculture; irrigation consumes 72% of Brazil's water supply. The World Bank report described how Brazil implemented steps to avoid water crises such as those that occurred in São Paulo, where deforestation, prolonged drought and insufficient storage resulted in São Paulo almost running out of water from 2014 to 2016.

The water supply problem in São Paulo is not the only region of Brazil facing water scarcity. Changing precipitation, hotter days and the fresh water demands of Brazil's fast-growing cities in regions that supply three-quarters of the nation's water supply have experienced some water supply shortages. Records of rainfall, water withdrawals for all purposes, and water storage capacity in Brazil's southeast have been kept since the 1960s, and in some basins, since the 1930s.

Analysis of this water storage capacity and rainfall amount data suggest that climate change and declining rainfall have contributed to the region's water scarcity, and are often the major contributing factors, but that they are not the only contributing factors. Other factors include the lack of investments in infrastructure works to improve supply systems, deforestation and other plant losses in the reservoir regions, urban sprawl and so on. Research in similar regions in the USA, the Czech Republic, Mexico, Spain and Nigeria reveal similar contributing factor combinations. The world bank researchers concluded that the possible future impacts that climate change combined with these other factors will have on water management, "are worrying and require immediate adaptations in supply systems to maintain water availability in the coming decades" (De Lima et al. 2018, 228).

More than 140 Brazilian cities were forced to require periodic water rationing during the country's recent 20-year drought, with some restricted to receiving water only three days a week. Research by the Intergovernment Panel on Climate change (IPCC) predicts continuing increased heating and extreme weather events for all of Brazil, caused by heat waves and declining rainfall in the North East and Amazon Basin.

Deforestation is considered to be a significant contributor to climate change and changes in precipitation patterns. Recognizing this, Brazil has successfully launched a program to curb deforestation that has resulted in an 82% reduction in deforestation of the Amazon Basin. Environmental protection regulations such as the Forestry Code, were developed to help protect natural resources in rural areas, and are recent successes following decades of the devastation of Brazil's forests.

Water policy in Brazil also includes environmental considerations (Petelet-Giraud et al. 2018). The policy in effect throughout the long drought was adopted in 1988 with the Federal Constitution. For the first time, it included user-pays principles and participatory management practices. All waterways that extend across state lines or into other countries or are boundaries with other countries, are controlled by the federal government. Groundwater resources and streams all within a single state remained the responsibility of the states.

Climate Change, Heat and Water Scarcity

In July of 2018, much the world was suffering under another record-breaking heat wave, with more extreme weather events on the way. Associated Press (AP) and Reuters (TRF). AP reported that weather scientist are unanimous in their belief that these weather extremes are an effect of climate change. Adding insult to injury, they reported that the killer temperatures and water scarcity around the globe are likely to be followed by more devasting flooding.

According to the AP, heat waves and forest fires "are setting all-time temperature records across the globe. Europe suffered its deadliest fire in more than a century, and one of nearly 90 large fires in the U.S. West … More heat records fell in parts of Massachusetts, Maine, Wyoming, Colorado, Oregon, New Mexico and Texas." The 2018 Redding, California fire destroyed more than 800 homes and left 50,000 people under mandatory evacuation orders Another two wildfires burning about 14 miles apart threatened more than 10,000 homes. Two Fall California forest fires resulted in nearly 80 deaths and loss of more than 14,000 homes. "More heat records fell in parts of Massachusetts, Maine, Wyoming, Colorado, Oregon, New Mexico and Texas. Meanwhile, heavy downpours were causing floods across the Eastern USA.

Elsewhere, the AP report added, the temperature in Japan hit 106 degrees on July 23, its hottest temperature ever. Northern Europe was suffering from climate change-induced heat records, where "normally chill Norway, Sweden and Finland all saw temperatures they have never seen before on any date, pushing past 90 degrees. At least 118 of these all-time heat records have been set or tied across the globe, according to the National Oceanic and Atmospheric Administration (NOAA). A devastating fire in Greece with at least 83 fatalities was said to be the deadliest fire in Europe since 1900, according to the International Disaster Database run by the Centre for the Research on the Epidemiology of Disasters in Brussels, Belgium.

The same week, Reuters (TRF) reported that the extreme heat was adding to the already critical water shortages in other parts of the globe. For example, in South Africa, droughts and resulting extreme water shortages are predicted to become far more common. More than 650 million people in 500 cities, among them Sao Paulo and Tehran, were seen as likely to experience reduced access to water by 2050. Many of the world's most populous cities have already periodically limited water use per person: Los Angeles 52.8 gallons a day, Melbourne 40.95 gallons, and Cape Town an extremely low 13.2 gallons per person per day.

Summary

Clean water is fast becoming an extremely scarce resource. The problem is not that we are running out of water; the 1% or so of all the freshwater on earth is not going away. The problem is we are using that water faster than nature can replenish it through the normal water cycle. That means that it will require more money and energy to obtain what is needed by city dwellers, farmers and industrialists. Climate change is increasing droughts in many arid and semi-arid parts of the world, including the Pacific Southwest, and increasing the number and severity of extreme rainstorms. People in areas of water scarcity can expect to see even more water scarcity. People living in areas that have traditionally enjoyed sufficient water will begin to see more devastating flooding. Declining water supplies mean more human suffering and increased risk of instability, violent conflict and migration. Many of the countries or regions that are most deeply affected by environmental changes are already impoverished. They often lack the resources necessary for implementation of the sound water management that will get them through the changes now underway.

This chapter looked at the ways that four areas with extreme water scarcity are working through their growing water scarcity. Sustainability case studies described different ways to go about planning for and implementing solutions to water scarcity problems and alternative affordable methods of approaching the need for achieving sustainability in water-scarce regions. A widening gap between water supply and demand is projected to occur in 2030 in four locations: India, China, South Africa and Brazil. Total water demand in these regions already exceeds current water requirements. The finite water supply for all the uses in all water-use sectors is the total of the renewable water resource, with a theoretical upper limit of all surface water and groundwater that can be withdrawn for all uses in the region.

References

Boccaletti, Guilio, Martin Studhtey and Marc van Olst. 2010. Confronting South Africa's water challenge. Accessed July 8, 2018 from www.mckinsey.com/business-functions/sustainability-and-resource-productivity/our-insights/confronting-south-africas-water-challenge.

Cho, Renee. 2011. How china is dealing with its water crisis. *State of the Planet* blog. Columbia University Earth Institute. Accessed April 14, 2014 from https://blogs.ei.columbia.edu/2011/05/05/how-china-is-dealing-with-its-water-crisis/.

Currell, Mathew J., and Dongmei Han. 2017. The global drain: Why China's water pollution problems should matter to the rest of the world. *Environment: Science and Policy for Sustainable Development* 59 (1): 16–29.

De Lima, N. Gabriela, Magda A. Lombardo, and V. Magaña. 2018. Urban water supply and the changes in the precipitation patterns in the metropolitan area of São Paulo, Brazil. *Applied Geography* 94: 223–229.

Fishman, Ram, Naresh Devineni and Swaiminathan Raman. 2015. Can improved agricultural water use efficiency save India's groundwater? Accessed April 12, 2018 from iopscience.iop.org/article/10.1088/1748-9326/10/8/084022.pdf.

Luthra, Sonia and Amrita Kundu. 2013. India's water crisis and cures: An interview with Kirit S. Parikha. The National Bureau of Asian Research (NBR). Accessed June 28, 2018 from http://www.nbr.org/research/activity.aspx?id=356.

McKinsey & Co. 2009. Charting our water future. 2030 Water Resources Group. *Sustainability & Resource Productivity* 18 (1): 1–18. Accessed December 15, 2017 from www.mckinsey.com/~/media/mckinsey/dotcom/client_service/sustainability/pdfs/charting%20our%20water%20future/charting_our_water_future_full_report_.ashx.

Petelet-Giraud, Emmanuelle, Lisa Cary, Paul Cary, Guillaume Bertrand, Armelle Goglio-Jacqumot, Ricardo Hirata, Luc Aquilina, Lincoln Muniz Alves, Veridiana Martins, Ana Maria Melo, Suzana Montenegro, Eliot Chatton, Melissa Franzen, and Axel Aurouet. 2018. Multi-layered water resources, management, and uses under the impacts of global changes in a southern coastal metropolis: When will it be already too late? Crossed analysis in Recife, NE Brazil. *Science of the Total Environment* 618: 645–657.

Rogers, Kyle. 2018. Day Zero: City braces for water supply drying up. *Municipal Sewer and Water*, January 31. Accessed February 3, 2018 from www.mswmag.com/online_exclusives/2018/01/day-zero-city-braces-for-water-supply-drying-up.

Swain, Ashok. 2017. Here's why the idea to link India's rivers will remain a pipe dream. *Outlook,* September 3. Accessed April 12, 2018 from https://www.outlookindia.com/website/story/heres-why-the-idea-to-link-indias-rivers-will-remain-a-pipe-dream/301157.

SWP (Sustainable Water Partnership). 2018. What is water security? Accessed August 7, 2018 from www.swpwater.org/what-is-water-security/.

Tao, Tao, and Kunlun Xin. 2014. A sustainable plan for China's drinking water. *Nature* 511: 527–529.

Udimal, Thomas B., Zhuang Jincai, Emmanuel C. Ayamba, and Samuel M. Owusu. 2017. China's water situation: The supply of water and the pattern of it uses. *International Journal of Sustainable Built Environment* 6 (2017): 491–500.

11

Pathways to Sustainable Cities

In conducting the research for this chapter, it soon became clear that what characteristics constitute a sustainable city depends upon who is giving the answer. If your concern is in public power, the definition focuses on the sustainable supply of affordable electricity for all residents. However, if you are a public transportation official, a sustainable city is one that ensures that all residents have access to affordable public transportation. If your concern is ensuring that both public and private housing is available for everyone at affordable prices your definition focuses on housing. If your concern is drinking water, your definition is on ways to ensure that your city or community has enough clean, reliable and safe water for all sectors. If it is wastewater, your concern is that your sewage system never fails and discharges only clean, safe water that can be recycled for other users downstream. The chapter begins with a brief example of why it is difficult to define sustainable cites. It then examines the evolution of a planning framework for managing water for achieving and sustaining sustainable cities. It will close with a discussion of some of the actions that water suppliers and municipal leaders are initiating for achieving and supporting sustainable cities.

What Are Sustainable Cities?

In their review of nine each French and English professional research papers and news articles related to the sustainability see changes topic, Hamman et al. (2017, 336) reached four conclusions indicating the difficulty of arriving at a consensus definition for what constitutes a sustainable city:

D. E. McNabb, *Global Pathways to Water Sustainability*,
https://doi.org/10.1007/978-3-030-04085-7_11

1. There is no agreement on what constitutes a 'sustainable city.'
2. The different literature approaches (conceptual, institutional, practical or geographic) were not mutually exclusive.
3. The 'sustainable city' concept appears in the French and U.S. journals analyzed as a genuinely political theme, with a wide range of actors and institutions acting as filters, promoters or detractors.
4. In the contexts within which the terms 'sustainable' and 'development' were found, they were most often associated with political, economic and social issues. However, the themes of what constitutes a sustainable city were constantly changing and were not fixed, in both content and procedures.

Overall, the researchers found four synonyms used for sustainable cities: The green city (green infrastructure, tree-line streets, parks); the city of short distances (housing and urban transit); the just city (social or economic inequality); and, most commonly seen, the participatory city (participatory democracy, interaction between stakeholders, etc.). In French texts, prominent sustainability themes in the literature included in-city nature (green infrastructure) concerns, public transport and urban farming. In English language texts, the major sustainable city topics mentioned included public transport, energy, climate and natural resources. Mentions of water and sanitation were conspicuously absent.

This brief introduction makes it clear that planning for these and other critical urban services is done by the people who manage those services, generally without consultation with planners or managers of related public service inputs. A similar planning issue exists with the mix of different roles of the federal, state and local oversight agencies of each individual public service. However, at times, it appears that a movement exists calling for more cooperation among agencies. The following federal agency example shows that cooperation in planning for and supporting local efforts at achieving sustainability in some infrastructure development programs among government agencies can happen.

This HUD-DOT-EPA Partnership for Sustainable Communities set out to ensure that affordable housing and transportation needs were met while promoting more inclusive neighborhoods. According to Regina C. Gray, a social science analyst with HUD, the partnership established clear goals for reducing energy consumption and protecting the environment, with the following six guiding principles first introduced in June of 2009. Together, these form the basis of a sustainable city; none of these goals can be maintained for long without water and sanitation services (Gray 2015).

1. Provide more transportation choices.
2. Promote equitable, affordable housing.
3. Enhance economic competitiveness.
4. Support existing communities.
5. Coordinate and leverage federal policies and investment.
6. Value communities and all neighborhoods.

Although the principles went far in establishing a roadmap for future urban development, the cooperative program for America's sustainable cities did not include water or wastewater facilities. All urban areas must recognize that sustainable water supplies and the infrastructure necessary to distribute water are together the one asset that can make future urban living possible. Sustainable cities must also have an adequate and reliable wastewater collection and treatment before discharge. Meanwhile, planning for the infrastructure needed for water and wastewater is still largely managed by engineering departments, with little awareness of how their actions influence and are influenced by other departments, agencies or public served. This chapter touches upon the cooperative management of all aspects and forms of water that is necessary for achieving and maintaining sustainable urban growth.

The Dublin Warning

The UN International Conference on Water and the Environment held in Dublin, Ireland in 1992 issued a warning that the emerging urban water situation was fast becoming critical, and most of the world's cities needed to change their business-as-usual attitudes with regard to water. The unbridled over-exploitation and pollution of global water resources, coupled with urbanization in the world's poorest nations was unsustainable and was fast outpacing those cities' ability to provide clean, safe water and sanitation services. The following paragraph was included in the end of conference Dublin Statement:

> The sustainability of urban growth is threatened by curtailment of the copious supplies of cheap water, as a result of the depletion and degradation caused by past profligacy. After a generation or more of excessive water use and reckless discharge of municipal and industrial wastes, the situation in the majority of the world's major cities is appalling and getting worse. As water scarcity and pollution force development of ever more distant sources, marginal costs of meeting fresh demands are growing rapidly. Future guaranteed

supplies must be based on appropriate water charges and discharge controls. Residual contamination of land and water can no longer be seen as a reasonable trade-off for the jobs and prosperity brought by industrial growth. (ICWE 1992)

How Sustainable Are U.S. Cities?

The third annual report from the Sustainable Development Solutions Network (SDSN) continued its ranking of the 100 most populous metropolitan areas in the USA on the Sustainable Development Goals (SDGs) that were agreed upon by 100 of the UN's members in 2015 (Espey et al. 2018). Approximately 85.5% of all residents of the USA live in cities as measured by metropolitan statistical areas (MSAs); roughly 66% of the total live in the 10 largest MSAs.

The report's authors concluded that none of the 100 cities analyzed were on track to achieve the agreed upon SDGs. Each city was analyzed on 44 factors associated with sustainable development challenges in environmental, social and economic factors that hinder achieving their objectives. "The study highlights widespread inequality in urban areas and that gender, age, race and income all impact a person's likelihood of realizing education, healthcare and economic opportunities." Having achieved 68.57% of its DSCs, The San Jose-Sunnyvale-Santa Clara MSA was the best performing community. The Seattle-Tacoma-Bellevue metropolitan area was second on the list, with 63.52% of its goals reached. Seattle has accomplished 100% of their objectives in both energy and water sustainability in their green architecture goals. Table 11.2 lists the water sustainability categories and actions taken by Seattle.

Table 11.1 lists the percentages for the top 10 cities; almost two-thirds of the 100 cities analyzed are less than half-way there; 12 cities had achieved less than 50% of their Sustainable Development Solution (SDS) goals. The two cities with the lowest SDS achievement goals were Wichita, Kansas with 47.88% and Dallas-Fort Worth-Arlington, Texas with 47.44%. Great progress was reached on the access to clean water and sanitation category, with 10 cities scoring 98.95% or higher on this factor. Five cities had perfect 100% scores: Manchester-Nashua, NH; Palm Bay-Melbourne-Titusville, FL; Charleston-North Charleston, SC; Seattle-Tacoma-Bellevue, WA; and Springfield, MA. Table 11.2 lists the water sustainability categories and actions taken by the Seattle MSA.

Table 11.1 The top ten U.S. cities and sustainability index rating on a scale of 1–100%

Metropolitan statistical area	Rank	2017 achievement index score
San Jose-Sunnyvale-Santa Clara, CA	1	68.57
Seattle-Tacoma-Bellevue, WA	2	63.52
Boston-Cambridge-Newton, MA	3	61.86
San Francisco-Oakland-Hayward, CA	4	61.70
Madison, WI	5	61.23
Manchester-Nashua, RH	6	60.76
Albany-Schenectady-Troy, NY	7	60.44
Bridgeport-Stamford-Norwalk, CT	8	60.16
Oxnard-Thousand Oaks, Ventura, CA	9	60.10
Provo-Orem, UT	10	56.56

Source Esprey et al. (2018)

Table 11.2 Water sustainable city items and activities for Seattle, Washington

Green building categories	Examples of actions taken
Solar heating of water	Solar energy
Water conservation program	Stormwater management plan
Installation of water-efficient toilets	Toilet installation rebates
More efficient urban irrigation equipment	Rebates for sprinkling timer upgrades
More water-efficient kitchen appliances	Water-efficient commercial dish washers
More water-efficient medical equipment	Water-efficient sterilization systems
More efficient laundry equipment	Water-efficient commercial washers
Other water-efficiency technologies applied	Other water-efficient technologies
Rainwater harvesting for water reuse	City RainWise program
Land close to urban creeks reforestation	Protect Our Waters program

Source From material in Ribeiro et al. (2018)

Evolution of Urban Water Sustainability Management

Cooperation between water providers over water resource management issues is becoming more common and this is indicative of the idea that an urban area can only be sustainable if its water and wastewater service is reliable and managed on an integrated basis. As a result, water resource planning is often done at the catchment level. A catchment or catchment-basin includes the total area drained by a single river or other type of water body. A catchment may extend across national boundaries. The integrated catchment management concept promises multiple benefits for all the partner water sector participants and water-use sectors in the catchment, as well as the all towns and cities in the region.

Careful planning for meeting the water infrastructure needs in the world's cities should be based on precise estimates of where, when and how much water main replacement will be needed to ensure utilities can support the continuing growth in urban population while repairing and replacing their ancient and failing existing infrastructure (AWWA 2012). There are more than 51,000 community water utilities operating in the USA. The EPA and Department of Agriculture jointly identified a list of key management tasks necessary for achieving sustainability: a consistently quality product that included infrastructure stability, customer satisfaction, and financial viability. Since the last decades of the last century, both the EPA and Department of Agriculture have endorsed adoption of a variety of management tools to serve as guides to help small and rural utilities maintain sustainable operations. In addition to the original WRM model, two additional water management models now exist for assisting water providers' plan for water supply security. These are Integrated Water Resource Management (IWRM) and Integrated Urban Water Management (IUWM).

Sustainability Planning at the Catchment Level

The integrated basin or catchment-level approach began in the 1990s with the adoption of an integrated catchment-based water resource planning model that grew out of the 1970s' Water Resources Management (WRM) initiative supported by UNESCO (Saveniji and Van der Aaag 2008; Mees et al. 2017). The Dublin Statement that came out of the UN's 1992 International Conference on Water and the Environment and the list of sustainability objectives (Agenda 21) evolved into what was to become the IWRM framework for how water resources should be managed on the basin and aquifer levels in order for urban water supplies to be sustainable. The section on sustainable urban areas in the Dublin Statement warned the decades of disregard for the over-exploitation of finite freshwater supplies could no longer be allowed to continue.

The Global Water Partnership, one of the IWRM's strongest supporters, defined sustainability as "a process which promotes the coordinated development and management of water, land and related resources in order to maximize economic and social welfare in an equitable manner without compromising the sustainability of critical ecosystems" (GWP 2000).

Integrating planning for all classes of water, its resources and uses was seen as a way to ensure better management of all water classes supplied to residents of the cities in the basin. It would include planning for actions for reducing flood risk, controlling pollution, protecting drinking water resources, improving biodiversity and improving the health of and recreation for citizens in the region's communities. Integrated planning at the catchment level was also considered to be more cost-effective than tackling single-issue urban area problems. Adoption of the IWRM approach has been slow, despite its recognized advantages. Some authors consider this slow pace to be a result of the very large basin level scope of the required planning (Mees et al. 2017; Chidammodzi and Muhandiki 2017).

Over the last half of the twentieth century, fast-growing cities in developing nations have found it increasingly difficult to manage water supply resources and maintain the infrastructure needed to supply the large numbers of rural families migrating to their cities (Hukka and Katko 2015). With climate change, these cities face longer and more severe conditions of increasing water scarcity, extreme storms and flooding, along with the poor water quality that has already become debilitating. In many of these regions, population growth, urbanization and inability to fund needed infrastructure development are exacerbating the fresh water and sanitation deficits (Studart and Gallagher 2018).

Integrated Urban Water Management (IUWM)

Although basin-level water resource planning was developed as an aid for planning for and monitoring regional or catchment-wide water supplies for all sectors in a basin, it was seen as too broad in scope to help with sustainability planning for cities, even though it has particular relevance for the water suppliers managing the supplies for residents of urban areas. What was needed was an urban planning model that would do much the same, but on a local city or community basis. Thus, an IUWM approach was born. IUWN differs in scope from the basin-level resource management approach since it includes managing all water classes and estimating current and future urban demands for water and needs for wastewater services (Fig. 11.1). IUWM would also include adjusting water needs and scarcity estimates to include to reflect the probable effects of climate change (Furlong et al. 2016, 2017). Because it includes considering all classes of water, IUWM has become a tool for planning the water infrastructure needed to achieve sustainable cities (Kirshen et al. 2018).

By considering the importance of water-use efficiency and economic efficiency in terms of urban environments, IUWM includes planning for the infrastructure needed for all the classes of water without which an urban community will soon no longer be sustainable. It also recognizes that

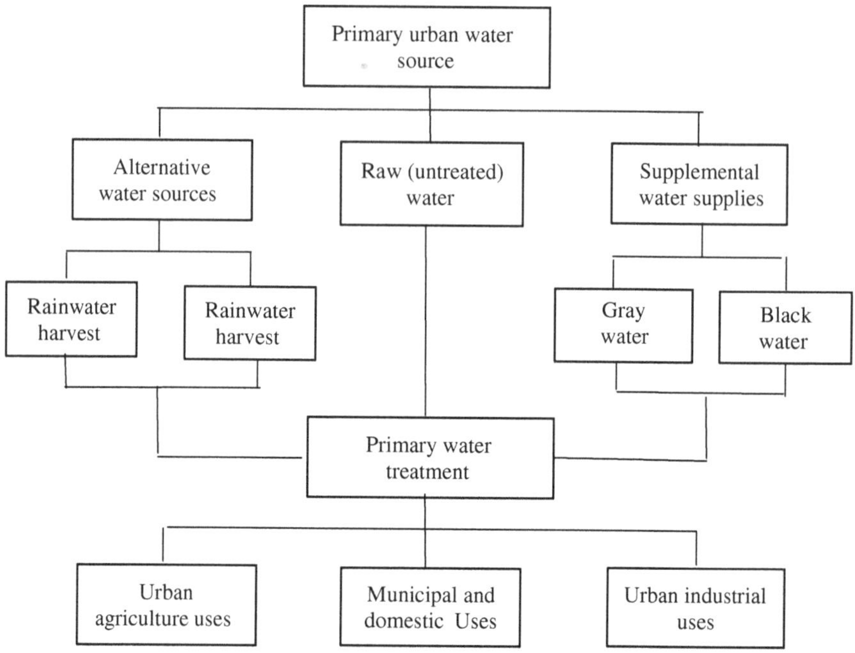

Fig. 11.1 Abbreviated model of the all-water scope of the IUWM system (*Source* The author)

different kinds of water are used for different purposes in many parts of the world, including the USA. Scarcity of the main source or sources of a community water supply has forced many regions to include alternative freshwater sources and supplemental treated reclaimed water.

Benefits of IUWM were described by the GWP as offering water managers better coordinated, responsive and sustainable resource management practice. It integrates water sources, water use sectors, water services and water management. Further beneficial features for water and wastewater infrastructure management include the following:

- Recognizes alternative water sources;
- Differentiates the qualities and potential uses of water sources;
- Views water storage, distribution, treatment, recycling, and disposal as part of the same resource management cycle;
- Seeks to protect, conserve and exploit water at its source;
- Accounts for non-urban users that are dependent on the same water source;

- Aligns formal institutions (organizations, legislation and policies) and informal practices (norms and conventions) that govern water in and for cities;
- Recognizes the relationships among water resources, land use and energy;
- Simultaneously pursues economic efficiency, social equity and environmental sustainability; and
- Encourages participation by all stakeholders (Bahri 2012).

Sustainability planning at the city level involves formal and informal collaboration among all water sector participants. This includes all those that collect, treat and distribute water, those that catch water in drainage systems, and those that use the water for various purposes, including parks and recreation departments, transportation districts, waste management service providers, public works departments, schools, colleges and universities, private developers and the civic organizations dealing with environmental stewardship, urban forestry and urban agriculture.

Municipal water and wastewater service suppliers are addressing the sustainability issue from many different directions, with activities linked closely to the characteristics of individual organizations, rates varying by use or by sector, political influences and urban planning philosophies (Herrick and Pratt 2012). Water utilities come in many different sizes, and have many different strengths and weaknesses. Their leaders chose models of sustainability they believe to be achievable; some are heavy investors in green infrastructure, whereas others invest in professional sports arenas. Some have a high tolerance for development while others focus on triple bottom line and life cycle costing. Water utilities employ sustainability actions based on their openness to innovative ways of dealing with external factors such as climate change, combined stormwater and sewage overflows, failing infrastructure and economic constraints. In 2016, the EPA and U.S. Department of Agriculture published a guide for managers of small water utilities to use when planning for sustainability.

Awareness of the need for greater investing in water and wastewater infrastructure has long existed. For example, a 2004 report to the U.S. Senate included with this warning: "Mounting evidence suggests that the integrity of the nation's drinking water and wastewater infrastructure is at risk without a concerted effort to improve the management of key assets—pipelines, treatment plants, and other facilities—and a significant investment in maintaining, rehabilitating, and replacing those assets".

What was needed was a management system that suppliers of water and wastewater service could use to effectively plan and budget for repairing, replacing and extending their critical infrastructure. This was achieved when a plan for achieving the goal of water sustainability for the world's cities was agreed upon at the UN, Sustainable Water Management Improves Tomorrows Cities Health (SWITCH) and several others conference on waster sustainability in Zaragoza, Spain.

SWITCH was a 5-year research partnership funded by the European Commission with a budget exceeding €20 million over the period 2006–2011. It involved an implementing consortium of 33 partners from 15 countries. SWITCH case studies focused on ways for advancing applications of concepts in sustainable urban water management or IUWM. The theme of the 5-year project was identifying the contribution water management principle for the 'city of the future.' The studies supported by SWITCH sought to "challenge existing paradigms and to find and promote more sustainable alternatives to the conventional ways of managing urban water. SWITCH also set out to do things differently by carrying out action-orientated research in cities that was more demand-led reflecting the expressed needs of cities" (SWITCH 2011, 3). In addition to the many city studies and conferences promoted by SWITCH was an urban sustainability project toolkit on 35 ways that cities can address their own sustainability challenges (a list of all SWITCH publications is available online at www.switchurbanwater.eu/pubs.php). The SWITCH project supported research on the following themes:

- Water management in cities in four continents and at various stages of development;
- All aspects of the water cycle (freshwater, wastewater, stormwater and water systems);
- A wide range of climatic, socio-economic and institutional situations;
- Social, economic and environmental perspectives as they relate to water and city growth;
- Develop relevant scales for measuring progress on water goals at household to city levels;
- Clarify water's role as part of urban planning and the built environment; and
- Project current water management circumstances to the City of the Future.

Table 11.3 Technologies and practices implemented by cities for water sustainability

Category	Discussion and examples
Green infrastructure	Green roofs, permeable pavements, rain gardens, rainwater harvesting, stormwater harvesting planted swales, urban tree-shaded lanes, and other programs
Upgrading utility facilities	Energy efficient and onsite renewable energy production, green product purchases, wastewater recycling for reuse, sustainable landscaping, water use restrictions, energy conservation
Land use practices	Land use conservation, rebuilding wetlands, improving riparian vegetation for stream cooling, new flood control methods, improving steam flow by dam removals
Water conservation	District water conservation policies, encouraging low-flow toilets, smart meters for leak detection, reusing treated wastewater for urban irrigation
Transportation equipment	Fuel efficient fleet, including hybrid and electric-power vehicles, use of biofuels for energy
New treatment facilities	Movement to tertiary treatment of wastewater for agriculture irrigation and potable level for recharging groundwater
Other programs	Decentralized systems, point-of-use and point-of-entry treatment systems, biosolids management programs, smart metering for energy and water use, asset management planning for infrastructure repair and replacement

Source From material in Herrick and Pratt (2012)

What Cities Are Doing to Achieve Sustinability

Herrick and Pratt (2012) selected a variety of technologies and practices from a larger list of the activities that cities are adopting to achieve the status of a sustainable city. Among the major actions listed were green infrastructure and new and improved treatment facilities. Table 11.3 lists seven categories of sustainability actions.

Achieving Sustainability in Exeter, New Hampshire

In 2016, the population of the section of the town of Exeter, New Hampshire was 14,690. The town's water and wastewater services are provided by the Public Works Department. The water system provides drinking water and fire protection to the town. Exeter relies on a combination of surface water and groundwater to meet the community's water supply needs. The primary water supply source is the Exeter River, which flows through the center of town, where it was dammed to serve as a run-of-river reservoir. The dam was removed in 2016 to improve the flow of the river, particularly as a means of reducing the number of city floods associated with heavier and or more frequently occurring heavy rainstorms.

Beyond the falls, the river becomes the Squamscott River, which empties into Great Bay and the Gulf of Maine. Water is pumped to the town's water treatment plant before being distributed or sent to one or more reservoirs. Other water sources are the Exeter Reservoir, which is fed by Dearborn Brook. Exeter's groundwater sources include Skinner Springs, and three wells. Wastewater is treated before being discharged into the Squamscott River downstream of the town. The wastewater treatment plan was upgraded and moved in 2016 to a higher location in order to avoiding stormwater flood damage. Water quality sampling of the Exeter River in 2012 had indicated extensive water quality impairment at ten stretches of the river, including five within the town of Exeter.

A team of nine researchers conducted a detailed analysis of the Exeter water and wastewater facilities to determine the potential impact from expected extreme weather conditions. The study followed a One Water, IUWM framework that focused on all parts of the water cycle and full range of present and future water demands. Among the sustainability improvement options suggested in the report were protecting, restoring and expanding tidal and freshwater wetlands that over the years had been filled in for construction of the town's wastewater treatment plant and for parks and other recreational uses. Expanding the range of banned developments in the region's floodplains would increase flood management while also increasing aquifer recharging (Box 11.1).

Box 11.1 Cities ranked by their preparedness to survive extreme weather events

Eleven European cities were among the top 12 the 2017 Arcadis global report on the water sustainability index of 50 cities around the globe. The index rates are based on cities' efforts to achieve sustainability when threatened with extreme weather events and to cope with unexpected water shortages. Toronto, Canada was the only city in North America to be ranked in the top ten.

The index examines city's water sustainability preparedness on three factors: resiliency, efficiency and quality. Resiliency is measured on a city's ability to bounce back after being hit with water-related disasters, and their current vulnerability to extreme weather events. Efficiency is measured by the amount of water lost to leaking infrastructure, use of metering, degree of water reuse, sustainable supply coverage, and water rates. Water quality is measured by the health of residents, sanitation infrastructure, water pollution and effects on the ecosystems and environment.

Among the ten most water sustainable were the European cities of Rotterdam, Copenhagen, Berlin, Brussels, Frankfurt, Birmingham and Manchester, Toronto, Canada and Sydney Australia. Washington, DC, ranked at 13 on the list, was the highest ranked U.S. city. Other U.S. cities in ranked above 20 were New York, Houston, Boston, Philadelphia, Dallas and Chicago. Cities in the bottom of the 50 that were rated included New Delhi, Mumbai, Manila, Jakarta, Nairobi, Johannesburg, Rio De Janeiro, Mexico City, Jeddah and Doha. *Source* Hill (2017)

These actions alone would improve Exeter's total urban water management, but they would not eliminate flooding, reverse growing water pollution or increase groundwater supplies to the point of sustainability. Planting more trees and other riparian vegetation has been recommended to enhance the riverside environment and cool the river water during the warmer summer months. To further mitigate flood damage from the increasing number and severity of extreme weather events, structural actions were also possible. These included raising the height of roadways and structures in the flood-threatened areas. Low impact development such as parks, green spaces and vegetative buffers would result in better control over development on the flood plain.

Improvement of the water supply could be achieved by the two major actions that are immediately available to Exeter: area-wide water conservation, augmented by greater natural increases in groundwater recharging through flood control catchments. Currently, rainwater and stormwater are discharged directly into the Exeter River. If this practice were replaced by rainwater and stormwater harvesting, the water supply could be significantly improved. Water reuse and recycled water, while increasingly common in many locations around the globe was not considered feasible at the time of the study due to the perceived high cost of potable level treatment.

Summary

Cities cannot be sustainable without efficient and reliable supplies of water and energy. While some power can be generated on-site with either wind or solar energy, a sustainable water supply for all water user sectors must be withdrawn for either a surface or a groundwater source. Overpopulation and climate change-influenced changes in precipitation patterns are rapidly depleting primary water sources in many parts of the globe. As a result, cities have had to find new or supplemental water sources. Among these supplemental sources are water conservation in all water-use sectors, repairing and replacing water infrastructure to eliminate water main leaks, reusing once-used municipal water, harvesting rainwater, collecting and storing stormwater for urban landscaping and recharging aquifers, and urban green architecture such as water-friendly buildings, permeable pavements and green areas.

A set of urban sustainability goals were universally adopted by members of the UN in 2015, Progress toward achieving the SDGs has been monitored each year since the program's adoption. A report on the portion of 100 U.S. MSAs indicated that while much progress has been made since 2015, more than 62 cities are less than halfway toward implementing the programs and policies needed by a sustainable city. Even as cities do approach meeting their sustainability goals, climate change and the continuing migration to urban areas is making overcoming their sustainable city challenges more difficult.

References

AWWA (American Water Works Association). 2012. Buried no longer: Confronting America's water infrastructure challenge. Accessed May 21, 2018 from www.awwa.org/portals/0/files/legreg/documents/buriednolonger.pdf.

Bahri, Akiça. 2012. Integrated urban water management. TEC Background Papers No. 16, Global Water Partnership, Technical Committee. Accessed October 24, 2017 from www.gwp.org/globalassets/global/toolbox/publications/background-papers-integrated-urban-water-management-2012.pdf.

Chidammodzi, C.L., and V.S. Muhandiki. 2017. Water resources management and integrated water resources management implementation in Malawi: Status and implications for lake basin management. *Lakes and Reservoirs: Research and Management* 22: 101–114.

Espey, Jessica, Hayden Dahmm, and Laurie Manderino. 2018. Leaving no U.S. city behind: The U.S. cities sustainable development goals index. United Nations Sustainable Development Solutions Network. Accessed August 5, 2018. Available at http://unsdsn.org/wp-content/uploads/2018/06/US-Cities-Index-Report.pdf.

Furlong, Casey, Ryan Brotchie, Robert Considine, Greg Finlayson, and Lachian Guthrie. 2017. Key concepts for integrated urban water management infrastructure planning: Lessons from Melbourne. *Utilities Policy* 45: 84–96.

Furlong, Casey, Saman De Silva, Lachlan Guthrie, and Robert Considine. 2016. Developing a water infrastructure planning network for the complex modern planning environment. *Utilities Policy* 38: 1–10.

Gray, Regina C. 2015. Guest Editor's introduction: Building a research agenda for creating sustainable and inclusive communities for all. *Cityscape: A Journal of Policy Development and Research* 17(2): 2–15.

GWP (Global Water Partnership). 2000. Integrated water resources management. TAC Background Papers (4). Accessed August 4, 2018 from www.gwp.org/globalassets/global/toolbox/publications/background-papers/04-integrated-water-resources-management-2000-english.pdf.

Hamman, Philippe, Virginie Anquetin, and Céline Monicolle. 2017. Contemporary meanings of the 'Sustainable City': A comparative review of the French and English language literature. *Sustainable Development* 25 (4): 336–355.

Herrick, Charles, and Joanna Pratt. 2012. Sustainability in the water sector: Enabling lasting change through leadership and cultural transformation. *Nature and Culture* 7 (3): 285–313.

Hill, Christopher. 2017. Sustainable cities water index: Harnessing water for long-term success. Accessed September 3, 2018 from www.arcadis.com/en/united-states/our-perspectives/2016/sustainable-cities-water-indexharnessing-water-for-long-term-success/.

Hukka, Jarmo J., and Tapio S. Katko. 2015. Resilient asset management and governance for deteriorating water service infrastructure. *Procedia of Economics and Finance* 21: 112–119.

ICWE (International Conference on Water and the Environment). 1992. The Dublin statement on water and the environment. Accessed August 4, 2018 from https://www.gdrc.org/uem/water/dublin-statement.html.

Kirshen, Paul, Semra Aytur, Jory Hecht, Andrew Walker, David Burdick, Stephen Jones, Neil Fennessey, Renee Boudeau, and Lorilee Mather. 2018. Integrated urban water management applied to adaptation to climate change. *Urban Climate* 24: 246–263.

Mees, Hannelore, Cathy Suykens, and Ann Crabbe. 2017. Evaluating conditions for integrated water resource management at sub-basin scale: A comparison of the Flemish Sub-Basin boards and Walloon River contracts. *Environmental Policy and Governance* 27: 59–73.

Ribeiro, João M.P., Stephane L. Bocasanta, Risa O. Ávila, Mica Magoto, Ana V. Jonck, George M. Gabriel, José B.S.O. de Andrade Guerra. 2018. The adoption of strategies for sustainable cities: A comparative study between Seattle and Florianópolis legislation for energy and water efficiency in buildings. *Journal of Cleaner Production* 197: 366–378.

Savenije, H.H.G., and P. Van der Zaag. 2008. Integrated water resources management: Concepts and issues. *Physics and Chemistry of the Earth* 33: 290–297.

Studart, Rogerio, and Kevin Gallagher. 2018. Guaranteeing sustainable infrastructure. *International Economics* 155 (2018): 84–91.

SWITCH (Sustainable Waster Management Improves Tomorrows Cities Health). 2006. Sustainable water management in the city of the future. Accessed October 11, 2017 from www.switchurbanwater.eu/outputs/pdfs/SWITCH-Final-Report.pdf.

12

Pathways to Sustainable Agriculture

Water for all uses will become increasingly scarce during the rest of this century. High on the list of the difficulties that agriculture faces include continued inefficient water use, overdrafts of already-stressed groundwater aquifers, and agriculture runoff that results in surface water quality degradation. Ensuring that the water needed to meet the food needs of the world's fast-growing population is being made more difficult by the effects of climate change on rainwater. The global demand for agricultural water is predicted to increase by about 20% by 2050 (UN 2014a). Many regions of the world already suffer under conditions of water scarcity. Population growth in poorer countries largely dependent on inefficient agriculture practices is further contributing to the food supply problem. Water scarcity and water source pollution are two of the biggest barriers to achieving a secure and reliable food supply for the growing population.

Globally, agriculture already accounts for about 70% of world's water withdrawals. Most of that water is used for irrigation. The runoff from agriculture can result in water pollution from the fertilizers, pesticides, animal wastes, and in some regions, human waste. Inefficient use of the freshwater withdrawals for crop irrigation and livestock raising ends up as nonpoint pollution of both surface water and groundwater sources (Dilekli et al. 2018).

D. E. McNabb, *Global Pathways to Water Sustainability*,
https://doi.org/10.1007/978-3-030-04085-7_12

Agriculture Irrigation Methods

The three main methods of agriculture irrigation currently in use are flood or furrow, sprinkler, drip, and micro-irrigation (ATS Irrigation 2018). Subsurface application of irrigation water is a relatively new method that is sometimes listed independently or included as one of the micro-irrigation methods. In flood irrigation, water is allowed to flow across a flat field and allowed to permeate into the soil.

In the furrow irrigation method, water flows down furrows that run between raised rows of field crops. It is collected at the end of the row and either used later or discarded. Water to the furrows is often supplied by an aboveground PVC pipe with an opening at each furrow. The opening of the gate can be varied to adjust the flow to the furrow. In some cases, a large diameter thin wall poly pipe with a hole is punched to divert water to each furrow is installed.

There are four standard types of agriculture sprinkler systems: hand-moved pipe systems, solid set systems, center pivots of other mechanical movement systems, and hose reel or traveling gun systems. Hand-moved lines are sections of aluminum pipe that are attached end to end in a section installed on the surface of a field. One end of the set of pipes is attached to the water supply and may include a provision for the insertion of fertilizer and pest-control materials. The set of pipes is moved by hand to another position after a set period of time. Solid set systems are sections of PVC pipe permanently installed underground with risers rising above ground, each with a sprinkler head attached.

Center pivots and related systems are a row of rigid pipes with sprinklers attached above or below the pipe, often with a larger sprinkler operating at the end of the moving pipe. In a center pivot system, the set of pipe revolves around a circular field. Hose reel or traveling gun systems consist of one large sprinkler attached to a hose on a rack attached to a cart. The cart is pulled down the field, unreeling the hose. When the sprinkler is in place at the end of the field, water flowing through the hose operates the reel and rolls the hose back on its spool as it moves down the field.

The drip irrigation method uses small diameter PVC tubing with holes for nozzles or 'emitters' set along the length of the pipe to apply water to a small area directly to the root zone of a crop. The emitters can be installed into the tubing by hand to water a specific tree or plant or set by a factory on at previously set distances apart. Drip tape is a type of drip irrigation with drip emitters installed from 6 to 12 inches apart in a very thin tube that comes in coils or rolls. Drip tape is usually used to irrigate vegetable crops and gardens but can be buried in order to irrigate crops such as cotton or corn.

Micro irrigation—the method with the best promise for achieving sustainability—uses small, low pressure and low volume sprinklers called micro-sprinklers on systems that irrigate orchards and vineyards. In most cases, the micro-sprinkler is installed at the base of plant to water only that plant or tree. In some cases, the micro-sprinkler is installed between two trees. Water can be provided to the micro-sprinklers from an aboveground polyethylene hose-like pipe, or by a PVC pipe buried below ground.

Irrigation Water Demand

Irrigation water demand varies from season to season and from crop to crop, as well as with soil characteristics and local climate conditions. Researchers agree that under the existing variability of conditions and farming methods, meeting the projected 60% increase in global food demand will require the expansion of arable land and increased plowing of the soil and application of agrochemicals, energy and water. This additional farmland, additional tillage, and use of chemical fertilizers are estimated to account for 70% of the loss of the earth's biodiversity by 2050 (Levidow et al. 2014). However, meeting the food needs of the world's growing population will require more land, more tillage, more water and probably more chemical fertilizers and pesticides (UN 2014b). A better way to use the land and water is needed.

Pathways to Sustainable Agriculture

Water is a fundamental resource necessary for sustainable agriculture. The approaches taken to achieve sustainability can be the integrated pathways that require transformational changes in the way agriculture is managed, or they can be farm-specific technologies that are relatively easy to implement. Two broad management pathways are reshaping the way agriculture is practiced today and are helping farmers around the world prepare for supplying the food that will be needed to meet the 60% increase in foodstuffs that will be needed by 2050 by the growing world population; one is Climate Smart Agriculture (CSA) and the second is Conservation Agriculture (CA).

Box 12.1 Water's role in achieving sustainable agriculture

"By 2050, the proportion of the population facing stressed water supplies is expected to increase by 500 percent and the number facing full water scarcity is expected to increase by 800 percent. Climate variability is expected to impact many major crops, cutting productivity in key food crops like irrigated rice by as much as 27 percent, rainfed wheat by 25 percent, and rainfed [corn] by 15 percent. Therefore, farmers around the globe must be involved in making crops more resistant to stress and cropping techniques and be more water efficient. There are things that farmers can do to help reduce water use. Safe drinking water systems and adequate sanitation that effectively disposes of human waste will be essential to ensure cities and towns grow sustainably."

The Farming First coalition advocates this six-point action plan for enhancing sustainable development through agriculture:

1. Safeguard natural resources; be drought prepared; plant drought-resistant crops
2. Share knowledge of water-efficient technologies; use best practices
3. Build local capacity; invest in infrastructure, irrigation and water storage systems
4. Protect harvests; minimize pre- and post-harvest losses; support higher quality crops
5. Enable access to markets; build better storage; reward water use innovation
6. Prioritize research imperatives; invest in R&D; promote farmer–researcher partnerships

Source Farming First (2013)

CSA is described by the United Nations Food and Agriculture Organization (FAO) as an approach to farming that addresses the interlinked challenges of food security and climate change. CA is a change in the traditional practices of soil preparation, ground cover and crop rotation. Box 12.1 describes water's role in sustainable agriculture. The technological advances are designed to increase the efficiency and effectiveness of the way that water is used in the world's large and small farms. Reducing water losses in the distribution network and by evaporation from the way water is applied are technical objectives that do not require large investments in infrastructure. Efficient irrigation technologies such as low-volume drip methods and micro-sprinkler irrigation systems are already applied in most farming regions.

External Forces Driving Change

Decades of recorded weather observations clearly reveal that changes in the world's climate are resulting in rising temperatures and in the increased frequency and severity of extreme weather events, including flooding and soil erosion from severe rainstorms. These adverse natural events are resulting in negative impacts on food crops, livestock, forestry, fisheries and aquaculture—all of which are necessary for food security and are important contributors to the economies of both developing and developed nations. Hence, efforts to ensure food security must include programs to strengthen farmers' resiliency and capability to survive these shocks to their environment. For this to happen, a transformational shift to more productive and resilient agriculture must replace existing business-as-usual operations. This is particularly critical for the small subsistence farms of the world. The shift must take place in the way land, water, fertilizers and pesticides are managed, and resources must be used more efficiently and sustainably (FAO 2013). These changes are addressed in CSA.

Climate-Smart Agriculture

Climate change and population growth will mean that the needed 60% increase in food supplies may not be possible if the agriculture sector continues to operate with a business-as-usual model. The agriculture sector is already struggling to meet food supply demands in many parts of Africa and Southeast Asia because of water scarcity and low water quality.

The Agriculture and Food Security research center at The Netherlands' Wageningen University explained the climate change-related environment difficulties that face the food sector in the twenty first century: "Many of the resources needed for sustainable food security are already stretched; the food security challenges are huge. At the same time, climate change is already negatively impacting agricultural production globally and locally. Climate risks to cropping, livestock and fisheries are expected to increase in coming decades, particularly in low-income countries where adaptive capacity is weaker. Impacts on agriculture threaten both food security and agriculture's pivotal role in rural livelihoods and broad-based development" (CCAFS 2014, online).

CSA was defined by FAO in 2010 as "an approach to developing the technical, policy and investment conditions to achieve sustainable agriculture development for food security under climate change." It contributes to the achievement of national food security and development goals with these three objectives:

1. Sustainably increase agricultural productivity while supporting equitable increases in farm incomes, food security and development.
2. Adapt and build the resilience of agricultural and food security systems to survive climate change at multiple levels.
3. Reduce greenhouse gas emissions from agriculture (including crops, livestock and fisheries).

Despite the UN's interest in ensuring the sustainable development contribution of CSA, it is important to remember that it not a set of practices that can be universally applied. Rather, CSA is an approach that requires different responses and different technologies to be regionally or locally applied as appropriate. CSA is an agriculture philosophy that applies to both on-farm and beyond-the-farm actions, technologies, policies, institutions and investments (CCAFS 2014). The CSA approach to reducing food insecurity, providing continued economic growth and reducing poverty includes the following features (FAO 2013):

- CSA addresses adaptation and builds agriculture's resilience to shocks;
- It sees mitigating climate change effects as a potential economic co-benefit of agriculture;
- It is a location-specific and knowledge-intensive approach;
- It is an integrated approach sponsoring synergies and reducing negative tradeoffs;
- It identifies and provides solutions for barriers to process adoption;
- CSA improves farmer livelihoods by enabling better access to services, knowledge and resources; and
- It integrates solutions for climate change impacts with traditional agricultural investments.

Conservation Agriculture

To meet future food needs with more efficient water-smart agriculture in the face of the changes occurring in weather, urban migration and population growth requires adoption of the second sustainability movement: CA. In the world's most arid regions, CA may be the solution to the problems that are in the way of agriculture meeting its growing production demands. One of the advantages of CA over other approaches is that it results in less labor, lower investment in tillage machinery, and in the reduction in applications of fertilizers and pesticides.

CA is not a new concept. The movement has its roots in the 1930s' Dust Bowl of the USA. A prolonged period of drought and excessive tillage of the soil accelerated evaporation of the soil moisture and subsequent soil erosion, with top soil from U.S. Great Plains states blown as far as the East Coast.

Quoting the FAO, Berger et al. (2010, 80), defined CA as: "A concept for resource-saving agricultural crop production that strives to achieve acceptable profits together with high and sustained production levels while concurrently conserving the environment. CA is based on enhancing natural biological processes above and below the ground. Interventions such as mechanical soil tillage are reduced to an absolute minimum, and the use of external inputs such as agrochemicals and nutrients of mineral or organic origin are applied at an optimum level and in a way and quantity that does not interfere with, or disrupt, the biological processes." The three guiding principles of CA are limiting tillage, maintaining soil cover, and diversified crop rotation.

Minimal mechanical soil disturbance (the reduced tillage principle) means that the plowed area should be less than 15 cm (6 inches) wide or 25% of the cropped area—whichever is lower. The most important benefits of reduced tillage are the significant reductions in soil erosion due to wind and water and less loss of soil moisture from evaporation. Other advantages include reduced fuel and labor requirements during planting and post-harvest seasons. A possible disadvantage is that reverting to traditional reliance on herbicides with reduced-tillage systems may be used for with some tillage systems. No tillage, or greatly reduced tillage, is necessary for achieving the benefits of CA but it alone is not enough to achieve sustainability goals. It must be accompanied by the two other CA practices: soil cover and diversified crop rotation (Berger et al. 2010).

Soil cover refers to a permanent organic ground cover retained over the soil, while reducing disruption of the soil surface as much as possible. In no-tillage or reduced-tillage practices, soil cover should be in place for more than 30% of the tilled area. Any amount lower than this would not be considered CA. Cover crops may be crop stubble or planted as a single, soil-building species or in mixtures of vegetation. The most common mixtures include a legume such as a cereal grain like rye, which additionally serves to protect the soil from winter erosion better than a pure stand of vetch. In addition to controlling erosion, a regular supply of decaying organic materials can increase soil fertility, suppress germination of weeds, and feed valuable soil organisms. Cover crops and decaying organic mulch can also reduce pollution by preventing runoff of nutrients and pesticides into surface water.

The main benefits of diversified crop rotation have been known for centuries. The system was known for its ability to improve crop yields, eliminate or reducing damage by plant pests, and for controlling soil erosion from wind and water damage over cleared fields. Among the earliest documented uses of a crop rotation system are the early written reports of the Romans, where a three-plant rotation system called 'food, feed, fallow' was followed. Roman farms were regularly divided into three plots. A food crop was grown in one plot each year, the second plot was planted in an animal feed crop such as oats, and the third plot was left undisturbed. Crops were then rotated the next year. Among the chief benefits cited for this system are two that improve water quality and conserve water supplies: reduced pollution of water and soil from leached nitrogen, and the lower requirement for irrigation due to retained soil moisture. Others include lower risk of flooding because soils with retained root structure may act like a sponge, allowing filtration replenishment of groundwater aquifers, and improved soil structure that enhances the water-holding capacity of soils (Slavikova 2018).

In CA with annual plants, the rotation should involve at least three different crops. An example is rotating from corn, to cotton and then to mustard. Mustard is a special crop that has a farm value ranging from about US$50–100 million dollars annually. Mustard is often grown in rotation with small grain crops and to suppress weed growth. Small grains include winter and spring wheat, winter and spring barley, oats and rye.

Agriculture water conservation and efficiency are of critical importance in the arid and semi-arid USA, sub-Saran Africa, North Africa and the Middle East, and sections of South America. The CA approach requires significantly less water use for crop irrigation due to the increased infiltration and enhanced water-holding capacity from crop residues left on the soil surface. Crop residue and mulches left on the soil and no-tillage seeding also protect the soil surface from extreme temperatures and greatly reduce surface evaporation. Reducing soil surface evaporation is particularly important in the American Southwest, in the arid Middle East and North Africa, and in tropical and sub-tropical climates, as in Latin America and parts of South America. The following 2016 statement was made by Agrihome Expressions, an agriculture support agency in Ghana regarding the water benefits for African farmers from the adoption of CA: "Conservation agriculture requires significantly less water use due to increased infiltration and enhanced water holding capacity from crop residues left on the soil surface. Mulches also protect the soil surface from extreme temperatures and generally reduce surface evaporation, which is particularly important in tropical and sub-tropical climates. In sub-Saharan Africa, as with other dryland regions, the benefits of CA are most salient during drought years, when the risk of total cop failure is significantly reduced due to enhanced water use efficiency" (Agrihome 2016).

World Adoption of Conservation Agriculture

No-till farming began with Department of Agriculture experiments in the mid-1930s as a way of eliminating the soil erosion from the severe dust storms that devastated the American Southern Great Plains states, further contributing to the economic losses of the Great Depression. Several years of drought had left the region's fields bone-dry. The traditional deep plowing method of preparing soil for crop seeding had removed all soil cover. High winds swept away the region's top soils from an area from Texas to Nebraska. The choking dust killed people and livestock and crops failed across the entire region. The losses drove many farming families—including this author's grandparents—on a desperate migration west in search of work and better living conditions. Many small Midwest farming communities simply disappeared. Those that remained were the pioneer farmers of what became no-till agriculture.

Changing weather patterns saw an end to the Dust Bowl and early guidance and intervention by the Department of Agriculture's soil conservation service (since 1994, the Natural Resources and Conservation Service) helped farmers learn how to keep their topsoil on their farms. World War II revitalized agriculture and kick-started the economy, finally putting an end to the Great Depression. However, by the 1960s, reoccurring drought and failed crops brought new interest to the method. New research in the benefits and pitfalls of CA began in the United Kingdom, the USA, Argentina, Chile, Brazil, Australia and in Africa. In the USA, a book by Shirly Phillips and Harry Young titled *No-Till Farming* appeared in 1973, giving another boost to the method. By the year 2000, no-till farming had become CA, and was being applied around the globe. Other names that have been used for the practice include no-tillage, direct-drilling, minimum-tillage and/or ridge-tilling (Baker et al. 2002; Hobbs 2006).

Table 12.1 Adoption of Conservation Agriculture (CA) methods by continent, 2011 data

Region	Area (in hectares)	Area[a] (in acres)	Total (%)	CA as a percentage of total arable cropland (%)
South America	55,464,100	137,054,775	45	57.3
North America	19,981,000	49,374,126	12	15.4
Australia and New Zealand	17,162,000	42,408,226	14	69.0
Russia and Ukraine	5,100,000	12,602,374	4	3.3
Asia	4,723,000	11,670,787	4	0.9
Europe	1,351,900	3,340,617	1	0.5
Africa	1,012,840	2,502,782	1	0.3
World	124,794,940	308,375,012	100	8.8

Source From UN FAO (2011) data in Kassam et al. (2014)
[a]Acre figures may not equal total due to rounding

Table 12.1 shows the amount of CA land use underway in all major regions and the total acreage devoted to CA in 2011. South America is the leading adopter of CA with more than 57% of its total arable cropland devoted to CA farming. With 69% of the total cropland under CA methods, Australia and New Zealand have the greatest percentage of their total cropland following CA. The data in Table 12.1 are presented in both hectares and acres. Europe and Africa were the slowest regions to adopt the principles of conservation farming, each with less than 1% of their total arable cropland under CA practices.

The Department of Agriculture conducts a Census of Agriculture every five years, with the data published some two years later. Data for the 2012 census was available in 2014; 2017 census data should be available in 2019 or 2020. In the 2012 census, the total amount of cropland of the 278,290 farms on which CA's no-till practices were used was 96,476,497 acres, with 347 acres being the average farm size using CA. At the same time, there were another 185,738 farms using CA methods other than no-till methods, such as crop rotation and soil coverage; this was taking place on 76,639,604 acres with an average farm size of 392 acres. In comparison, there were 450,692 farms with a total of 105,707,971 acres, and an average size of 261 acres, who used traditional tillage practices. The largest total acreage farmed without any CA methods was 33,324,426 acres, farmed by a total of 9324 farms.

CA Example in India

Results of a CA, low-tillage experiment in the farming region of the Odisha province in East central India were reported by Pradhan et al. (2018). Traditional agriculture in the rain-fed uplands of India suffer from low productivity due to poor soil fertility, high soil erosion, the effects of extreme weather events associated with climate change, plus the impact of population growth endemic to the Indian sub-continent. The research project described by Pradhan et al. occurred over the three-year period from 2011 to 2014. It included reduced tillage, inter-cropping and cover cropping practices designed for the region's maize-base crop systems. Progress was measured by crop yield, system productivity and profitability. No changes were found in any maize yields, including intercropping with cowpeas. although the maize–cow pea combination resulted in income from the second crops of cowpeas and horse gram. Mustard and horse gram were grown in same plots after the maize harvest and maize and cowpea intercropping (horse gram, or *Kulthi* in Hindi, is a lentil widely consumed as a staple diet in the southern parts of India) (Table 12.2).

Table 12.2 Indian conservation agriculture experiment results

Tillage	Cropping system	Residue (mulch) cover	Average yield (kg per ha)	Gross field benefits (US$ per ha)	Total variable costs (US$ per ka)	Net benefits (US$ per ha)
Conventional tillage	Maize	Fallow	4777	681	370	311
		Mustard	7186	1024	500	524
		Horse gram	6237	889	494	395
	Maize + cowpea	Fallow	7735	1102	474	628
		Mustard	10,585	1508	604	904
		Horse gram	9882	1408	698	810
Minimum tillage	Maize	Fallow	4260	607	338	219
		Mustard	6461	921	518	403
		Horse gram	5776	823	512	311
	Maize + cowpea	Fallow	7739	1103	440	663
		Mustard	10,731	1529	570	959
		Horse gram	9726	1386	564	822

Source After material in Pradhan et al. (2018)

Overall, the design of minimum tillage, maize–cowpea intercropping and mustard mulch left on the soil for moisture retention had significantly higher system productivity and net benefits over traditional practices of conventional tillage, soil maize cropping and no mustard resident retention. Mustard and horse gram grown after the intercropping and mustard residue yielded higher grain crops of 25% and 37%, respectively. The authors' interpretation of the research findings was that, although no universal strategy exists for ending food scarcity and rural poverty, the combination of the following practices will go far in helping to meet the growing need for food: reduced tillage, cross-cropping and selected crop residue left on the soil as mulch, reduced water loss, and improved farmer include by providing additional crops on the same plots.

Efficient Irrigation Water Use

As noted earlier, there are four main methods of agriculture irrigation: gravity, sprinkler, drip and microsprinkler, and subsurface. In the gravity method, water is brought or pumped to the field either by flooding the field or in furrows between crop rows. Sprinkler irrigation systems, as with the common central pivot systems, irrigate a crop by applying water by means of sprinklers attached above or below the main system pipe. Fertilizers and pest control chemicals can be included in the sprayed water. Drip and microsprinkler systems apply small quantities of water on or below the soil surface as drops, tiny streams, or miniature spray placed along a furrow between crop rows or buried under ground alongside plants. Water application methods include bubble, drip, subsurface drip, mist or spray systems. Subsurface systems include micro-irrigation systems in which water is applied directly to the root zone of plants by means of applicators (e.g., orifices, emitters, and porous tubing) placed below the ground surface. According to the U.S. Department of Agriculture, subsurface systems are considered to be one of the more advanced irrigation methods in use. Because evaporation is significantly reduced, these are considered to be the most efficient irrigation methods.

Drip and microsprinkler irrigation systems are reported to have the highest irrigation efficiency rating of 87.5–90%, compared with traditional sprinkler systems that range from 70 to 82.5%, depending on the type of system. Irrigation efficiency for most surface irrigation systems can range widely, with furrow and gravity systems having a reported 67.5–75.0% efficiency. Overall, micro-irrigation and subsurface systems are the most efficient way to irrigate. In water-scarce regions, drip and other micro methods have long been used to irrigate vegetable crops and fruit and nut orchards. On the island of Crete, for example, close to 82% of the olive orchards are irrigated with high-efficiency irrigation methods (Udias et al. 2018). Drip and microsprinkler irrigation are now the most commonly used micro method, largely because it is relatively inexpensive to install and is generally a more efficient way of getting water to a plant's root system. The traditional spray irrigation method loses more water irrigation water by evaporation.

Despite the fast-growing adoption of CA, a majority of the U.S. agriculture industry still uses traditional flood and furrow or sprinkler methods. In water-scarce California, for example, in 2010, flood and furrow irrigation accounted for 43% of all irrigated acres, and in 2015 these were still the predominant irrigation methods. The shift to drip and microsprinkler irrigation systems in the state is growing, however. The change may be due to mandated water access limitations during the recent prolonged drought. The acreage to which gravity (furrow flood) irrigation methods were applied declined by 36%, and the sprinkler method declined by 15%. During this same period, the drip and microsprinkler method increased by 150% over the same period, and the subsurface irrigation method increased by 380%.

Summary

The combined effects of climate change and population growth are placing a tremendous demand upon the world's finite supply of freshwater, which in turn is creating a necessary search for ways to conserve and augment water for agriculture. In this chapter, two complementary approaches to water sustainability were discussed: transformational changes to water management and the application of innovative methods for improving the efficiency of irrigation techniques.

The two pathways shaping the transformational changes in the planning and management of water supplies in the agriculture sector are CSA and CA. CSA incorporates the current and future effects of climate change on the supply and demand of water for agriculture. It includes these three objectives in its water plans: increasing agricultural productivity while supporting increases in farm incomes, food security and development; building resilience of agricultural and food security systems for surviving climate change; and reducing greenhouse gas emissions associated with all forms of agriculture production. The three guiding principles of CA are limiting tillage, maintaining soil cover, and diversified crop rotation. Limiting tillage reduces soil erosion and evaporation of soil moisture, reducing the need for crop irrigation. Soil cover also retains soil moisture, while also improving the soil quality. Crop rotation also improves soil quality while also reducing the need for added fertilizers and pesticides.

CA water application methods include bubble, drip, subsurface drip, mist, or spray systems. The main techniques for sustainable agriculture water use are drip and microsprinkler systems. These methods apply small quantities of water on or below the soil surface as drops, tiny streams, or miniature spray placed along a furrow between crop rows or buried under ground between furrows. Subsurface micro-irrigation systems are applied directly to the root zone of plants and are considered to be one of the more advanced irrigation methods in use. Because evaporation is significantly reduced, these are considered the most efficient irrigation methods. In water-scarce regions, drip and other micro methods are increasingly used to irrigate vegetable crops as well as in fruit and nut orchards.

References

Agrihome Expressions. 2016. Conservation agriculture (CA): Increase yields and improve farmlands. Accessed August 14, 2018 from http://agrihomegh.com/conservation-agriculture-benefits/.

ATS Irrigation, Inc. 2018. What are the different types of farm irrigation systems? Accessed August 15, 2018 from www.atsirrigation.com/what-are-the-different-types-of-farm-irrigation-systems/.

Baker, C.J., K.E. Saxton, and R.R. Ritchie. 2002. *No-Till Seeding: Science and Practice*, 2nd ed. Oxford, UK: CAB International.

Berger, Amélie, Theodor Friedrich, and Josef Kienzle. 2010. Conservation agriculture. In *Soils, Plant Growth and Crop Production*, ed. Willy H. Verheye, vol. 1, 79–105. Oxford: EOLSS Publishing.

CCAFS (Climate Change, Agriculture and Food Security). 2014. Climate smart agriculture. Wageningen University and Research, The Netherlands. Accessed August 11, 2018 from https://ccafs.cgiar.org/climate-smart-agriculture-0#.W28Vfc5KiUk.

Dilekli, Naci, Faye Duchin, and Ignacio Cazcarro. 2018. Restricting water withdrawals of the thermal power sector: An input-output analysis for the northeast of the United States. *Journal of Cleaner Production* 198: 258–268.

FAO (Food and Agriculture Organization of the United Nations). 2012. China: Geography, climate and population: Water report. Aquastat, Food and Agriculture Organization of the UN. Accessed December 9, 2017 from www.fao.org/nr/water/aquastat/countries_regions/CHN/.

Farming First. 2013. Enhancing sustainable development through agriculture: Applying the Farming First principles to increase water use efficiency. Accessed January 21, 2018 from https://farmingfirst.org/wordpress/wp-content/uploads/2010/10/FarmingFirst-Water-Paper.pdf.

Hobbs, Peter R. 2006. Conservation agriculture: What it is and why is it important for future sustainable food production? Paper presented at the international workshop on Increasing Wheat Yield Potential, Obregon, Mexico, March 20–24.

Kassam, Amir, Rolf Derpsch, and Theodor Friedrich. 2014. Global achievements in soil and water conservation: The case of Conservation Agriculture. *International Soil and Water Conservation Research* 2 (1): 5–13.

Levidow, Les, Daniele Zaccaria, Rodrigo Maia, Eduardo Vivas, Mladen Todorovic, and Alessendra Scardigno. 2014. Improving water-efficient irrigation: Prospects and difficulties of innovative practices. *Agricultural Water Management* 46: 4184–4194.

Pradhan, Piyush, Ajay Verma, and Vithlesh Kumar. 2018. Need of conservation agriculture in India: Sustainability. *International Journal of Current Microbilogy* 7 (1): 308–314.

Slavikova, Sara. 2018. 10 benefits of crop rotation in agriculture. Accessed August 14, 2018 from https://greentumble.com/10-benefits-of-crop-rotation/.

Udias, Angel, Marco Pastori, Anna Malago, Olga Vigiak, Nikolaos P. Nikolaidis, and Faycal Bouraoui. 2018. Identifying efficient agricultural irrigation strategies in Crete. *Science of the Total Environment* 633: 271–284.

UN (United Nations). 2014a. International decade for action 'Water for Life' 2005–2015. Accessed November 5, 2017 from http://www.un.org/waterforlife-decade/food_security.shtml.

UN (United Nations). 2014b. World water development report 2014: Water and energy. Accessed July 2018 from http://www.unesco.org/new/en/natural-sciences/environment/water/wwap/wwdr/2014-water-and-energy/.

13

Pathways to Water-Smart Power

In 2015, the global production of energy accounted for approximately 15% of all freshwater withdrawals. That percentage is expected to grow to 20% by 2035. As more developing nations industrialize, and their populations continue to grow, more people moving to cities will need power for lighting, heating and air-conditioning and everyone will use more water. To draw irrigation water from deeper and deeper water tables, farmers will need to install pumps that are more powerful and use more energy. Water and energy planners are finding it more and more difficult to achieve their global goal of a sustainable future. Moreover, the water and energy problems they face are tightly interconnected, and both are necessary for sustainable growth. By 2035, for example, the cost of energy could account for as much as 40% of the total operating cost of a water utility (Khalkahali et al. 2018). The interconnection of the two vital resources means that more water to supply the needs of a growing population must be found. More energy will be needed for the acquisition and treatment of freshwater and for the discharge of treated wastewater. The water–energy nexus means that both water and energy are needed for economy growth and for the support and quality of life.

The UN estimates that the cooling of water withdrawn for thermoelectric generation accounts for 43% of all freshwater withdrawals in the European Union and 50% of all freshwater withdrawals in the USA. Globally, close to 53 billion cubic meters of freshwater is consumed by thermoelectric production (water withdrawals denote water withdrawn from all sources; water consumption is water withdrawn but not discharged back to water bodies for future use). In addition to the need for water in all stages of the production

D. E. McNabb, *Global Pathways to Water Sustainability*,
https://doi.org/10.1007/978-3-030-04085-7_13

of power, water in very large amounts is also important for the capture, cleaning, processing, and shipping of the primary energy sources such as coal and natural gas that together fuel close to 90% of all power-generating facilities. Much of the water used for producing these fuels is either lost to evaporation, returned by deep wells to the earth, or polluted beyond affordable reuse. The demand for water and electrical energy are growing worldwide, while the supplies of both are impacted by weather and precipitation patterns brought on by global climate change (Liao et al. 2016; UN 2018).

The Water–Energy Nexus

Globally, thermoelectric generating plants produce most of the world's electricity. In the USA, it is as high as 89–90%, with the remaining 10% using either stored water for hydroelectric generation or renewable sources such as wind, sunlight, biofuels or other sources. Thermoelectric plants use coal, natural gas or nuclear power to generate power. These plants require large amounts of freshwater for producing steam in the boiler, for cooling and condensing the steam after it is used to power turbines for generating electricity, and for scrubbing flue gas to remove and collect potential air pollutants. The largest use of water in power plants is for cooling. The U.S. Geological Survey (USGS) estimates that approximately 195 billion gallons of water is used every day for cooling steam and the equipment used to generate power. Coal-fired plants in China, for example, consume as much as 11% of all water consumed by Chinese industry. In the USA, coal-fired power plants account for 39% of all freshwater withdrawals (Regucki et al. 2018).

The water–energy nexus is a common way of referring to the interrelationship of water and energy. It includes the water used to generate the steam for generating electricity and for cooling the steam used in the power generation process. It also includes the electricity used to collect, clean, move and store fresh water, and for collecting, treating and disposing of wastewater. Water is needed in all stages of electric energy conversion from primary energy fuels to the making of electricity. Power plants use one of two methods, or combinations of the two, for producing steam for generating electricity and for cooling and condensing the steam after its use in turbines. Thus, all power systems are potentially vulnerable to water scarcity and climate warming. Any decline in the accessible water for steam and cooling, or temperature increase in the rivers into which they discharge their partially cooled water, can significantly limit the generation of electricity (Rogers et al. 2013; Gjorgiev and Sansavini 2017). Federal regulations require the temperature

Table 13.1 Cooling methods used in power plants by fuel source, 2014

Fuel source cooling systems					
Primary energy source	Once-through	Re-circulating	Dry-cooling	Wet and dry hybrid cooling	Total cooling systems
Coal	398	368	4	1	771
Natural gas	197	422	51	4	674
Nuclear	50	44	0	0	94
Other	74	41	1	0	116
Totals	719	876	56	5	1655

Source EIA (2014)

of the discharged water to be below levels that can harm local stream life; this requires steam to be kept at lower levels, thereby reducing the efficiency of the generating process. Table 13.1 lists the distribution of the cooling systems used in the USA by the major primary energy source.

Until the 1970s, most of these plants burned coal for their primary energy source and used once-through cooling systems (EIA 2014). Generating facilities were constructed adjacent to water bodies—mostly rivers. Water was withdrawn from the nearby surface water sources, passed through the plant once, and then returned to the source or an alternative water body. However, coal-fired plants constructed after the 1970s are more likely to have installed recirculating, closed-loop cooling systems. Construction of these coal-burning plants peaked during the four-year period from 2000–2004. Since 2004, most of the 201 new plants that went online were predominantly the environmentally friendly natural gas burners that greatly reduced the discharging of CO_2 and other greenhouse gasses, together with closed-loop cooling systems.

By the end of the 1970s, most new generating plants had also turned to a waterless system known as dry cooling. Instead of using water to lower steam temperatures in either once-through or recirculating systems, the dry-cooling system uses large fans to blow air over the cooling water or uses cooling towers and cooling ponds and recycles the water. These systems have a penalty, however; its takes much of the electricity generated to power the cooling fans and, in the process, water is lost to evaporation. This makes them less desirable for large generating plants that require large amounts of steam, such as coal or nuclear plants. As of 2014, of the 1655 cooling systems operating in the USA, 875 (53%) were reusing water, either having it passed through a cooling tower or letting the heated water rest in a cooling pond before discharging or reusing it. During the same period, 719 systems (43%) did not reuse water, 56 systems had adopted the dry-cooling methods, and 5 were employing a hybrid system (Table 13.1).

However, it is important to note that, regardless of the primary fuel used, not all the water required for cooling is lost to evaporation. In coal-powered plants, almost all the water withdrawn for a power plant with a once-through cooling system is returned to the source from which it was taken or discharged in another water body. While plants with closed-loop systems withdraw far less water, they consume much of that water through evaporation during the cooling process. Once-through systems lose less water to evaporation but tend to discharge hotter water back to its source. During the hot summer months and low stream flows, heat pollution can be as much of a problem as water lost due to evaporation during cooling. When still-hot water is added to warmer source water, significant damage to the environment can occur and this has resulted in many extensive fish kills. Hotter water in the cooling chain also reduces the plant's generating efficiency.

While only a moderate percentage of the water used to cool all thermoelectric power plants is actually consumed, very large volumes are required for use. Water permits are required to withdraw large volumes of water from surface water and groundwater sources, and as competition for water supplies increases, it may be difficult or impossible to acquire new water permits. Moreover, the water withdrawn by the power plants may also be needed for municipalities, industry or agriculture. During the prolonged droughts that are occurring more often and lasting for longer periods, all these water users have had to compete for access to the limited water remaining. Restrictions occurred in both the American Southwest and Southeast during the first two decades of this century. As a result, many of the U.S. older coal-fired power plants that were constructed with once-through cooling systems are being replaced by newer natural gas combined cycle plants and with recirculating water-cooling systems. In areas of more extreme water scarcity such as Arizona, Nevada, California and Texas, some utilities are taking the path of using reclaimed wastewater for cooling, thereby leaving the limited supplies of freshwater for other sectors. In these regions, and as competition for a limited resource increases, alterative water resources have not only become more economically viable, they have become essential.

Box 13.1 Planning for a water-smart energy future

In light of the effect of global warming on a sustainable future, the following 2013 report of the Energy and Water in a Warming World Initiative group of the Union of Concerned Scientists included this warning on the critical need for integrating national, regional and local planning for all electricity generation and water supply:

"No single platform exists for sound, long-term decisions at the nexus of electricity and water, but those made in isolation will serve neither sector. Instead, actors across sectors and scales need to engage ...Understanding and addressing the water impact of our electricity choices is urgent business. Because most power sector decisions are long-lived, what we do in the near term commits us to risks or resiliencies for decades. We can untangle the production of electricity from the water supply, and we can build an electricity system that produces no carbon emissions. But we cannot wait, nor do either in isolation, without compromising both. For our climate—and for a secure supply of water and power—we must get this right."
Source Rogers et al. (2013, 5)

Water Restrictions Increasing

All thermoelectric power plants need water. However, climate change has made acquiring that needed water problematic for many plants. More days with very high temperatures and longer warm days in general, have reduced stream flows and warmed surface waters during the summer season when demand is greatest. Warmer input water and periodic water scarcity significantly reduces the efficiency of the power-generating equipment. The USA, Europe and other regions have enacted strict restrictions on how much power-generating plants can add to the streams and lakes. However, those restrictions are not always enforced. During the prolonged periods of drought over the last two decades, all water users faced mandated restrictions in their allowed withdrawals. In the USA and Europe, some power plants had to cut production and, in some cases, close production completely due to the lack of water. Box 13.1 is an excerpt from a 2014 report in which the authors explain the need for considering both water and energy issues in planning for future actions by both sectors.

Controlling Heated Water Discharges

Dilekli et al. (2018) studied the forced power output restrictions on power plants in 12 Northeast U.S. states and Washington, DC. to determine what operational changes would be necessary in order for plants to comply with strict enforcement of current water legislation. At the time of the study, New Jersey, Maryland and Vermont were the states experiencing the greatest required cuts in power production.

Although the majority of plants in the study area used coal-fired, once-through cooling technology, natural gas and closed-loop systems and nuclear power dry-cooling (tower) systems were also included in the study. The researchers compared operating costs (baseline) without any change in water use against costs for tower cooling and with required cuts in water availability. Costs would increase from US$50 million to US$180 million when water withdrawals were restricted. Additional costs for capital and labor for tower cooling were the main reasons for the increase. Moreover, tower cooling reduces the efficiency of power generation, further contributing to the cost increases. The increases depended upon whether restrictions in interstate power transmission were implemented. Sharing power with regions not forced to cut production limits the impact of the lost revenues. Current legislative restrictions have led to increases in operating costs. The researchers concluded that, despite the increases, the region has more than enough installed capacity to fully comply with the restrictions on water withdrawals—provided interstate transmission is allowed to continue.

Restrictions Occurring Globally

Europe has a long history of energy-related water management activities; for example, reductions in river flows throughout Europe have been recorded since the 1960s. Carrying those patterns forward, now with the added factor of climate change, suggests that declining annual summer flows will not only continue, they will decline at an even faster rate (Gjorgiev and Sansavini 2017). A flow rate study of 200 of Europe's rivers from 1948–2004 found that rivers with declining flows exceed those increased flow by a ratio of 2.5–1. Almost all of Europe's power-generating plants were constructed adjacent to these existing water bodies.

Plant operators and researchers are searching for ways to avoid these extremely costly forced curtailments. Some have turned to groundwater when surface water supplies run short or are too warm for optimal power generation. In China, where almost all thermoelectric plants in the interior of the country use coal and once-through cooling systems, the use of groundwater by Union of Concerned Scientists (UCS) power plants has been banned since 2004 (Liao et al. 2016). Plants on the coast in many nations are more likely to use seawater for cooling. The UCS have added these water and power problems to their concerns, sponsoring research on possible pathways for coping with the water–power nexus.

In 2015 in the EU, nuclear power generated 27% of the union's electricity. Another 27% was produced in coal-powered plants, 17% came from plants using natural gas, 2% from oil, and a final 27% from renewables (more than half of which was from hydro). The percentage contributed by nuclear energy is declining in many areas, including the USA, where it is expected to be zero in 2050.

Nuclear power generation remains a major enegy-sector contributor in Europe. According to the WNA (2018), the 128 nuclear reactors operating in 14 of the 28 EU member states account for over a quarter of the electricity generated in the whole of the EU. Half of the EU's nuclear electricity is produced in France. The 53 plants operating in three non-EU countries (Russia, Ukraine and Switzerland) account for about 17% of the electricity in the rest of Europe. Norway and Switzerland are part of the EU synchronous grid. In 2008, one European researcher included fossil fuels as a major contributor to the meeting of the EU's power supplies: "The future supply of electricity mainly depends on fossil fuel use, nuclear power, renewable resources and electricity efficiency" (Verbruggen 2018, 4037). However, by 2017, the EU's attitudes regarding fossil fuels had shifted to alternative paths for meeting the demand for electricity: "The European Union is stepping up its efforts towards phasing out fossil fuel subsidies. To reach the EU's climate and energy targets for 2030, annual investments of some €379 billion (US$400 billion) is required over the 2020–2030 period.

Water Scarcity and Power

Large portions of the world are experiencing water scarcity, undergoing more and longer extreme water events, and enduring water stress stemming from the combination of climate change-altered changes in precipitation patterns, population growth and urbanization. When power plants cannot get the water they need, they have few alternatives; they must cut power production, find new sources of water, or completely stop production altogether. In 2014, coal-fired plants accounted for 33% of the all U.S. carbon emissions. World leaders are being forced to search for and adopt new ways of planning that include the interconnections between their energy as well as their water resources. Replacing old coal-fired, once-through cooling system generating plants with natural gas fired plants with recirculating cooling systems is already underway. This has the added benefit of greatly reducing carbon emissions.

Adapting to Climate Change

Clearly, the world needs to become more efficient in the production of energy as well as the way it uses the water necessary for producing that energy. If power projects are to remain resilient and long lived, they must be ready to react to the changes brought about by climate change, including greater precipitation pattern variability and higher peak demand as traditional warmer months get even warmer, all the while continuing to reduce carbon emissions (Rogers et al. 2013). Among the pathways with greatest possible impact is replacing all once-through cooling systems with closed-loop systems, dry-cooling systems or a combination of the two. In 2014, there were a little more than 50 recirculating systems using natural gas for fuel instead of the older coal-fired plants with once-through cooling systems. As more plants are converted from coal to natural gas, more recirculating systems are also being installed. Much of the world is already on this path. A much smaller number of plants are converting to not using any water for cooling, relying instead upon dry-cooling systems that use cooling towers or cooling ponds.

A 2017 study reported on another way to avoid halting production due to once-through water that exceeded the permitted source water temperatures would be to cooperate with other plants on a river system (Gjorgiev and Sansavini 2017). The objective would be to avoid forced restrictions on water intakes or on discharges of heated water from power plants by programing up-river reservoir releases so that enough fresh water flow is available to cool the discharged water in downstream regions. Cooling water that is overheated due to very high ambient air temperatures will be mixed with newly released water. The system would include redirecting other, colder river water for maintaining stream temperature control and, hence, permit plants to continue to produce electricity while discharging water that is hotter than regulations allow. Figure 13.1 is a photo of a Tennessee Valley Authority (TVA) combined cycle generating facility located in Mississippi.

The TVA plant consists of three separate and distinct one-on-one generating units, each consisting of a single combustion turbine driving a generator rated for approximately 160 MW, a heat recovery steam generator (HRSG) equipped with duct burners and a steam turbine generator (STG) rated for approximately 118 MW. The plant is capable of generating a total of 765 MW (TVA, n.d.).

Fig. 13.1 Tennessee Valley Authority natural gas combined-cycle power plant showing steam evaporation during the cooling cycle (*Source* Public domain photo www.tva.gov)

Pathways for Water-Smart Power in the USA

Water is used to produce electricity by many methods other than burning fossil fuels. Two important types are hydropower and two types of solar power. Nuclear power is usually included in renewable statistics. Another increasingly visible method is generating power by wind energy. Other sources of energy often mentioned include biomass, geothermal and tidal motion. Hydropower has long been a reliable power source; for many years it was the only reliable source. There are two main types of hydroelectric generating operations: conventional and pumped storage. Most of the hydroelectric generators are the conventional type; these account for 7% of the total electricity generating capacity in the USA and from 6 to 7% of U.S. electricity generated each year. Half of U.S. hydroelectricity capacity is located in three states: Washington, California and Oregon. Washington, with 21.3 % of the total hydroelectric capacity, has by far the greatest capacity (EIA 2017).

Hydroelectric facilities face two types of restrictions: the amount of water they receive and store, and the demands for stored water by other sectors. Agriculture, municipalities, industries and fish migration can all restrict reservoir water release. The amount of power many hydro plants generate each year varies by the water available in dams and rivers, and how much they must release for other sector uses. As a result, hydroelectric facilities often do not run at full output. For example, U.S. hydroelectric capacity, which measures actual output as a percentage of total capacity, averages between 30 and 40%.

Two types of solar power are used for electricity generation: solar thermal plants and photovoltaic panels. Solar thermal plants use concentrated sunlight to heat water for traditional steam turbine power generation. Research and data from test installations reveal solar thermal methods are heavy users of water for traditional cooling purposes, often exceeding the water use of coal plants. While they are advantageous as a means of reducing carbon emissions, they are not a valuable transition from coal-burning thermal plants (Wu and Chen 2017). Photovoltaic systems, on the other hand, use no water for production or operation. The only water used is for an occasional washing of the cells. Solar panels are fast becoming a preferred renewable application for both individual household installation and commercial power suppliers. In 2018, for example, at the time of writing, the Arkansas Public Service Commission reported that the number of households that have installed solar panels is almost double the previous year and more than 40 times the number in 2007. Many new owners are selling their excess power back to the utility. The commission's data shows net metering in the state of Arkansas at the end of 2017 could churn out enough power—about 9 megawatts—for 1000 typical homes based on industry estimates. In 2015, the capacity of renewable power surpassed coal as the biggest potential source of global power generation. As of 2018, solar panels, wind and hydropower were the preferred environmentally friendly energy options.

Electricity Generation with Wind

In 2017, wind turbines in the USA were the source of nearly 6.3% of the total electricity generated by power utilities. The amount of electricity generated from wind has grown significantly since 2000. Electricity generation from wind in the USA increased from about 6 billion kilowatt hours (kWh) in 2000 to about 254 billion kWh in 2017. Figure 13.2 displays the growth of wind power capacity in the USA. New technologies have decreased the cost of producing electricity from wind, and growth in wind power has been encouraged by governments around the globe as well as suppliers of electricity.

A large body of research on new technologies for reducing both water withdrawals and carbon emissions from burning coal has appeared since the early 2000s. Much of the research was triggered by the costly requirement to reduce carbon emissions by coal-powered power generating plants and the increased availability and significant decline in price of domestic natural gas fuels made available by the fracturing process (Peer and Sanders 2018). Except for a new goal plant in Texas, all new U.S. power plants include advanced

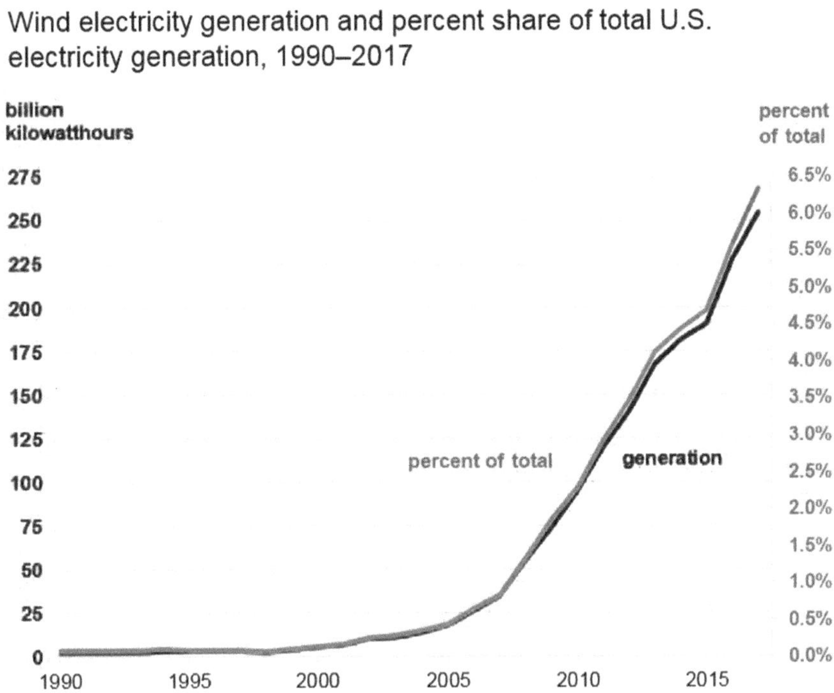

Fig. 13.2 Growth of wind power's share of U.S. electricity generation, 1990–2017 (*Source* US EIA)

technology natural gas combined cycle generating and either closed-loop, dry or a combination of the two cooling processes. Figure 13.3 is an EPA schematic drawing of the combined cycle process, showing that the heat used for power generation is redirected to the generation of steam for a second power generating cycle.

In their review of the literature on transitions underway in the U.S. power sector, Peer and Sanders focused on the shift from coal-powered generating facilities to natural gas fuels, with a smaller but substantial shift to renewables and a continued decline in nuclear powered plants. This trend is being matched with shifts from once-through cooling to the widespread adoption of closed-loop cooling systems, with a much smaller number of dry-cooling installations. One way that shortages in some areas can be relieved is to keep power generation withdrawals to a minimum by moving power plants to more water-sustainable regions. This "virtual water transfer" by long-range power transmission is being implemented in China (Liao et al. 2018). Among the water availability trends underway in the USA are:

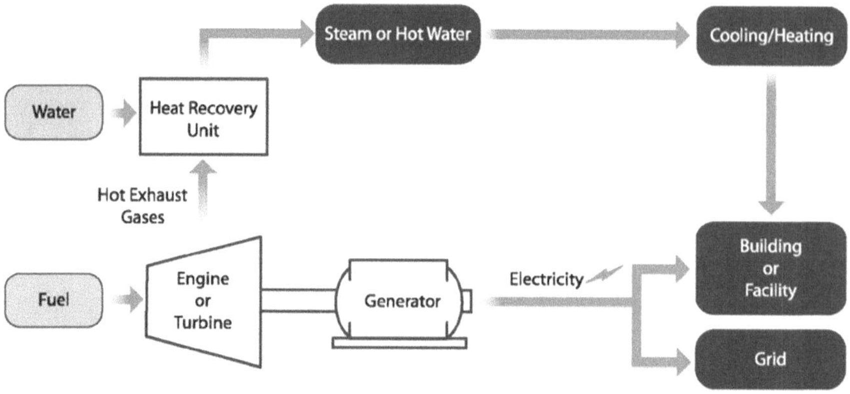

Fig. 13.3 Schematic of a simple heat and power combined-cycle system (*Source* EPA, n.d. www.epa.gov/chp/what-chp)

- A comparison by Gubert et al. (2012) of the water intensity of natural gas and coal extraction for electricity production and emissions controls at Texas fossil fuel power plants concluded that efficient benefits from switching from coal to natural gas combined cycle has the potential for a 60% reduction in annual freshwater consumption, despite the heavy use of water in hydraulic fracturing.
- A 2011 study by Stillwell et al. showed potential reductions of up to 700 million cubic meters of water withdrawals could result from converting once-through cooling systems in Texas power plants to alternatives such as recirculating cooling towers or dry-cooling methods; it would also contribute to increased stream flow and reduced water stress along the Texas Gulf.
- A 2014 study by Tidwell et al. concluded that transitioning all national thermoelectric plants from freshwater withdrawals and once-through systems to dry-cooling or wet-cooling using reclaimed water had the potential for achieving zero freshwater withdrawals in the eastern USA.
- A 2014 case study by Stillwell and Webber found that 60% of all Texas power plants are located within 25 miles of a reclaimed water source, making the switch from freshwater to reclaimed water for cooling in power plants economically feasible and able to improve existing water quality and help alleviate water availability concerns.

Summary

Water and power, both of which are essential to life, are closely interrelated. Water is necessary for the extraction of fossil fuels, for the steam that drives power generators and for cooling the operation. Power is needed for withdrawing water from surface water and groundwater sources, and for transmitting, treating and distributing freshwater to where it is needed and for treating and discharging wastewater. In all these activities, water and power production can have an impact on the quality and accessibility of water. Both water and power are also affected by climate changes underway around the globe. Water and power managers must work together to ensure both resources are available as the effects of climate change create even greater barriers to the sustainability of water and power.

Power's need for water will not be replaced at any time in the near future. However, innovative activities are underway for relieving the stress and mitigating the dangers brought about by climate change, population growth and industrialization. Among the trends contributing to the problems facing both water and power are conversions from coal to cleaner natural gas in thermoelectric generating plants. Changes from once-through freshwater withdrawals to closed-loop systems and dry methods for cooling operations are also beginning to make a significant reduction in the amount of freshwater lost to evaporation during the air and tower cooling process.

Mandated restrictions on used hot water discharges into freshwater sources are still a problem and are of particular concern in arid and semi-arid regions, including parts of Europe and the USA. Dry and air cooling and the use of alternative water sources are helping to alleviate this problem. Groundwater has become a solution that some power plants are turning to for supplementing declining stream flows. However, this is not a long-term solution. Many aquifers are the result of thousands of years of infiltration and cannot be recharged sufficiently. Aquifers in most of the developed world and arid regions of the world are already overexploited for agriculture or municipal use, or both. Changing precipitation patterns are helping to make groundwater supplies decline faster than they can be recharged.

Nuclear energy, wind power, solar power, hydro power, and a variety of renewables are among the many ways that electricity is generated around the globe. Until the early twentieth century hydro and wind power were almost the exclusive methods of renewable power generation. Major dam building and reservoir development began in earnest in the 1930s.

References

Dilekli, Naci, Faye Duchin, and Ignacio Cazcarro. 2018. Restricting water withdrawals of the thermal power sector: An input-output analysis for the northeast of the United States. *Journal of Cleaner Production* 198: 258–268.

EIA (U.S. Energy Information Administration). 2014. Many newer power plants have cooling systems that reuse water. Accessed July 20 2018 from www.eia.gov/todayinenergy/detail.php?id=14971.

EIA (U.S. Energy Information Administration). 2017. Hydroelectric generators are among the United States' oldest power plants. Accessed July 25, 2018 from www.eia.gov/todayinenergy/detail.php?id=30312.

Gjorgiev, Blaže, and Giovanni Sansavini. 2017. Water-energy nexus: Impact on electrical energy conversion and mitigation by smart water resources management. *Energy Conversion and Management* 148: 1114–1126.

Grubert, Emily A., Fred C. Beach, and Michael E. Webber. 2012. Can switching fuels save water? A life cycle quantification of freshwater consumption for Texas coal- and natural gas-fired electricity. *Environmental Research Letters* 7 (4): 1–11.

Khalkhali, Masoumeh, Kirk Westphal, and Weiwei Mo. 2018. The water-energy nexus at water supply and its implications on the integrated water and energy management. *Science of the Total Environment* 636: 1257–1267.

Liao, Xiawei, Jim W. Hall, and Nick Eyre. 2016. Water use in China's thermoelectric power sector. *Global Climate Change* 41: 142–152.

Liao, Xiawei, Xu Zhaob, Jim W. Hall, and Dabo Guan. 2018. Categorizing virtual water transfers through China's electric power sector. *Applied Energy* 226: 252–260.

Peer, Rebecca A.M., and Kelly T. Sanders. 2018. The water consequence of a transitioning US power sector. *Applied Energy* 210: 613–622.

Reguckia, Pawel, Marek Lewkowicza, Renata Krzyżyńskab, and Hussam Jouharac. 2018. Numerical study of water flow rates in power plant cooling systems. *Thermal Science and Engineering Progress* 7: 27–32.

Rogers, John, Kristen Averyt, Steve Clemmer, Michelle Davis, Francisco Flores-Lopez, Peter Frumhoff, Doug Kenney, Jordan Macknick, Nadia Madden, James Meldrum, Jonothan Overpeck, Sandra Sattler, Erika Spanger-Siegfried, and David Yates. 2013. Water-smart power: Strengthening the U.S. electricity system in a warming world. Cambridge, MA: Union of Concerned Scientists. Available at www.ucsusa.org/sites/default/files/legacy/assets/documents/clean_energy/Water-Smart-Power-Full-Report.pdf.

TVA (Tennessee Valley Authority). n.d. Caledonia combined cycle plant. Accessed July 24, 2018 from www.tva.gov/Energy/Our-Power-System/Natural-Gas/Caledonia-Combined-Cycle-Plant.

UN (United Nations). 2018. 2018 World Water Development Report. Accessed July 23, 2018 from http://unesdoc.unesco.org/images/0026/002614/261424e.pdf.

Verbruggen, Aviel. 2018. Renewable and nuclear power: A common future? *Energy Policy* 36: 4036–4037.

WNA (World Nuclear Association). 2018. Nuclear power in the European Union. Accessed July 24, 2018 from www.world-nuclear.org/information-library/country-profiles/others/european-union.aspx.

Wu, X. D., and C.Q. Chen. 2017. Energy and water nexus in power generation: The surprisingly high amount of industrial water use induced by solar power infrastructure in China. *Applied Energy* 195: 125–136.

14

Paths to Alternative and Supplemental Water Supplies

The fundamental task of water management involves the balancing of water supply with water demand. Today the task is made more difficult due to barriers such as climate change, population growth and urbanization that have strained the resource to the point where many water suppliers have been forced to seek ways to close the growing gap between water demand and water supply. Water demand projections are determined as unconstrained water requirements under an unchanged water policy regime. Water supply refers not to the raw resource, but rather to the amount of raw (untreated) water that is accessible by a demand center, and that can be reliably and sustainably delivered within the constraints of an environment or finite base (McKinsey and Company 2009). This definition of water supply includes by inference the facilities and infrastructure to make it safe and to deliver it to where it is needed.

This chapter describes examples of the successes and failures of nations in their efforts to achieve a sustainable supply of clean, safe water. The focus is not on water demand, but on supply, particularly to areas of the world where supplies are under sustainability stress.

The examples chosen are experiencing high to extreme water stress in the face of natural and human-caused barriers to achieving water sustainability. Supply gaps examined include Russia, China, India, South Africa and Jordan. Present and future water crises exist in many other parts of the world as well. The problems of supplying an adequate, reliable supply of freshwater for all purposes has become a global concern. Closing the gap might be easier if more unpolluted freshwater was accessible. However, the water we

D. E. McNabb, *Global Pathways to Water Sustainability*,
https://doi.org/10.1007/978-3-030-04085-7_14

possess is the water we are going to get. Therefore, the search for alternative sources has to focus on other ways to augment our existing supplies. For relatively wealthy countries with access to sea or brackish water, desalination is a path toward greater water security. Wealthy countries like Israel and Saudi Arabia are already heavily invested in desalination facilities. For poorer, less economically stable nations and those without easy access to seawater, recycled water may be the best way to augment their resources. However, many technical, social and institutional barriers bar the pathway to acceptance of this supply alternative (Bichi et al. 2018). As a result, to achieve water sustainability, many countries will have to rely on conservation and greater efficiency in such areas as energy production and agriculture. Still others must find ways to increase their investments in the acquisition, storage and treatment infrastructure needed to capture stormwater and urban runoff. Green architecture is another way of collecting rainwater for recharging groundwater aquifers that is currently underway in many urban centers. Much progress has been made in capturing alternative water supplies but, as the examples included in this chapter reveal, much work still remains.

To be sure that water researchers and planners are talking about the same things when they discuss the supply gap of water scarcity, water stress and the nation's water risk, agreement of their definitions became necessary late in the last century. Representatives of the world's major water management organizations have reached agreement on definitions of water scarcity, water stress and water risk. Those definitions, published by the Pacific Institute (2014), are:

- Water scarcity: "the volumetric abundance, or lack thereof, of water supply. This is typically calculated as a ratio of human water consumption to available water supply in a given area."
- Water stress: "the ability, or lack thereof, to meet human and ecological demand for water. Compared to scarcity, 'water stress' is a more inclusive and broader concept. It considers several physical aspects related to water resources, including water scarcity, but also water quality, environmental flows, and the accessibility of water."
- Water risk: "the probability of an entity experiencing a deleterious water-related event. Water risk is felt differently by every sector of society and the organizations within them and thus is defined and interpreted differently (even when they experience the same degree of water scarcity or water stress)."

Alternative Water Supplies in Russia

With 10% of the world's total supply, the Russian Federation (RF) is the second largest water-rich nation on the globe, led only by Brazil (Sheychuk 2013). However, the distribution of Russia's vast water resources is not even, nor is it always easily accessible. Moreover, its water infrastructure needs significant repair, replacement and upgrading. Moreover, only 15% of the country's water resources are located in the most populous and industrially developed section, west of the Ural Mountains. In total, 70–80% of the population live where most of the industry is located—in this part of the country. It is, therefore, where most of the public supply of water exists.

The Russian water system includes more than 30,000 water reservoirs and smaller impoundments. Most of these are small- and medium-sized dams built for reservoirs to collect and store water needed for agriculture. Moreover, many of those dams have gone without maintenance for more than 30 or more years and need improvement. The year 2020 will be the final year of Russia's eight-year water resource development plan, in which it was determined that an estimated of 12.8 billion rubles (US$193 million) per year for ten years would be needed. However, the global recession and subsequent drop in oil income meant that few of the plan's goals would be achieved.

In a study of Russia's future water supplies and wastewater services under four different scenarios, researchers at the National Research University in Moscow concluded that the need for greater investments in water conservation and wastewater infrastructure investments are likely to become critical in the near future (Proskuryakova et al. 2018). Different socioeconomic scenarios and alternative outcomes were developed based on their effect on (1) the sustainability of existing water systems; (2) domestic and industry water use; and (3) new water products and services. The four alternative outcomes were: (1) a near perfect future that permits national economic growth; (2) problem conservation (economic stagnations); (3) losses and accidents (economic depression); and (4) national priority (manual operation). Outcomes 2 and 3 were determined to be the most likely.

An estimated one-fifth of Russia's total population does not have adequate access to public water supply services, while 60% of the total water supply is not sufficiently treated. More than 60% of the water supply systems must operate with deteriorated infrastructure. Owing to these conditions, large portions of the urban populations may have to depend on bottled water or install domestic water filter systems. Most of Russia's water is tied up in the permafrost that covers the northern European region and central and eastern Siberia. The UN Food and Agriculture Organization (FAO) reports that there are 120,000 rivers longer than 10 kilometers (16.01 miles) in Siberia, with roughly 71% of the total drinking water available in the northern regions (FAO 2016).

Many of Russia's surface water sources remain heavily polluted, from either untreated industrial, domestic or agricultural discharges, or from a combination of two or more sources. Clean and treated water that meets international standards for human consumption is only available for 61% of the total domestic water demand. The great distances from some water sources and demand centers results in periodic dry-year water supply deficits.

Management of Russia's current water system began in 2004 with the passage of a new water code. The code included the establishment of the Federal Water Resources Agency for overseeing all water resources, the adoption of regulations for access to and payment for water, and the management of all water-related property. A revised water code followed in 2006 that gave the agency jurisdiction over surface water runoff (particularly flood control), replenishment of groundwater aquifers, the prediction of future water problems, and reaction-planning for and mediating damage from extreme water events. The agency now has these three main responsibilities:

- To ensure that all public and industry water needs are met now and in the future;
- To ensure that all hydraulic installations (dams, treatment facilities, pipelines, etc.) are safe in all conditions;
- To ensure that human activities benefit from government actions to prevent and reduce damage from floods and other extreme water events.

Since its founding in 2003, nearly three-quarters of the Russian Water Agency's finances have gone to flood defense activities. Rain and stormwater harvesting appears to still be a secondary concern. Floods have long been among the most frequent forms of natural disasters encountered in Russia. Every year, an average of 50,000 square kilometers (16,305 square miles) of land is subjected to flooding. Of the 400,000 square kilometers (154,441 square miles) of land that is considered to be flood plain, 150,000 km^2 (nearly 58,000 square miles) is considered to be susceptible to catastrophic flooding.

Most Likely Outcome Scenario Planning

Water planning in Russia includes a system of potential outcome scenarios. The scenario with the most damaging water outcome foresees an annual growth rate of only 1%, with a commensurate decline in water and wastewater infrastructure investments. However, as industrial production declines, less wastewater treatment will be needed. The reduction in investments will exacerbate the gradual deterioration of water and wastewater facilities, resulting in lower drinking water quality and more wastewater pollution being discharged. As water quality deteriorates with reductions in the availability of infrastructure funding in public sector water services, an increase in private sector participation is likely to occur, predominantly in urban areas with populations greater than 200,000. This market could grow to 35–40% of the total domestic water supply.

The losses-and-accidents scenario includes a flat or negative growth in GDP. In the depression-like scenario, many businesses are predicted to close, resulting in greatly reduced water withdrawals for manufacturing businesses, mining and paper production. Ostensibly desirable, the reduction in water consumption will result in less income for water supply services, further reducing the available funds for infrastructure repair and replacement. It is also likely that the Russian government will reduce domestic water rates, again reducing income from water supply organizations income and accelerating the deterioration of fixed assets.

The drive for achieving water sustainability is blocked by a host of daunting barriers, including population growth, urbanization, climate change, relatively inefficient agriculture, harmful energy choices, and the universal need to acquire the capital needed to install, repair and upgrade the country's often ignored water distribution systems. Achieving global sustainability for water clearly poses a serious challenge. However, if managed efficiently and fairly, Russia can succeed in its critical role in strengthening the sustainable social, economic and environmental systems that are being assailed in the light of fast and unforeseeable changes (Grimaldos 2015).

Closing the Water Gap in China

Water contributes significantly to the economy of China, which is a heavy user of water for agriculture, power generation, textiles manufacture and mining, and for its growing urban population. China is a nation with heavy to extreme water stress. Nearly 80% of all water supplies are from surface resources. However, demand has resulted in a growing reliance upon groundwater as well. Much of this groundwater is withdrawn from aquifers in Northeast China, where little rainfall occurs. This is resulting in in rapidly dropping water tables.

The climate of most of South China is a product of the East Asia Monsoon. The Himalayan mountains and high plateaus block most of the precipitation from the Indian ocean from penetrating far into Northern China and the Northwest of Asia. In winter, much of China is subject to dry cold air masses from Siberia.

Agriculture Water Use

Agriculture, including livestock raising, accounts for nearly 65% of all water withdrawals in China. However, the FAO reports that only 45% of the total withdrawals for agriculture are actually used by crops and livestock, the rest is lost due to low efficiency of the irrigation system. Industry, which includes power generation, accounts for a little more than 23% and all domestic uses account for a little more than 12%. Most of this water is from surface water sources, including the more than 50,000 rivers and 2300 permanent lakes. Close to two dozen desalination projects in china's coastal areas have been developed since the late 1950s. China's agriculture is produced in three major irrigation zones:

- The Northwest region and middle portions of the Huang river basin, where irrigation is absolutely necessary. Rainfall is less than 15.7 inches (400 millimeters) per year.
- The Northeast and Hangh Huai Hai plain, where irrigation is usually necessary. Rainfall ranges from 15–40 inches per year (400–1000 millimeters).
- The middle and lower reaches of the Chang, Zhu, and Min revers and parts of Southwest China, where irrigation is supplementary, then usually only for upland crops. Rainfall exceeds roughly 40 inches per year (1000 millimeters).

Portions of China's Southwest and Pacific Coast areas enjoy more than adequate precipitation with annual averages that approach 80 inches per year. Rainfall exceeds 40 inches per year in areas south of the Yangtze River. It is less than one inch in the Tarim River basin in China's Northwest and the Qaidam River basin in the west. Overall, China's average annual rainfall is a little more than 25 inches.

Water shortages occur mostly in the dry norther regions. The north depends on irrigation and demands are high and increasing. Rapid industrial growth and urbanization are contributing to the region's water stress. Nearly half of the 640 cities faced water shortages as early as 2000, with 100 cities having to cope with extreme water scarcity (FAO 2012).

Water supply stress in China already exists in many parts of the nation. The portion of the land facing high and extreme water stress has increased from 28% in 2001 to 30% in 2010. Part of this is because of relatively low water supply in many areas, and part is because of excessive demand for all water-use sectors. New research suggests that climate change is adding to the needs of the growing population, in turn adding to the already high and extreme stress levels (Wang et al. 2017).

Southern China is highly dependent upon irrigated agriculture. This region produces close to three-quarters of the nation's cereals and more than 90% of its cotton, other fruits, vegetables and other crops. Meeting the food needs of the increased population will be a challenge due to the difficulty of adding productive farmland and the already limited water. In addition, droughts occur every year from region to region, affecting roughly 13% of the total farming area. Crop damage also occurs from monsoon-produced flooding.

Similar to what has happened in many areas with high to extreme water stress, more farmers and municipalities in China have augmented limited surface water supplies with more and more groundwater. Total water withdrawals by source in 2005 consisted of 80% surface water, 18% groundwater, 0.002% desalinated water, and 2.4% recycled treated wastewater. Most of the surface water withdrawals occur in southern China; more than two thirds of groundwater withdrawals occur in northern China. Excessive pumping of groundwater has resulted in a rapid depletion of groundwater aquifers. In the Hai river basin, in 2009, the water table in deep wells had dropped 90 meters (nearly 300 feet), and in shallow groundwater tables as much as 50 meters (164 feet). In Beijing, groundwater tables in some wells have dropped from 100 meters (328 feet) to as much as 300 meters (984 feet).

China adopted the Integrated Water Resources Management (IWRM) approach in 2002, amending its 1988 water law to provide a legal basis for water resources and demand management. Water productivity, water pollution protection, demand management and ecosystem health are all included in policies and planning.

Transporting Water to North China

China has long suffered from a shortage of freshwater. This shortage is focused on the arid northwestern provinces, home to more than 40% of the population and where the largest cities and industrial bases are located. Its total water resources per capita are only 25% of the world's average (Zhao et al. 2017). China has elected to attack its northern water supply gap by transferring large quantities of surface water from the country's water-rich south to its water-scarce north. The South–North Water Diversion Project (SNWD), also translated as the South-to-North Water Transfer Project), is part of a multi-billion-dollar mega-project to bring Southern China's water north to Beijing and the arid Northwest. It is reported to be the largest project of its kind in the world. The diversion concept has been a dream of China's leaders since it was first mentioned by Mao Zedong in 1952 as an answer to China's water problems. When completed, the project will tie together China's four main surfacewater sources: the Yangtze, Yellow, Huihe and Haihe rivers.

Construction on the first segment, the central route began in 2003, with the first diverted water reaching the capital in 2014 after a 800-mile journey from the upper Han River, the largest tributary of the Yangtze River to Beijing and Tianjin. Completion of the first segment was delayed for a variety of reasons, including acquiring the property and the rising cost of construction materials (Yu et al. 2018). Originally estimated to cost US$62 billion, by 2014, that amount had grown to US$79 billion. This was the first of three stages.

The other two segments are the eastern route through the Grand Canal and the western route, which transfers water from three tributaries of the Yangtze River to deliver water to China's western provinces.

The Water Gap in India

Dr. Kirit S. Parikh, chairman of India's Integrated Research for Action for Development, was quoted in a 2013 interview as identifying India's water supply gap as being the result of three social, economic and environmental factors. The first is the nation's continuing population growth. This means there is just not enough water available at all times for all the people. The second is the past political unwillingness to invest in urban water treatment infrastructure. Investments in water infrastructure has not kept pace with the water demands of India's fast-growing urbanization. The third is a result of excessive groundwater withdrawals by India's farmers and pollution caused by inefficient agriculture irrigation runoff (Luthra and Kundu 2013). Without other available supplies, farmers and communities turn to uncontrolled groundwater exploitation.

Primary Water Sources

India has a tropical monsoon climate, with total annual average rainfall more than 46 inches (1170 millimeters) per year. However, the distribution of that rainfall varies greatly from region to region and often from year to year; droughts in varying severity are expected to occur four years out of every ten. During the period from June to September, 70–95% of the annual precipitation comes with the warm, humid air of the southwest monsoon blowing from the Indian Ocean. The two main primary sources of water in India are monsoon rainfall and glacial snowmelt in the Himalayan mountain range during the hottest parts of the year. Close to 80% of river flow occurs during the several months of the southwest monsoon.

Along India's west coast and the sub-Himalayan regions, average annual rainfall can exceed 98 inches (2500 millimeters). Little rain falls during the winter months as the northeast monsoon brings dry, cold air from Central Asia and the Himalayan mountains. Excessive withdrawals of surface water and groundwater cause water tables in many areas to fall as much as a meter (3.2 feet) every year. Both surface water and ground water sources are heavily polluted with organic and hazardous pollutants from generally poor sanitation and little attention is paid to solid waste removal, in much of the country.

India's Limited Usable Water

The total amount of usable water in India has been estimated to be between 700 billion to 1.2 trillion cubic meters (bcm) annually. With a population of 1.2 billion according to the 2011 census, India has only 1000 cubic meters of water per person, Researchers agree that a country is considered to have a water supply gap if it has less than 1700 cubic meters per person per year. Insufficient and delayed investment in urban water-treatment facilities means there is not enough clean, safe drinking water and there is insufficient access to sanitation facilities. Water in most rivers in India is largely not fit for drinking, and many are even unfit for bathing. Dwindling groundwater supplies are caused by the over-extraction for agricultural irrigation. Groundwater is considered an open-access resource, and anyone can pump water from under his or her own land. Agriculture accounts for 90% of water use, with only 5% left for increasing urban domestic use and 5% for India's growing industrial sector. India's primary water source for all sectors is groundwater. But with the increasing population, economic growth and water consumption, groundwater sources are depleting faster than they can be recharged.

The world's second most populated nation in 2017, by 2022, India was expected to surpass China to become the most populous nation on Earth. The population will continue to grow, reaching 1.5 billion in 2030 and 1.7 billion in 2050. Most of the food for India's population is grown in the country. The primary source of supply for all water withdrawn in India is groundwater from overexploited aquifers. As the population continues to increase, the country's aquifers are rapidly drying up.

Surface water withdrawals from the largest rivers in India, including the Ganges, Krishna, Brahmaputra, Godavari and Mahanadi, are limited by seasonal conditions and the high level of pollution. Flow in these rivers varies significantly from season to season. Flooding is a common problem in 12% of the nation, with little or no efforts in place to harvest stormwater.

Population growth in India with unprecedented urbanization is making it extremely difficult for the nation to keep pace with the growing need for additional improvements to both the availability and quality of municipal drinking water. The growing populations continue to stress the water and wastewater systems. Despite the many quality improvements made to the water sector, many freshwater resources are heavily polluted with both biological and agricultural or industrial chemical pollutants. More than 21% of the diseases in India are water related and only 33% of the population has access to effective sanitation facilities (Snyder 2017).

2016 Water Crisis

In 2016, India was facing what has been described as its worst water crisis in generations. The drought was considered to have been brought on by several years of below normal rainfall. That weather-caused crisis has been said to be further exacerbated by a variety of human-caused activities, including insufficient or ill-advised political actions, and years of inability to coordinate water management among the states and federal governments. Excessive groundwater withdrawals are considered to be a result of the government policy of providing free electricity to farmers for withdrawing irrigation water. A highly critical report in a digital global business publication described the crisis and underlying mismanagement thus:

> Centuries of mismanagement, political and institutional incompetence, indifference at central state and municipal levels, a steadily increasing population that will reach an estimated 1.7 billion by 2050, a rapidly mushrooming middle class demanding an increasingly protein-rich diet that requires significantly more water to produce—together these are leading the country towards disaster. On top of that there is an absence of serious and sustained attempts at the central or state levels to manage water quantity and quality, a lack of implementation of existing laws and regulations, and pervasive corruption and poor adoption rates of new and cost-effective technologies ... Despite this sad state of affairs, there are no signs of politicians waking up to the situation, or those willing to take hard but essential political decisions. Actions are mostly cosmetic. (Biswas et al. 2017)

India's Urban Water Supply Gap

India's continuing economic development has clearly provided an opportunity for a better life for more of the country's already large and fast-growing population. Every year, thousands of young men and women leave the dominant agricultural sector and relocate to urban areas where job opportunities are more available. Although India is still among the least urbanized states of the world with only 28% of the population living in towns and cities, huge population shift is causing a serious infrastructure crisis for the nation's cities and towns. It is estimated that by 2050 more than 900 million people will be living in India's urban areas.

Urbanization has always been associated with economic, social and political progress. However, in many developing areas in India, it is also contributing to serious socio-economic problems. Some of these are air pollution, urban sprawl and overcrowding, lack of adequate housing, unemployment and under-employment, inadequate education systems, urban slums and squatter settlements, missing or unsafe drinking water, missing or insufficient sanitation facilities including sewer collection and treatment facilities, inadequate transportation infrastructure, and overtaxed solid waste disposal. Water scarcity is a major problem in India's urban areas. Most of the water sources are contaminated by sewage and agricultural runoff. With open defecation and other unsanitary practices, waterborne disease is a major source of communicable disease in India. As of 2018, efforts to close the still-widening water supply gap were too little, too late.

The Supply Gap in the Middle East

The Middle Eastern nation of Jordan is among the world's most water-stressed regions. In 2017, the fast-growing population of Jordan, a land-locked, water-scarce, arid nation, was 10.3 million, including the more than 1.3 million refugees from Iraq, Syria and Palestine housed in camps in Jordan. Water-stress projections to 2040 put five of the six most water-stressed nations in the world in the Middle East.

Most of Jordan's surfacewater supplies are shared with neighboring countries of Israel, Syria and Saudi Arabia. The larger neighbors have greater control over the distribution of Jordan's surfacewater resources. Jordon has three major rivers: the Jordan, the Zarqa and the Yarmouk. Most of the raw water in these rivers is saline and hence—unless desalinated—is unfit for human use or for agriculture (Hadadin et al. 2010).

The use of these surface water resources already exceeds supply. The lack of control and lost share has forced Jordan to turn to the unsustainable over-drawing of groundwater from inland aquifers, lowering local water tables and lowering water quality. In 2013, Jordan's population was 8.1 million. By 2040, Jordan's population is projected to be more than 11 million, and it is projected to be the 14th most water-stressed nation on the globe. Water stress is ranked on a 5-point scale. Jordan's ranking was 4.86. The water program in Box 14.1 describes some of what Jordan is doing or is contemplating in order to close its water supply gap.

Box 14.1 Water and wastewater projects in Jordan

With the objective of improving how water resources are utilized and wastewater facilities are managed over the next 25 years, the USAID is assisting Jordan Water and Wastewater Infrastructure Project (WIP) is responding to the improvements needed in the Middle Eastern nation's water sector. Project objectives are to: (1) deliver critically needed water and wastewater infrastructure facilities, and (2) provide training and capacity building to the Water authority of Jordan (WAJ) and water companies throughout Jordan.

Launched in May 2010, USAID's five-year investment program involves the delivery of improvements in Amman, Zarqa, Ma'an, Tafileh, Jerash, and Irbid water/wastewater facilities. In the short-term, the WIP was to implement an improved and more efficient water sector over the next five years. In the long-term, a comprehensive and fully responsive water strategy will be developed through the planning horizon of 2035. Project activities include:

- Studying and preparing water/wastewater master plans, implementing feasibility studies, and preparing tender documents for urgent infrastructure needs.
- Selecting sites for new water/wastewater facilities and designing facilities that meet requirements such as reuse standards for treated wastewater.
- Incorporating staff from WAJ and water companies into project activities for on-the-job training, improving their ability to plan and design water/wastewater facilities effectively and sustainably.
- Providing quality designs and construction supervision, including the testing and start-up of new/upgraded facilities.

Source USAID (2017)

The Jordan water and wastewater project is an example of the type of water and wastewater infrastructure aid provided by government agencies. The project included developing projections on population, water demand and wastewater flow, preparing geographic information system (GIS) mapping, data collection, and hydraulic models for master plans and feasibility studies of water and wastewater systems. Other needed infrastructure improvements included completed designs and construction bids for Amman water projects and wastewater treatment facilities. In addition to training operating personnel, the aid program included designs, surveys and geotechnical investigations for additional water and wastewater infrastructure improvements.

For both Africa and the Middle East, the problems of a gap between the finite traditional water supply and the water demanded for human consumption, industry, energy production and agriculture is wide and growing wider every year.

Summary

Global water scarcity and water stress are targets of diverse efforts to close the gap and achieve sustainability of the resource for all sectors. This chapter looked at examples of the water supply gap and the efforts of nations to achieve a sustainable supply of clean, safe water. After defining water scarcity, water stress and water risk, the discussion began with the water supply in Russia and a review of water scarcity problems and some solutions implemented or planned in the countries experiences high water stress, including China, India and Jordan. The examples chosen are experiencing high to extreme water stress in the face of natural and human-caused barriers to achieving water sustainability.

Russia has the second largest amount of freshwater in the world, second only to Brazil. However, both Russia and Brazil experience water scarcity in some regions of their large landmasses. Much of Russia's water supply is contained east of the Ural Mountain, and much of that remains frozen in permafrost, while some 70% of Russia's agriculture, industry and municipalities are located west of the Ural Mountains. More than 10 million Russians do not have access to safe drinking water; almost 30% of the nation's water infrastructure is in need of immediate repair and replacement; and more than half of its surfacewater sources are heavily polluted.

China's water scarcity problem is a result of two factors. One is that most of its water is in the water-rich south, while most of its population and industry is in the arid north. The second is the very large demand for water needed for irrigated agriculture to feed its more than 1.4 billion people. China's major effort to close its water supply gap is the mega-development of transfers from the water-rich South to the water-scarce North.

The water supply gap in India, the world's second most populous nation, is largely a result of water demand by the 1.34 billion people. Most of its surfacewater is over-subscribed or too heavily polluted for inefficiently irrigated agriculture, which now must depend on groundwater. As a result, aquifers are being depleted faster than they can be replenished. India is also undergoing a rapid growth in urbanization, where the existing infrastructure to meet the water demand is incapable of fulfilling that demand.

Water-scarce nations in Jordan and the Middle East in general are seeing their water gap widening by a variety of factors. One reason is that precipitation patterns are changing due to climate change, resulting in less precipitation and faster evaporation due to more extremely hot days and nights. Another reason is the growing population that results in a greater demand for water for agriculture, along with migration to cites that cannot provide the needed water or wastewater facilities to meet the demand.

References

Bichai, Françoise, Arani Kajenthira Grindle, and Sharmila L. Murphy. 2018. Addressing barriers in the water-recycling innovation system to reach water security in arid countries. *Journal of Cleaner Production* 171: S97–S109.

Biswas, Asit K., Cecilia Tortajada, and Udisha Saklan. 2017. India is facing the worst water crisis in generations. *Quartz India*. Accessed December 13, 2017 from http://qz.com/931878/india-is-facing-the-worst-water-crisis-in-generations.

FAO (Food and Agriculture Organization of the United Nations). 2012. China: Geography, climate and population: Water report. Aquastat, Food and Agriculture Organization of the UN. Accessed December 9, 2017 from www.fao.org/nr/water/aquastat/countries_regions/CHN/.

FAO (Food and Agriculture Organization of the United Nations). 2016. Russian Federation: Geography, climate and population. Food and Agriculture Organization of the UN, Aquastat. Accessed January 22, 2018 from www.fao.org/nr/water/aquastat/countries_regions/RUS/print1.stm.

Grimaldos, Angel Simón. 2015. Water for a sustainable future: New paradigm, new vision. In *Water and Sustainable Development*, ed. Ramiro A. Lopera, 68–73. United Nations Water for Life. Accessed December 7, 2017 from www.un.org/waterforlifedecade/pdf/wm-iii-eng.pdf.

Hadadin, Nidal, Qaqish Maher, Emad Adawwi, and Ahmed Bdour. 2010. Water shortage in Jordan: Sustainable solutions. *Desalination* 250: 197–202.

Luthra, Sonia, and Amrita Kundu. 2013. India's water crisis and cures: An Interview with Kirit S. Parikha. The National Bureau of Asian Research (NBR). Accessed June 28, 2018 from http://www.nbr.org/research/activity.aspx?id=356.

McKinsey & Co. 2009. Charting our water future. 2030 Water Resources Group. *Sustainability & Resource Productivity* 18 (1): 1–18. Accessed December 15, 2017 from www.mckinsey.com/~/media/mckinsey/dotcom/client_service/sustainability/pdfs/charting%20our%20water%20future/charting_our_water_future_full_report_.ashx.

Pacific Institute. 2014. Defining water scarcity, water stress and water risk: It's not just semantics. *Pacific Instituter Insights*, February 4. Accessed June 30, 2018 from http://pacinst.org/water-definitions/.

Proskuryakova, Liliana N., Oxcan Saritas, and Sergey Sivaev. 2018. Global water trends and future scenarios for sustainable development. *Journal of Cleaner Production* 170: 867–879.

Shevchuk, Anatoly. 2013. Water resource management, water ecosystem problems. In *Sustainable Development in Russia*, ed. Sergei Bobylev and Renat Perelet, 144–153. Berlin: Russian-German Exchange.

Snyder, Shannyn. 2017. Water in crisis: India. The Water Project. Concord, New Hampshire. Accessed December 9, 2017 from https://thewaterproject.org/water-crisis/water-in-crisis-India.

USAID (United States Agency for International Development). 2017. Water and wastewater project. Accessed November 23, 2017 from www.usaid.gov/jordan/fact-sheets/water-and-wastewater-infrastructure-project.

Wang, Jiao, Linjun Zhong, and Charles Iceland. 2017. China's water stress in on the rise. World Resources Institute, Washington, DC. Accessed December 8, 2017 from www.wri.org/blog/2017/01/chinas-water-stress-rise.

Yu, Min, Chaoran Wang, Yi Liu, Gustaf Olsson, and Chunyan Wang. 2018. Sustainability of mega water diversion projects: Experience and lessons from China. *Science of the Total Environment* 619–620: 721–731.

Zhao, Zhen-Yu., Jian Zuo, and George Zillante. 2017. Transformation of water resource management: A case study of the South-to-North Water Diversion project. *Journal of Cleaner Production* 163: 136–145.

15

Pathways to Supplemental Water Resources

In 1950, there were 2.55 billion people in the world, and world leaders were worrying about how to feed the growing population. A few others were beginning to worry about the damage being done to the environment, the world's freshwater resources, and the pollution of the atmosphere. At the start of the new century, fifty years later, the world's population had nearly tripled to 6.15 billion, and leaders were worrying about how to deal with all the waste and pollution that the billions of people were generating. In just another ten years, the world added nearly another billion people to reach 6.9 billion. By 2020, another ten years later, the population is expected to reach 7.8 billion, and world leaders are now worrying about how climate change is going to impact the world's finite water supply. They are also concerned with what appeared to be an unstoppable global migration to cities by so many of the world's people. World population will reach 10 billion in 2055. Additional water is absolutely necessary in most urban areas to supplement the existing supplies of surface water and groundwater. This chapter looks at some of the things urban water suppliers are doing to add nontraditional water sources to supplement their existing water supplies.

Three supplemental water solutions have been tested by water managers in both water-scarce and water-rich areas. Supplemental water supplies include harvested rainwater, collected and stored stormwater, and municipal graywater to collect both rainwater and stormwater, while also adding green spaces to the urban environment. Urban planners are adopting green infrastructure to also control minor flooding while capturing this supplemental water for municipal nonpotable uses.

D. E. McNabb, *Global Pathways to Water Sustainability*,
https://doi.org/10.1007/978-3-030-04085-7_15

Rainwater harvesting is collecting precipitation from the roofs of building and other urban structures and storing it in cisterns and downspout rain barrels until it is needed for domestic and public green apace irrigation. Stormwater harvesting is more complex in that large amounts of precipitation or snowmelt that is now captured in combined sanitary systems must be redirected to single-purpose storm drain systems. Stormwater management also includes the infrastructure, such as runoff basins and levies, for flood control.

Graywater, another available resource, is collected from non-toilet uses such as bathing, dishwashing and laundry uses by domestic and commercial water users. The problem with collecting graywater is that it needs a separate domestic collection system to separate it from blackwater. Once collected, graywater may be given some treatment before being used for agricultural and urban green space irrigation of schoolgrounds and parks (Oppenheimer et al. 2017).

The harvesting of rainwater, stormwater, normal urban water runoff, and the reuse of municipal graywater are now integral components of far-sighted sustainable urban water management (SUWM) programs. The world's fast-growing cities are the focus of much of the interest in developing these alternative water resources. Using larger proportions of cities' nontraditionally sourced water is increasingly critical to achieving the world's sustainable development goals (IPS 2017; Li and Bergen 2018).

Harvesting Rainwater

As more land is converted to urban and suburban areas, the natural flow of rainwater is altered forever. In direct rainwater collection systems, high water volumes and urban area pollutants are transported into rivers, lakes and estuaries, degrading water quality and riparian habitat in most if not all urban stream systems. Untreated runoff water is generally directed directly into urban area rivers and streams. During extreme storm events, the water that is collected in combined water and sewer systems that are unable to cope with extreme precipitation has long been absorbed into the sewage stream. The result too often is an overflowing of mixed raw sewage and polluted stormwater into the urban environment. The following benefits of harvesting rainwater were selected from a list published by Chris Maxwell-Gaines (2018) for *Innovative Water Solutions*:

- Rainwater harvesting is socially acceptable and environmentally responsible; it promotes self-sufficiency and helps conserve water.
- It reduces municipal water consumption and it can be used as a main source of water or as a back-up source to wells and municipal water.
- It can be helpful source supply during emergencies.
- Rainwater is the best water source for landscape irrigation due to the lack of chlorine and other treatment chemicals.
- Rainwater harvesting can reduce the frequency and intensity of flooding around a house or property.
- Rainwater harvesting can reduce stormwater pollution, lessening its impact on local watersheds.

Rainwater harvesting has long been a major water supply source in many arid and semiarid regions of the world. With the use of microfilters to improve the quality and cleanliness of rainwater that must be stored for long time periods, rainwater is the sole water supply source for more than 1 million people in Australia, for example. Rainwater harvesting systems also supply drinking water for some 15% of the rural population in France, and more than 2 million people in Brazil. Similar rainwater harvesting systems also serve millions of people in Africa, India and rural areas of the USA (Dos Santos and de Farias 2017).

Rainwater Harvesting in Washington State

The state departments of ecology, with health departments, oversee aspects of water use and water rights in the state. Some states require a permit for collecting and using rainwater, provided it is used on the same property where it was harvested, that it is collected from an existing structure (no special structure for rainwater harvesting is allowed), and that it does not affect existing water rights. Local and individual counties rules apply for new structures in which a rain harvesting system is planned. An example of how a developer incorporated rain harvesting elements in a new Seattle, Washington office building is described in Box 15.1.

Box 15.1 Rainwater harvesting in a Seattle office building

A new 52,000 square feet office building in Seattle, Washington for the Bullitt Foundation was the first to be designated a "Living Building" by the United States Green Building Council (USGBC). To be so designated, a building must comply with the following rules: it must generate its own energy from renewable, non-toxic materials, collect and treat its own water, and operate

efficiently. The building uses treated rainwater as its sole potable water source and captures its own graywater for landscape irrigation and groundwater recharge. The water treatment system uses a system of ultrafiltration, activated carbon filters, and ultraviolet disinfection. The system has storage capacity of 56,000 gallons of potable water. Graywater is captured and treated using a wetland system on the building's roof, with a capacity of 500 treated gallons per day. Treated water is pumped into bioswales for groundwater recharging.
Source Oppenheimer et al. (2017)

Global population growth and the migration of people from rural areas to cities in water-rich and water-poor countries has resulted in an insufficient supply of fresh water for millions of people. In the fast-growing cities in semi-arid regions of otherwise water-rich Brazil, for example, water scarcity is experienced approximately 41% of the time. The scarcity problem everywhere is exacerbated by changes in traditional precipitation patterns and the warming effects associated with climate change. Longer dry seasons and an increasing number of extreme rainfall events has left the water collection, storage and distribution infrastructures of these cities unable to meet the growing water demand. However, where it has been implemented, harvested rainwater has been shown to provide a viable supplemental supply for on-site non-potable uses, saving from 25 to 40% or more treated freshwater for domestic potable use.

Harvesting Stormwater

Heavy urban runoff is the movement of stormwater from impermeable sources such as buildings, parking lots and paved streets and highways to streams, rivers and lakes. The flow of rain and melting snow water through urban areas picks up a variety of pollutants, including fertilizers and pesticides from private lawns and urban green infrastructure areas such as schools and parks. This polluted water then flows through city drains that may join with city sewage systems or independently discharge the collected water to streams and other water bodies. Although considered a major flooding problem for many cities, stormwater has become a viable addition to urban area water supplies, as the following descriptions of stormwater harvesting in two of the largest cities in water-scarce California attest.

Stormwater Harvesting in California

The city and county of San Francisco, California implemented a non-potable water program in 2015 as a planning too for identifying and relaxing existing regulatory barriers to rain and stormwater harvesting and other green infrastructure actions. The first step was the adoption of a non-potable water ordinance addition to the health code. The new rule allows the collection, treatment and use of nonpotable water, including graywater, rainwater, stormwater, drainage water and wastewater for a range of water uses in commercial, residential and mixed uses developments (Oppenheimer et al. 2017). Follow-on additions have expanded the allowed uses of the supplemental water sources. In 2015, for example, new buildings over 250,000 square feet were required to use graywater, rainwater and/or foundation drainage water for toilet and urinal flushing and irrigation. The program also provides grants for the voluntary collection and treatment of alternative supplemental water sources.

Extreme weather events are predicted to increase in many areas, including the USA, where they are already appearing to increase in severity. This change in weather is a result of climate change. A growing number of urban centers are implementing green infrastructure into their urban design as a method of supplementing their water supplies while also controlling flooding (EPA 2018). The objective is the implementation of SUWM. Many urban planners are now including stormwater into their sustainability plans.

The city of Los Angeles Department of Water and Power (LADWP) is an example of how one of the largest metropolitan areas in the USA has begun to capture stormwater to supplement its supply of potable water. The LADWP imports large amounts of its surface water supply from the San Francisco Bay Delta, the Owens River Valley, and the Colorado River. LADWP has committed to reducing this volume of imported water by 50% by 2024. One way it plans to accomplish this is to use more stormwater, thereby increasing its supply reliability and reducing potable water use. As part of its stormwater master plan, some stormwater is captured in the basin and used for recharging regional aquifers. Traditionally, most of the rain that the Los Angeles basin receives has been captured in storm drains and then collected in large concrete culverts that carry it and surface runoff for discharge into the Pacific Ocean. A Stormwater Capture Master Plan (SCMP) was developed by the city to determine whether stormwater could be captured to supplement the local water supply. Deemed possible, the next step was to determine the most cost-effective method of making it happen, including a strategy for additional stormwater capture by 2035 (Oppenheimer et al. 2017; CASQA 2018).

The LADWP and its water and wastewater service partners harvest close to 29,000 acre-feet of stormwater annually for recharging local water aquifers. An additional 35,000 acre-feet per year is directed to surface ponds where it is naturally absorbed into the soil and into local aquifers. Research has shown that an additional 68,000–114,000 acre-feet per year could be captured in Los Angeles through a suite of projects, programs and policies over the next 20 years.

Stormwater Harvesting Problems

Stormwater can often accumulate too quickly to be controlled. As a result, this runoff is a major cause of floods in urban areas. Urban stormwater runoff carries with it the litter and pathogens that can accumulate in urban areas, including such pollutants as plastics and other trash, vehicle fuels and lubricants, industrial sediments, pet excreta, winter de-icing products such as road salt or de-icing chemicals, fertilizers and pesticides and other pollutants, which eventually ends up in the oceans of the world.

Many older sewer systems combine their systems for collecting urban runoff with their sanitary sewage systems. During heavy rainfall and sudden storms, these combined systems often cause sewer overflows. A combined sewer overflow (CSO) is defined as a mix of approximately 80% stormwater and 20% sewage. These combined systems remain common in many urban areas. For example, the EPA reports there are 151 National Pollutant Discharge Elimination Systems permits that authorize discharges from CSOs to water bodies in the Great Lakes Basin. In mid-June 2018, heavy rains caused combined sanitary and storm sewers to overflow into Milwaukee's rivers and Lake Michigan. The increasing number and severity of extreme water events associated with climate change is expected to result in more discharges of untreated sewage into the basin's water bodies.

Another example of the damage that can occur from combined sewage overflows occurred in Portland, Oregon in April of 2018. The state's environmental protection agency reported that prolonged heavy rains caused Portland's CSO system to overflow, discharging raw sewage into the Willamette River from several different points. The public was advised to not use water from the river, or to have contact with the water, due to increased bacteria in the water. In 2011, Portland completed a $1.4 billion 20-year project to reduce if not eliminate such overflows. Since then, CSOs to the Willamette River have declined by 94% and discharges into a slough adjoining the Columbia river have declined by 99%. Before the CSO mitigation project, overflows to the Willamette River from multiple outfalls occurred an average of 50 times a year, with some instances lasting days; overflows now only occur an average of four times per winter season, and once every three summers.

Using Graywater

The reuse of what had been considered a waste problem, a liability and a major draw on municipal water supplies to be an avoided is today in many parts of the world a valued resource and a way of supplementing existing freshwater resources, and a way for reducing water scarcity. Treated wastewater is now regularly used for irrigating gardens and public parks, some farm crops, orchards and vineyards, in the generation of electric power, in cooling systems, for wetland preservation, and is increasingly used for recharging and augmenting depleted aquifers. Treated domestic graywater is also gaining acceptance as a replacement for freshwater in toilet flushing, cooling and other uses in commercial facilities. Water treated to potable standards for reuse has also been approved in many arid-area communities.

Graywater (also spelled greywater or gray water) is the relatively clean wastewater from home kitchens, laundries, bathroom sinks and bathing facilities in residences and office and commercial buildings. Water used for toilets or for industrial purposes is not included. It can be collected at individual single-family residences or apartments and stored for short periods before being used for nonpotable purposes such as landscape irrigation. Graywater is considered a "a great source of irrigation water, which in many locations in the Southwest is 30 to 50 percent of the overall irrigation water use. Graywater will usually contain sodium, calcium, magnesium, nitrogen, potassium and other salt compounds. It will also have varying amounts of household chemicals, oils, makeup greases, nutrients and chemicals that can largely be managed by the types of products used within a household. Many of the nutrients in gray water are exactly the same as the fertilizers we buy at a local store to fertilize our landscapes" (Pushard 2018). The EPA recommends sanitizing graywater before use if stored for longer periods.

The permitted use of graywater for on-site landscape irrigation is becoming more popular in arid and semi-arid regions such as the American Southwest, Australia, and the Middle East. The International Plumbing Code (IPC) and the Uniform Plumbing Code (UPC) have adopted standards detailing how graywater systems should be installed, allowing state and local governments to adopt rules in their own building codes.

Arizona and New Mexico in the arid southwestern USA are the leading states in permitting graywater use (Pushard 2018). Neither New Mexico or Arizona make installing a graywater reuse system difficult. Installing a graywater system does not require a permit for private residential gray water systems that discharge fewer than 250 gallons per day for use onsite for gardening, composting or landscaping irrigation. The few restrictions that do exist include: graywater cannot be stored for more than 24 hours; it cannot be used in aboveground spray-irrigation systems; and the water should not contain hazardous chemicals derived from activities such as cleaning car parts, washing greasy or oily rags, or disposing of waste solutions from home photo labs or other home occupational activities. In New Mexico, the Skywater residential development, for example, requires the installation of graywater systems with any new home, and the city of Santa Fe promotes and offers rebates of US$150–$300 for the installation of 'Laundry to Landscape' graywater systems. In Tucson, Arizona, the city requires the installation of indoor plumbing for a graywater system at the time of construction.

Arizona and New Mexico were among the first states to adopt statewide codes to allow these systems at new residential developments. Other states are still debating whether to adopt state-wide rules for graywater systems, permitting its use local jurisdictions and not in others. San Francisco city and county permit installation in new residential and office construction. San Francisco's non-potable water regulations now includes greywater systems in new office buildings.

Other states permit graywater's limited use but have made it difficult to get a permit. Still other states are still lagging in this area of water reuse and continue to require it to be disposed of in traditional sewage systems. The infrastructure necessary to separate graywater from blackwater is expensive and therefore only recommended for new construction.

Green Infrastructure

Green infrastructure such as roadside retention ponds collects and holds stormwater and other urban runoff where it is then naturally filtered and absorbed into the soil. Also, rainwater harvesting with green infrastructure is becoming more common in many locations. These approaches include capturing rainwater that falls on urban roofs and holding it for later use for landscape irrigation or for other uses such as fire suppression. Rain gardens can capture runoff and clean it of pollutants before releasing it; household rain barrels can catch roof runoff for garden irrigation use instead of allowing it to flow into an overloaded storm sewer system. By reducing flooding, green infrastructure can help protect water quality, public health, and the environment. Green infrastructure reduces and treats stormwater at its source, providing environmental, social and economic benefits that far exceed the cost of its installation (Table 15.1).

Table 15.1 Major green infrastructure design models and techniques

	Technique	Description
1.	Downspout disconnect	Reroute rooftop drainage pipes from draining rainwater into the storm sewer into rain barrels, cisterns, or permeable areas; beneficial in cities with combined sewer systems
2.	Rainwater harvesting	Systems to collect and store rainfall for later use; they slow and reduce runoff and provide a source of water; can be particularly valuable in arid regions
3.	Rain gardens	Bioretention or bioinfiltration cells; shallow, vegetated basins that collect and absorb rooftop, sidewalk and street runoff
4.	Planter boxes	Urban rain gardens with vertical walls and open or closed bottoms to collect and absorb sidewalk, parking lot, and street runoff; ideal for space-limited sites
5.	Bioswales	Vegetated, mulched channels that provide runoff treatment, movement and retention, to slow, infiltrate and filter stormwater; good when placed by streets and parking lots
6.	Permeable pavement	Pervious concrete, porous asphalt, or permeable interlocking pavers to allow infiltrate, treat and/or store rainwater; good where land values are high and flooding or icing is a problem
7.	Green streets and alleys	Green infrastructure elements integrated into design to store, infiltrate, and evapotranspire stormwater; include permeable pavement, bioswales, planter boxes, and trees in the design
8.	Green parking lots	Green infrastructure integrated into parking lot designs; includes permeable pavements rain gardens and bioswales
9.	Green roofs	Roofs are covered with growing vegetation that enables rainfall infiltration and evapotranspiration of stored water
10.	Urban tree canopies	Rows of trees to reduce and slow stormwater by intercepting precipitation in their leaves and branches; add greenspaces
11.	Land conservation	Natural areas that collect stormwater, protect open spaces and sensitive natural areas in and near to a city and provide recreational opportunities for city residents

Source US EPA (2018)

Table 15.2 Example alternative green infrastructure solutions to water quality problems

Water issues and solutions			
Water pollution	Biological control	Water temperature control	Erosion control
Reforestation	Reforestation	Reforestation	Reforestation
Forest conservation	Forest conservation	Forest conservation	Forest conservation
Riparian buffers	Riparian buffers	Riparian buffers	Riparian buffers
Reconnect rivers to floodplains	Reconnect rivers to floodplains	Reconnect rivers to floodplains	Reconnect rivers to floodplains
Wetlands restoration	Wetlands restoration	Wetlands restoration	Urban green spaces
Wetlands conservation	Wetlands conservation	Wetlands conservation	Water way shading
Construct wetlands	Construct wetlands	Construct wetlands	

Source From material in UN Water (2018)

Green infrastructure incorporates a variety of urban design models and techniques for collecting, storing and putting stormwater and rainwater to beneficial use for people and businesses in urban areas. The idea has only recently received widespread attention, although the techniques used have been around for decades and, in many cases, centuries. The underlying objective is to protect and preserve urban area from the ravages of extreme weather events while providing city dwellers more accessible green areas and lowering summer temperatures. Table 15.2 lists sample green infrastructure methods for water resolving water quality issues.

Early Adopters

Portland, Oregon and Seattle, Washington, along with California, Arizona and New Mexico were early adopters of green infrastructure. Both Portland and Seattle endure winters with heavy and prolonged rainfall. Portland sits about ten miles above the junction of two major Northwest rivers, the Willamette and the Columbia. Seattle is situated on the coast of an arm of the Pacific Ocean.

Many severe winter rain storms commonly pass through the Columbia River Gorge from eastern Oregon and Washington, bringing with them heavy precipitation to Portland. The city uses green infrastructure to capture and hold stormwater to keep it out of its more than 80-year old combined rainwater and sewer system. Portland's urban runoff is first treated and then discharged at several points of the Willamette River, which then flows into the Columbia River. However, as the 2018 Portland CSO described above attests, extreme weather events can still result in costly and harmful damage.

The city of Seattle sits between Puget Sound and the large freshwater Lake Washington. Both Portland and the Seattle-Tacoma metropolitan area average about 37 inches of rain each year, less than many large U.S. cities, including Miami, Houston, Atlanta, New York City, Boston, Washington, DC, Miami, and Philadelphia, to name a few. However, Seattle has been receiving more rain than in the past, and much of that rain comes in heavy downpours. Rainfall and polluted stormwater runs into Puget Sound, reducing water quality. In 2013, Seattle's city council and the mayor directed city departments to work together to achieve Seattle's goal of annually eliminating 700 million gallons of stormwater runoff by 2026 with the use of green infrastructure.

Global Green Infrastructure Adoption

Dams and the pumping of collected ground water have been traditional solutions for flood prevention in addition to their roles in water supply acquisition and storage. Today, however, a variety of lower-cost and ecosystem-friendly green infrastructure solutions have been successfully implemented for flood control. Table 15.3 lists some of these approaches along with traditional flood control methods.

In 2018, researchers at the University of Copenhagen, Denmark, reported on a study of green infrastructure in five large urban centers: Singapore, Berlin, Philadelphia, Melbourne and the Sino-Singapore Tianjin Eco-City in China (Li and Bergen 2018). The five cities were selected from a group of green infrastructure innovators that included San Francisco, Portland, Vancouver, Canada, Bristol, Brussels, Stockholm, Masdar City (a planned development in Abu Dhabi, United Arab Emirates) and the Songdo International Business District (a new South Korean city built on 1500 acres of filled land on Inchon's waterfront). The five cities were selected because of their commitment to SUWM concepts and the availability of data and the regions' willingness to share that data. Selected characteristics of the five cases are shown in Table 15.4.

Table 15.3 Example alternative green infrastructure solutions for mitigating flooding

Flood mitigation solutions			
Riverine flood control	Urban stormwater runoff control	Coastal storm and flood control	Traditional infrastructure solutions
Reforestation	Reforestation	Protecting mangrove barriers	Dams and levees
Forest conservation	Green spaces for bioretention and infiltration	Restoring old mangrove barriers	Sea wall construction
Riparian buffers	Riparian buffers	Conservation of coastal marshes	Surface collection and transshipment of stormwater
Wetlands restoration	Rainwater harvesting	Conservation of coastal sand dunes	
Wetlands conservation	Stormwater harvesting	Wetlands conservation	Sea wall construction
Construct wetlands	Permeable pavements	Construct wetlands	
Establish flood bypass routes			
Permeable pavements			

Source From material in UN Water (2018)

Table 15.4 Selected characteristics of the five case studies in green infrastructure

City	2018 population	Annual rainfall (inches)	City water problems	Example green infrastructure
Berlin	3.6 million	22.5	Pollution buildup in closed-water cycle; difficulty meeting EU water directive	Reduce combined sewer overflows; process rainwater for supply; add green roofs, green facades
Melbourne	4.8 million	23.7	Pollutant buildup in surface water; the impact of climate change on patterns of precipitation	Rainwater harvesting for irrigation; increase permeability surface; use sub-catchments to avoid flooding

(continued)

Table 15.4 (continued)

City	2018 population	Annual rainfall (inches)	City water problems	Example green infrastructure
Philadelphia	1.3 million	41.5	Low flow in rivers, high polluted urban runoff; changes in precipitation patterns	Reduce fees for stormwater absorption in permeable surfaces; larger green infrastructure projects
Singapore	5.8 million	94.5	Increasing population and water demand; limited catchment area. Water is imported from Malaysia	Collect and store stormwater for supply and drainage; increase livability, eliminate imported water; eco districts and green buildings
Tianjin Eco-City	70,000 (planned for 30,000 when initiated in 2008)	23.6	Saline local aquifer, drinking water imported from Tianjin main city; high demand for green spaces	Artificial green areas irrigated by reclaimed water; collect stormwater runoff from paved areas

Source After material in Li and Bergen (2018)

Green Infrastructure in Washington, DC

Between 700 and 800 older U.S. cities have combined sewer and stormwater drainage systems. The increasing incidence of extreme weather events taking place today has forced cities from Honolulu to the east coast USA to discharge tons of CSO into the country's streams and rivers to avoid flooded streets, homes and businesses. The system in Washington, DC is one of those cities, being forced more and more often to discharge CSOs into local surface waters. Box 15.2 describes its early efforts to reverse that trend.

Box 15.2 Washington, DC adds green infrastructure to planned drain overflow control plan

The combined sanitary sewer system of Washington, DC was constructed in the late 1800s. Today, approximately one-third of DC is still served by a combined sewer and stormwater drainage system. In normal periods, the system carries all the sewage to the Blue Plains Advanced Wastewater Treatment facility. In heavy rainfall years, however, capacity at the treatment plant cannot accept all the combined water. To avoid flood streets and basements, the overload of mixed stormwater and raw sewage is discharged into local rivers and creeks.

The district received approval of its plan (the Clean Rivers Project) to control the combined over flows. The project originally included graywater infrastructure and large tunnels to capture and store combined sewage and stormwater. However, in 2017, the project received EPA authorization to include installation of green infrastructure (GI) in the project. GI was defined as "an approach to managing stormwater runoff that takes advantage of natural processes to slow down, reduce and in some cases reuse stormwater to keep it from overwhelming sewer systems and polluting waterways ... [and] to mimic the natural environment through the use of plants, trees and other measures."

DC Water has spent more than US$15 million since then to design and build demonstration projects to show that GI is capable of managing combined sewer and stormwater overflows (CSOs) in the District. Included in the project are bioretention planters between curbs and sidewalks, pervious pavement in parking lanes and alleys. The project has won two awards; the Government Finance Officer Association's Award of Excellence, and an award by *Bond Buyer Magazine* for the innovative pay-for-performance bond program.

Source Appel et al. (2017)

Recycled Water as a Supplemental Supply

Following longer and more commonly occurring droughts, several of southern California's water large water utilities have turned to water reuse as a viable alternative source of water to supplement their costly and limited sources of imported water (Maya 2017). The installed new infrastructure to prepare this alternative supply are advanced water treatment plants (AWTs), and the treatment procedures follow the same path: microfiltration, reversed osmosis, and ultraviolet light ad chemical treatment to potable-safe levels. Most of the final water is then returned to local acquirers as the key component in regional Groundwater Replenishment Systems.

Orange County Water District (OCWD) and the Metropolitan Water District of Southern California in Los Angeles County are leaders in this reuse of municipal wastewater. Both area water districts began recharging aquifers in the 1970s. The OCWD began accepting treated wastewater in 1975 from the Orange County Sanitation District (OCSD), which serves some 100 municipalities in the country. After final treatment, which includes screening for 400 contaminants, the freshwater-quality water is blended with deep well freshwater and injected underground, where it serves as a barrier to ocean water intrusion and as part of local municipal distribution. The 1975 facility was the first in the world to use reverse osmosis to purify wastewater. Reverse osmosis is now commonly used in treatment of wastewater to permeable status, the method uses semipermeable membranes without defined pores to remove microbiological pathogens and organic chemicals, including emerging concern pollutants such as hormones, pharmaceuticals, endocrine disruptors and personal care products (Bernados 2018).

Subsequent OCWD expansions included a US$143 million, 30 million gallons per day (MGD) final treatment facility in 2015. OCWD's treatment capacity goal is 130 MGD of fully treated water for reuse by 2013. The Metropolitan Water District of Southern California (Metropolitan) is a cooperative of fourteen cities, eleven municipal water districts and one county water authority that provides water to 19 billion people for an average of 2 billion gallons of water a day. The district was created in 1928, primarily to build and operate the Colorado River Aqueduct, which today provides the largest proportion of the 55% of Metropolitans water that is imported. The district also imports water supplies from near San Francisco, California via the 444-mile California Aqueduct. Metropolitan operates nine reservoirs and five water treatment plants. Four of these treatment plants are among the ten largest plants in the U.S. It provides treated water to parts of Los Angeles, Orange, San Diego, Riverside, San Bernardino and Ventura counties. The district operates the Regional Recycled Water Program with the Sanitation District of Los Angeles. A new 500,000 million-gallon demonstration wastewater treatment plant began operating in 2018. Eventual expansion to a 150 million gallon per day, US$2.7 billion treatment facility is planned by the cooperation.

Green Infrastructure in Philadelphia

In 2009, the newly elected mayor of Philadelphia, Michael Nutter, created the Mayor's Office of Sustainability, setting into motion the nation's first major test of actions to mitigate the effects of climate change on the ability of the city to control combined stormwater and sewage overflows (Fitzgerald and Laufer 2017). The green stormwater management plan called for cross-department cooperation among the several city departments with a stake in stormwater and wastewater management. Other city departments that had to collaborate in implementing the plan included planning, transportation, streets, power, finance, economic development, public health and education.

Although Philadelphia's stormwater management plan was the first submitted to the EPA to depend almost entirely on green infrastructure, this was not the city's first experience with stormwater control. Like other cities, Philadelphia was required by the Clean Water Act of 1994 to eliminate storm overflows of its combined stormwater and sewage system. In 1999, the city established an Office of Watersheds (OOW) in the Philadelphia Water Department (PWD) to oversee implementation of the city's CSO program. The OOW'S *Green City, Clean Waters* plan called for a 25-year, US$2.48 billion program for controlling CSOs through green infrastructure in close to 35% of the city infrastructure programs, which are included in Box 15.3.

Box 15.3 Philadelphia's stormwater control green infrastructure programs

Green streets: Green streets emphasize the capture of stormwater runoff from public rights-of-ways, such as streets. Various green stormwater infrastructure improvements include stormwater tree trenches, planters, and bump-outs, or pervious pavement.
Green schools: Schools can educate the local community about green stormwater infrastructure, while installing their own rain gardens, green roofs, pervious pavement, tree trenches, and rain barrels.
Green public facilities: The full array of green stormwater infrastructure practices that can be implemented at public facilities, include rain gardens, green roofs, pervious pavement, stormwater tree trenches, rain barrels, and cisterns.
Green parking: Green stormwater infrastructure practices that can be used to manage stormwater in parking lots, include vegetated strips and swales, rain gardens, infiltration beds and trenches and pervious pavements.
Green parks: Draining the nearby impervious areas to the open spaces enhances the visual appearance and the amenities at parks, in addition to managing stormwater runoff.
Green industry, business, commerce, and institutions: Stormwater management regulations for development and redevelopment and the parcel-based billing for stormwater management services provide incentives for private entities to install green stormwater infrastructure.
Green alleys, driveways, and walkways: These areas present an opportunity to retrofit, to allow infiltration, or to redesign to include diverted rooftop runoff to green stormwater infrastructure at the end of an alley and within the public right-of-way.
Green homes: Green homes projects include installing rain barrels and/or connecting rain leaders to rain gardens or flow-through planters. Others could include reducing the amount of impervious pavement, planting trees or building green roofs.
Source Philadelphia Water Department (2018)

Summary

Climate change is increasingly associated with significant alterations in water patterns. Many already water-stressed regions are experiencing even greater water scarcity, while others are seeing more and longer periods of extreme rainfall events. These effects of climate change, along with the increasing demand for water by a growing population, the migration to urban areas, and the water needed for more intensive agriculture are stimulating a global search for alternative and supplemental water supplies. Reusing water, harvesting stormwater and using green infrastructure to capture stormwater have been shown to be successful means for supplementing scarce freshwater supplies.

Controlling and capturing stormwater provides two supply benefits. Better controls of stormwater can help limit the damage to life and property from flooding, and captured stormwater and urban runoff can be an affordable source of water to supplement the limited supplies of freshwater.

Urban water system managers must deal with a variety of competing operational issues, ranging from protecting public health to providing critical water service to consumer, industrial and other sector customers. They must meet domestic demand with little, if any, say in area development policies, accept and process the liquid wastes of industries, hospitals and institutions, all the while adhering to a requirement for maintaining environmental quality. While urban water system managers see their existing supplies evaporating, they search for new and supplemental sources of supply. In that search, accepting the potential for the reuse of graywater, rainwater and stormwater supplies is becoming essential.

References

Appel, Todd, Bethany Bezak, and John Lisle. 2017. DC water green infrastructure financing: Pay for success can help utilities pursue innovate solution. *AWWA Journal* 109 (10): 28–31.

Bernados, Brian. 2018. Reverse osmosis for direct portable reuse in California. *American Water Works Association Journal* 110 (1): 28–36.

CASQA (California Stormwater Quality Association). 2018. The Los Angeles stormwater capture master plan: Harvesting stormwater for municipal supply. Accessed July 6, 2018 from www.casqa.org/asca/los-angeles-stormwater-capture-master-plan-harvesting-local-stormwater-municipal-supply.

Dos Santos, Sylvana M., and Maria Mariah M.W.E.C. de Farias. 2017. Potential for rainwater harvesting in a dry climate: Assessments in a semiarid region in northeast Brazil. *Journal of Cleaner Production* 164: 1007–1015.

EPA (Environmental Protection Agency). 2018. Green infrastructure. U.S. Environmental Protection Agency. Accessed July 3, 2018 from https://www.epa.gov/green-infrastructure/what-green-infrastructure#main-content.

Fitzgerald, Joan, and Joshua Laufer. 2017. Governing green stormwater infrastructure: The Philadelphia experience. *Local Environment* 22 (2): 256–268.

IPS (Inter Press Service). 2017. World water week: "Water and waste: Reduce and reuse", August 22. Accessed February 24, 2018 from http://www.ipsnews.net/2017/08/2017-world-water-week-water-waste-reduce-reuse/.

Li, Liu, and Jensen M. Bergen. 2018. Green infrastructure for sustainable urban water management: Practices of five forerunner cities. *Cities* 74: 126–133.

Maxwell-Gaines, Chris. 2018. What is rainwater harvesting? Innovative Water Solutions. Accessed July 6, 2018 from www.watercache.com/faqs/rainwater-harvesting-defined.

Maya, Anna. 2017. Securing Southern California's water future. *Water World* 33 (11): 10–13, 39–40.

Oppenheimer, Joan, Allegra Da Silva, Zita L.T. Yu, Mark Hanna, and Ken Susilo. 2017. Capturing alternative water sources to supplement drinking water supply. *Journal of the American Water Works Association* 109 (1): 18–25.

Pushard, Doug. 2018. Gray water: The new green? *Green Fire News*, April 4. Accessed July 7, 2018 from https://greenfiretimes.com/2018/04/gray-water-the-new-green/.

PWD (Philadelphia Water Department). 2018. Green stormwater management programs. Accessed August 8, 2018 from www.phillywatersheds.org/what_were_doing/green_infrastructure/programs.

UN Water. 2018. Nature-based solutions for water. United Nations World Water Development Report 2018. Accessed July 16, 2018 from unesdoc.unesco.org/images/0026/002614/261424e.pdf.

16

Alternative Sources of Water Supply

Achieving a sustainable supply of clean, safe water is not a problem just for developing nations; water stress and scarcity is a critical concern in the life of people in nations throughout most of the globe. Supplying water for a rapidly growing population and urbanization, providing irrigation water needed to grow the food to ensure an adequate supply of food, halting the indiscriminate pollution of surface and groundwater resources, while funding the new and replacement infrastructure needed to meet demands of a changing climate are problems for everyone. Water and wastewater service suppliers in many regions are struggling to be sure they have an adequate and sustainable water supply for their current and future customers. This chapter looks at some of the alternative pathways that some districts are taking to be certain that theirs is a sustainable water supply that meets present and anticipated future demand.

Complications in arriving at sound constant values for supply and demand make it difficult to accurately predict future needs. Even defining the terms supply and demand is difficult. The McKinsey 2009 report on water's future, for example, put forward a definition of supply as the total renewable water resource that can be available with the existing infrastructure and natural constraints, such as periodic droughts. However, that amount is limited by what can be withdrawn from the natural hydrologic system. The total amount of surface and groundwater that meets the requirement of being "accessible, reliable, and an environmentally sustainable supply" is limited by these external constraints. At best, forecasting future supply can only be an estimate.

D. E. McNabb, *Global Pathways to Water Sustainability*,
https://doi.org/10.1007/978-3-030-04085-7_16

Table 16.1 Considerations for selected alternative water sources

Alternative Source	Water quality concerns	Potential treatment	Potential alternative uses	Considerations
Rainwater	Suspended solids, pathogens	Filtration, possible sedimentation and disinfection	Irrigation, toilet and urinal flushing	Minimal treatment for irrigation use
Stormwater	Sediments, organics, pathogens	Filtration, possible sedimentation, disinfection and biological treatment	Irrigation, cooling tower makeup, industrial uses	Foundation drain water can be reused like stormwater
Onsite reclaimed wastewater	Pathogens, sediments, organics, solids, hardness	Filtration, disinfection, and biological treatment	Irrigation, cooling tower use, industrial use	Regulatory control over reclaimed water for irrigation
Graywater	Pathogens, sediments, organics, dissolved solids	Possible sedimentation and biological treatment	Toilet and urinal flushing, landscape irrigation	Subsurface irrigation appropriate for selected crops
Recycled blackwater	Pathogens, sediments, organics, dissolved solids	Requires expensive tertiary treatment; Government approval needed if used for fruit and nut crop irrigation	Growing use of tertiary treated sewage water for recharging aquifers; some potable use for food and beverages	Tertiary treatment becoming more cost effective; significant growth in agriculture use in arid countries
Desalinated water	Standard water purification practices on desalinated output	Desalinated and brackish water needs expensive equipment treatment	Increasingly used to supplement existing limited freshwater supplies	Common in some U.S. coastal cities, Middle East and China

Sources FEMP and other government sources

Defining demand is less problematic. The McKinsey researchers defined future demand as the amount of unrestricted water withdrawn from surface water and groundwater resources and unconventional sources as recycled and desalinated water, without considering returns to the sources.

The combined effects of climate change, population growth, urbanization and inefficient water use has brought many areas of the world to a position of extreme water scarcity. As a result, water supply managers are searching for accessible new and different approaches of augmenting their traditional supplies of surface water and groundwater water. This chapter looks at three alternative approaches in the search for alternative potable water supplies, beginning with the recycling treated water for potable and nonpotable uses, followed by the increasing investments being made in desalination of ocean and brackish water, and closing with a discussion of the mega-project diversion of existing surfacewater sources.

Organizations, industries, government agencies and institutions and municipal water utilities are increasingly looking for supplies of fresh water from alternative water resources. Alternative sources are sustainable supplies of water that are not from a new surface water or groundwater source in amounts sufficient to serve as a reliable alternative or supplemental water supply. The U.S. Federal power agency has provided an information document that describes the features and problems of the chief potential alternative water resources. Abridged data from the FEMP report and other government sources such as the EPA and state reports are included in Table 16.1.

Recycled Water as an Alternative Resource

The supply of freshwater is increasingly under stress in many parts of the world. Climate change is changing precipitation patterns, resulting in even less precipitation in already arid locations and more precipitation in others. Air and semi-arid regions such as the Middle East and the American Southwest are particularly in need of additional supplies of water. It has been the case for several decades that, after tertiary sanitation treatment some recycled water has been accepted as a substitute for drinking water for landscape watering and for irrigating agricultural crops not directly intended for human consumption. The use this recycled of water to augment scarce surface and groundwater supplies has gained more and more acceptance. Moreover, advances in treatment technologies have reduced the cost to the point where recycled water is in many instances cheaper than natural supplies (McNabb 2017).

Wastewater has typically been classified into two major categories according to its source: domestic (municipal) wastewater, and industrial wastewater. There are two types of domestic wastewater: graywater and blackwater. Graywater is used water from sinks, bathing facilities and laundry activities. Blackwater is wastewater (sewage) from domestic and public toilets. Treatment and purification that removes the unwanted components is necessary before the effluent can be discharged. Untreated blackwater may contain feces, urine and used toilet paper. Because body wastes may contain intestinal disease organisms, it is sanitized and biosolids are removed before the final clean effluent is discharged into a water source from which it originated, or into surface ponds or wetlands where it is allowed to evaporate or be absorbed into the earth.

Industrial blackwater is water used in industrial and commercial operations, including water that has been used for agricultural irrigation. Industrial wastewater that is used in industrial processes can contain such pollutants as residual acids, plating metals and toxic chemicals. It requires complex and expensive treatment before being discharged. However, industrial wastewater that has only been used for cooling purposes, as in thermoelectric power generation, is far less polluted and is often reused for the same purpose. Industrial processes generate many different pollutants in their wastewater. Separate processes are needed to treat each type of contaminant before the water is discharged or reused. After primary and/or secondary treatment, some highly polluted industrial wastewater may be sent for further treatment or entered into the usual municipal wastewater treatment system.

Among the innovative approaches now underway are rainwater and stormwater harvesting, collecting and reusing greywater for non-food agricultural and landscape irrigation, intensive treatment of domestic wastewater for further filtration through natural filtering and recharging of potable water aquifers, and desalination of ocean and brackish estuary waters (Oppenheimer et al. 2017). Four areas with extreme water scarcity were described in a comparative study of alternative affordable methods closing the gap between water supply and demand that is projected to occur by 2030: China, India, South Africa and São Paulo, Brazil. The study defined total water demand as "the projected water requirements if efficiency is unchanged and the policy environment is static. This demand is measured as the actual withdrawals from surface water, groundwater or nonconventional sources." Water supply was defined as the total renewable water resource (as represented by a theoretical upper limit) of all surface and groundwater that can be withdrawn for all uses in the basin.

Analysis of surface water supply of São Paulo was conducted on four suppy-influencing factors: accessibility, environmental requirements (ecosystem maintenance), reliability, and a collection of other relevant phenomena that could include water transfers from adjacent basins and the use of nonconventional sources. A similar set of variables were used for determining groundwater supplies, with an emphasis on an underground supply that is accessible by affordable pumping, is near a demand center (a farming center or urban area, for example), and is sustainable (an aquifer that is regularly recharged).

Different Supply Gap Solutions

The 2030 Water Resources Group (WRG) is a private sector organization formed to study and report on the water supply problem and the estimated cost of achieving water sustainability, with funding by the World Bank. Their report, *Charting Our Water Future*, reviewed future gaps between water demand and water supplies in four areas: China, India, South Africa, and the São Paulo state of Brazil. Since the 2009 release of the report, extended periods of drought, continuing population growth and urbanization have exacerbated the supply problems facing the four areas.

The WRG analysis included four estimates of: (1) the impact of economic and social growth on the nation's water resources; (2) costs and requirements of meeting future water demand; (3) comparisons of different growth scenarios; and (4) the implications for stakeholders. Recommendations included a mix of solutions to increase water supply and to improve water productivity. The group concluded that combinations of existing low-cost remediation approaches could eliminate up to 100% of the predicted 2030 demand–supply gap in countries studied. However, ten years after the release of that report, it can be seen that the actions taken were not extensive enough to come even near to closing the regions' supply–demand gaps.

The sustainability pathways that countries adopt must be based on some combination of solutions to three underlying water supply and demand factors: (1) means of expanding supply; (2) increasing the efficiency and productivity of existing water uses; and (3) reducing demand by shifting the economy towards less water-intensive activities. Different combinations of these three options will determine which pathway nations will follow in achieving their sustainable water supplies. Each pathway is expected to have different costs and time frames for completing the processes that will bring about their distinct sustainability package.

Varying the country's economic outputs, the water intensity of its economy, and its reliance on supply-side infrastructure are all potential levers for achieving water security. The critical question to be answered is: what is the optimal mix for this country, its people and economic status? (McKinsey and Company 2009). In the beginning of arriving at the right mix answer to this question, these four classes of technical measures, broadly covering increases in supply and different types of efficiency and productivity characteristics, may be analyzed:

1. Agricultural productivity measures that could improve both the efficiency of water use in irrigation and crop yields on both rain-fed and irrigated lands.
2. Industrial efficiency measures.
3. Domestic and municipal efficiency measures.
4. Supply measures that increase accessible, reliable and environmentally sustainable supply.

The analysis followed cost-effective ways to close the 2030 supply–demand gap projected for the four countries using only two of the three approaches: increasing water supply and improving water productivity. The research assumed that the two solutions together could likely close 100% of the 2030 supply–demand gap.

Annual capital costs to achieve this goal for the four countries were estimated to be US$19 billion in 2030. Expanded to cover the entire globe, the annual estimated costs were US$50 billion for integrated supply and productivity solutions for closing the projected water gap. This was thought to be one-quarter of the roughly US$200 billion that would be required for expansion of water supply solutions alone.

The cost curve developed for each of the four regions led the researchers to conclude that "different local costs and impacts of otherwise similar measures, [indicate] that the stand alone potential for technical measures to close a water supply–demand gap is unique for each country" (McKinsey and Company 2009, 73):

- India's cost-effective solution rests squarely on improving the water productivity of agriculture;
- China's solution focuses on better managing the rapidly growing demand from industrial and urban uses;
- São Paulo's solution would depend heavily on industrial as well as municipal and domestic efficiency;
- South Africa's least-cost solution would involve a combination of supply and industrial efficiency levers.

Closing the Gap in India

Although neither on the 2008 list of the top most water-stressed countries, nor included among the nations on the projected 2040 water-stressed list, India's huge population justifies its inclusion in the McKinsey Group shortlist of regions facing a water crisis. India's population growth, climate change-based disruption in precipitation patterns, and uncontrolled urbanization together put the water supply issue high on the list of concerns of the Indian government. India is on the path to becoming the world's most populous country in 2022, making it highly dependent upon groundwater for irrigated agriculture, both for export and to feed its growing population.

India's water difficulties have been blamed on three factors. First, the country does not have enough raw water for its existing population, let alone for keeping up with the growing demand from population growth. Global water planners consider a country to be under water stress if it has access to less than 1700 cubic meters of water per person. India has less than 1000 cubic meters of water per person.

Second, much of India's available water supply is of very low quality. This is largely a result of a failure to keep up with needed investments in water and wastewater infrastructure, and agriculture and industrial pollution. The water in most of India's rivers is unsafe for human consumption, with little or no treatment facilities available. Water pollution is the result of agriculture fertilizers, animal wastes, and insufficient treatment of human wastes and untreated industrial water discharges.

The third contributing factor is the depletion of groundwater by the nation's agriculture sector. Farmers are pumping out more water than can be recharged, with existing irrigation methods polluting water supply resources (Luthra and Kundu 2013). Of the three factors, improving agriculture methods is considered to the be the most cost-effective short-term path to improving water supplies.

Agriculture is the major contributor to the India's GDP. More than half of all workers in India are employed in the agriculture sector. In 2016, India exported close to US$39 billion in raw agricultural products and more than 4.4 million tons of milled rice. Rice and cotton, two of India's most valuable crops, are also two of the largest users of water. In 2014, irrigation accounted for some 84% of total available water, industry took 12% and the domestic sectors consumed about 4% of the total available water. Nearly 55% of all agriculture water used was groundwater.

For centuries, Indian farmers depended on rainwater for irrigation. The annual monsoon was dependable and sufficient for farmers' needs. Little planting survived the annual dry season. However, traditional agriculture began to change beginning with the advent of higher crop yields made possible by the green revolution during the late 1960s. New high-yield crop varieties such as dwarf wheat and rice, together with chemical fertilizers, chemical pesticides and controlled irrigation revolutionized agriculture around the globe.

Farmers with holdings of just several acres account for more than 40% of India's food grains, nearly all of which is grown in the dry season and irrigated with groundwater. To take advantage of the increased economic opportunity, new crops were planted during the dry season and irrigated with groundwater. Some districts still provide free or low-cost electricity for irrigation pumps. Subsidized power has contributed to the excessive withdrawals of groundwater. Water tables have dropped across India by as much as 25 feet to more than 50 feet since 2009 due to excessive pumping of groundwater for irrigation.

Pathways to Alternative Water Supply

India is concentrating on two pathways for closing its widening water supply gap: one, the more cost-effectinve path of greater agriculture groundwater irrigation efficiency and two, a very costly, large-scale surface water collection and distribution system known as the National River Linking Project (Luthra and Kundo 2013; Swain 2017). Improving agriculture efficiency includes shifting to new, less water intensive crops such as more low-water use barley and less high-water use wheat, and shifting from open channel row irrigation to controlled drip and sprinkler irrigation.

During India's monsoon wet season, more than three feet of rain may fall. Little of that rainfall is harvested; floods that spread agriculture and human waste pollution are common in many areas. Conversely, little precipitation occurs during the dry season, and without irrigation most crops cannot survive. Farmers typically based their planting on estimates of available water. If less rainwater or surface water is available, Indian farmers tend to increase the amount of land depending on groundwater for irrigation. There is little public demand for government construction of more and higher dams to store surplus monsoon water, and even less for increasing the cost of electricity to pump groundwater. Meanwhile, research has shown that increasing water efficiency does not reduce demand for groundwater. Rather, farmers take advantage of the now extra water to increase planting in less productive soils, thus balancing out any gains (Fishman et al. 2015).

The Linking River Project

In 1980, the Indian government announced a plan to connect the 14 major rivers from the Himalayas with the 16 rivers lower in the peninsula. Called the Indian National Perspective Water Development Plan, the idea was to supplement low water in some rivers with excess water from rivers with too much water, thereby reducing flood damage while taking steps to overcome the persistent water shortages in some area.

Although the project was looked upon as a good plan, the linking was never completely implemented. Pointing to the 22-year delay in completing the project, in 2002, the Indian Supreme Court informed the government that it should take steps to go back to work and compete the project within a reasonable time period (Swain 2017). In the now almost 40 years since it was announced, some progress has taken place. However, it is still nowhere near completion. At 2018 prices, the completed project has been estimated to cost somewhere between US$125 billion and $200 billion, money the Indian government has other pressing uses for.

Closing the Supply Gap in China

The lack of comprehensive regulation and control of non-point source pollution has resulted in China now facing what has been described as critical shortage of water for all usages (Tao and Xin 2014; Udimal et al. 2017). In 2014, the annual supply of freshwater was short by an average of 40 billion tons of water. As the nation continues to population continues to grow and the nation to urbanize, this supply–demand gap will continue to grow every year. Many aquifers can no longer meet water withdrawals. Analysts believe this largely exists because nearly half of the country's water sources are polluted.

Both shallow and deep aquifers are contaminated. The major sources of pollution in both surface and groundwater is caused by unregulated sewage and municipal wastewater, industrial wastewater discharge, and agriculture fertilizers and pesticides. A 2011 study of 800 wells in nine provinces and autonomous regions, including Beijing and Guangzhou, found nearly 79% of the water failed to meet government health standards. An estimated more than 200 million Chinese were forced to rely on unsafe raw water.

In 2014, the Chinese central government announced it was initiating a major war on water pollution. The goals of the program were (1), to ensure that at least 70% of the surface water quality in seven of the country's major river basis would meet Class III standards (middle of the quality levels), and that no more than 15% of the groundwater would be rate Cal V. Achieving the plan goals was to be based on the following ten steps:

1. Take (central government) control of and reduce pollutant discharges.
2. Promote transformation of the economic structure to lower the intensity of pollution.
3. Focus on protection of water resources through water saving (conservation).
4. Strengthen support for (water) science and technology.
5. Allow market mechanisms to impact water and pollution levies.
6. Strict environmental law enforcement and supervision.
7. Strengthen the management of the overall hydrologic cycle.
8. Ensure security of water for ecological and environmental purposes.
9. Confirm and implement the responsibilities for all parties involved.
10. Strengthen public participation and social supervision.

China's Use of Unconventional Water

Eliminating water pollution will make China's water safer, but protection alone will not close the supply–demand water gap. Alternative sources are needed as well. China is looking to increase augmenting its water supplies with reused or reclaimed water (Udimal et al. 2017). China's reserves of water are dropping each year; between 2000 and 2009, water reserves declined 13% as the country suffered through a series of prolonged droughts. Precipitation in China is also unevenly distributed. In normal years, Southern provinces receive close to 80 inches a year, while the north and west only receive from 8–16 inches of annual rain or snow.

The Giant Infrastructure Project

China has a giant infrastructure project underway to move excess water from the south to the arid north and west. Starting in 2002, the US$62 billion South-to-North Water Diversion Project (SNWDP) is the largest such water diversion scheme ever attempted. When completed in 2050, it will link the Yangtze, Yellow, Huaihe and Haihe rivers, annually diverting 44.8 billion cubic meters of freshwater from the south to the north (Cho 2011).

The project is not without its critics. The south is becoming drier as climate change is melting Himalayan glaciers and lowering precipitation continue. Southern farmers and cities are afraid they will soon not have enough water to share with the north.

China is also working with the energy sector to reduce its heavy use of water. It has become a global leader in the production of solar panel technology, investing in wind power and adding desalination facilities. Coal mining and use in power plants are heavy users of water. Chine is replacing old coal-powered electricity generation facilities with nuclear plants, and is substituting freshwater with seawater for cooling.

Summary

The desired payoff that will result from the implementation and accomplishment of these sustainability solutions are the global health and economic benefits that safe, adequate and sustainable water supplies provides for citizens and sectors. Clean water is necessary for healthy people in sustainable environments, increased prosperity for citizens and governments, equitable treatment for all sectors and every person, enhanced education and economic opportunities for all people and for resilient communities, and the salvage of distressed ecosystems and protection of all others for this and future generations.

Public acceptance of the reuse of water to augment domestic water supplies in arid and semi-arid regions of the country is increasing as droughts occur more often, last for longer periods, and temperatures become hotter. Greater acceptance of the idea that recycled water, the source of which can be stormwater or treated wastewater, is a valuable resource available for augmenting scarce surface and groundwater potable supplies, will be necessary in the near future if the world is going to reach water sustainability.

Alternative sources of supply are difficult and often extremely costly to acquire. Traditional farming methods and crops used in the world's largest nations are often difficult to change. When any new water is available, farmers are quick to take advantage of the supply, further adding to the over-use of surface water and the depletion of existing aquifers.

References

Cho, Renee. 2011. How china is dealing with its water crisis. *State of the Planet* blog. Columbia University Earth Institute. Accessed April 14, 2014 from https://blogs.ei.columbia.edu/2011/05/05/how-china-is-dealing-with-its-water-crisis/.

Fishman, Ram, Naresh Devineni, and Swaiminathan Raman. 2015. Can improved agricultural water use efficiency save India's groundwater? Accessed April 12, 2018 from https://iopscience.iop.org/article/10.1088/1748-9326/10/8/084022.pdf.

Luthra, Sonia, and Amrita Kundu. 2013. India's water crisis and cures: An interview with Kirit S. Parikha. The National Bureau of Asian Research (NBR). Accessed June 28, 2018 from http://www.nbr.org/research/activity.aspx?id=356.

McKinsey & Co. 2009. Charting our water future. 2030 Water Resources Group. *Sustainability & Resource Productivity* 18 (1): 1–18. Accessed December 15, 2017 from www.mckinsey.com/~/media/mckinsey/dotcom/client_service/sustainability/pdfs/charting%20our%20water%20future/charting_our_water_future_full_report_.ashx.

McNabb, David E. 2017. *Managing the Water Resource in an Era of Climate Change*. New York: Palgrave.

Oppenheimer, Joan, Allegra Da Silva, Zita L.T. Yu, Mark Hanna, and Ken Susilo. 2017. Capturing alternative water sources to supplement drinking water supply. *Journal of the American Water Works Association* 109 (1): 18–25.

Swain, Ashok. 2017. Here's why the idea to link India's rivers will remain a pipe dream. *Outlook,* September 3. Accessed April 12, 2018 from https://www.outlookindia.com/website/story/heres-why-the-idea-to-link-indias-rivers-will-remain-a-pipe-dream/301157.

Tao, Tao, and Kunlun Xin. 2014. A sustainable plan for China's drinking water. *Nature* 511: 527–529.

Udimal, Thomas B., Zhuang Jincai, Emmanuel C. Ayamba, and Samuel M. Owusu. 2017. China's water situation: The supply of water and the pattern of it uses. *International Journal of Sustainable Built Environment* 6 (2017): 491–500.

Part IV

Innovative Water Management

17

Managing Assets for Water Sustainability

Government and civil engineers looking at water's 'big picture' refer to a reliable infrastructure as the essential framework upon which sustainability will rest. But utilities and other water providers approach this problem from a point of view that is closer to home: how can we best manage our many assets for sustaining the desired amounts of high-quality water and water-related services to our customers? After the Second World War, water managers in Australia and New Zealand came up with a way to help guide local water utilities plan and operate their systems for attaining maximum reliability. They called the system Asset Management Planning.

Water system managers in the USA and other public utilities around the globe have adopted this approach, and many U.S. federal and state oversight agencies now require utilities to develop their own asset management plans. This example from New Jersey explains how states have adopted asset management plans as a utility requirement:

> To ensure drinking water and wastewater systems are adequately maintained and operated to continually and reliably meet customer service expectations as well as comply with applicable permit conditions, NJDEP rules generally require drinking water and wastewater utilities to demonstrate that they have adequate facilities, and equipment, and that they regularly perform operation and maintenance to meet the conditions in their permits. This includes conducting an inventory of system assets, providing adequate

D. E. McNabb, *Global Pathways to Water Sustainability*,
https://doi.org/10.1007/978-3-030-04085-7_17

> staffing and training, performing preventative maintenance, and demonstrating adequate funding. In order to meet these requirements, a utility must identify its needs and costs, and develop long-range financial plans. The current best management practice for ensuring such financial planning is through the development and implementation of an Asset Management Plan. (NJDEP 2016)

To maintain sustainable growth with system sustainability, water systems have had to adopt new ways for coping with rapidly changing conditions and for managing limited resources astutely. Regulatory and civic organizations have come up with many of the tools needed by system managers. Integrated water management and integrated urban water management are examples of these systems. Large and small water systems must invest large sums on infrastructure repair and replacement to retain operations. Many nations are also requiring or encouraging large and small systems to adopt asset management programs. Many states in the USA now require medium and large water and wastewater systems to submit asset management plans. To aid managers of small systems in the asset management planning processes, many state agencies have published a guide to help small systems develop their own asset plans. Some states require systems to submit an asset management plan before they can participate in the state water infrastructure resolving fund.

The asset management planning program is based on the desired level of service (LOS) each water system provides its ratepayers. Good asset management ensures that the infrastructure requirements to meet those operational goals are met at the at the lowest possible life cycle cost commensurate with required sustainability. Accordingly, "Water systems that take care of their assets are more likely to ensure safe and reliable drinking water for their customers. Lowest life cycle cost refers to the best appropriate cost for rehabilitating, repairing or replacing an asset. Asset management is implemented through an asset management program and typically includes a written asset management plan" (WSDOH 2009). Benefits that can accrue to a water system from preparing and following an asset management plan include the following:

- Longer asset life through systematic and more efficient decisions for rehabilitating, repairing or replacing water infrastructure.
- Improved ability for reliably and sustainably meeting customer service demands.
- Ability to set rates on the basis of sound financial planning for repairing and replacing critical assets.

- Better budgeting by focusing on operations designed to ensure safe, reliable and sustainable water supply.
- Ensuring the system meets the LOS it promises its customers while also meeting all regulatory requirements.
- Improved ability to successfully respond to emergencies.
- Improved systems security.

Maintaining the Commitment to Change

The effects of climate change, with the escalating costs of energy, labor and supplies are among the newest challenges water systems face in their commitment to providing safe and reliable water supplies and treatment services. As these challenges converge, it has become clear to many utilities that their operations must focus not just on meeting current needs. Instead, they must now plan for achieving and maintaining a sustainable ability for to continuously meet increasing demands under these rapidly changing environmental conditions. This demands that everyone involved in these systems retain a commitment to operating under a new management model. One of the most important aspects of asset management is that it cannot be something that the [water] utility does on the side as one of its many activities. Rather, it must be the way the utility does business. "Asset management thinking must underlie every activity, every action, and every decision that the utility undertakes … if there is insufficient acceptance of asset management, the plan will be much less successful and may not succeed at all" (NMEFC 2006, 4).

In planning for sustainability, water system managers must also plan for functioning in a manner that ensures their operations do not entail taking shortcuts around their responsibility as stewards of the environment. Water systems managers in many parts of the world are confronted with formidable environmental challenges as they work to maintain and often expand their systems. Getting to the goal of sustainability involves a commitment to change in the way systems are governed and managed, but it also includes protecting the environment for the benefit of future generations.

Asset management plans require planning for acquiring, treating and supplying clean and safe freshwater to all sector customers, collecting and treating wastewater discharged back into the environment, and managing stormwater runoff and help avoid or lessen flood damage. These problems have intensified as pressure from population growth and development have increased along with more and more severe infrastructure failures. For water systems, the twenty first century is the era of adjusting to a constantly increasing pressure on water supplies and resilience in the face of impending natural disasters.

Management according to a comprehensive asset management plan opens the pathway to sustainability. The chief benefit of operating a system according to asset management principles is, and always should be, to maintain reliable service in the face of uncontrollable change. Successful implementation of asset management will go far in a water system in terms of achieving and maintaining long-term sustainability, even when confronted with more and more severe natural disasters.

Box 17.1 Communicating the need for a change

"All organizations undergo change at different times in their life cycle. Your water and sewer service is no exception. Our operations team has indicated a desire to explore other career paths before the end of this year. We are currently in the process of planning for the resulting new staff structure that will take place as a result of that change. We do not expect the change to occur immediately, or even soon, but we have been alerted to the pending change and are preparing for a smooth transition. To be better able to make the transition with little or no new costs, we have begun examining the strengths and weaknesses of a different operational management structures.

Changes in governance typically occur with every election as commissioners serve their terms and are replaced by new commissioners, each of whom may come to the position with little knowledge of the responsibilities and processes of the utility's operations. This approach results in the loss of institutional knowledge. As of this January, none of the original commissioners who designed the present operational management system remains on the board of commissioners. Now, after ten years of operations, we newly elected commissioners must learn for ourselves which management structure is right for the utility considering today's changing environment and rapid growth.

We have begun our analysis by investigating and evaluating the number and types of management problems and benefits that result from a change of operations management structures. Seeing how other utilities have solved this problem is the first step in our making sure your utility maintains sustainable and secure water and wastewater service here at your water service. We will keep you posted on our progress."

Source The author

A Better Way to Do Business

Before beginning planning for the transformation process to asset management and sustainability, commissioners. administrators and managers must identify and adjust for any disparity between sustainability goals and persistent attitudes against change among staff and ratepayers. Transformational change cannot begin until the causes of the disparity are identified, and a strategy designed for bringing the attitudes of all stakeholders to a point where the new policy and new way of operating will be accepted. Communicating the need for change to ratepayers is necessary. Box 17.1 is an example of how a small water district began informing its customers of an impending change in the way the system was governed and the growing need to plan for major infrastructure repair, refurbishment and replacement. Much of the initial infrastructure had been installed with greater concern for the equipment's cost than for maximizing the available life span without failures.

Planning for a transformational change such as shifting from a business-as-usual approach to following a prepared schedule for repairing, refurbishing and replacing large sections of the small water and wastewater system's assets on a life cycle schedule requires the support of ratepayers. Such a system begins with an internal analysis to determine the readiness of the utility's ratepayers to accept—and pay for—the change. Rates are already high in districts where customers are asked to pay the full cost of supplying their clean, safe water and wastewater services rates. Getting the political overseers—the councilmen or elected district commissioners—of all public utilities and the systems' ratepayers to agree to the increases necessary for sustaining those services is often particularly difficult in a small public service organization. Before any rate increase can be adopted, an analysis is needed to determine whether the community is ready and willing to accept the increase.

Investing in the considerable time and effort for acquiring support for the high cost of replacing worn out or damaged critical infrastructure in order to maintain reliable operations may be difficult to arrange, but it is necessary. Internal support is also necessary. The willingness to accept and pay for change is seldom a widespread characteristic among a system's administrators or its ratepayers. Often, water system administrators and managers will have very different attitudes toward making an infrastructure investment before a breakdown occurs compared with the attitudes of others. Even some senior managers and the employees charged with implementing and overseeing the change may balk at the idea. The philosophy of "if it works, don't change it" exists everywhere.

Adopting Asset Management

During the presidency of George W. Bush, the General Accounting Office (GAO)—now the Government Accountability Office—was directed by the chairman of the Committee on Natural Resources and Environment of the U.S. Senate to conduct a study on a new management tool called asset management. The study was to focus on whether asset management could help water and wastewater utilities deal with their infrastructure problems. GAO was to determine (1) the potential benefits of comprehensive asset management for drinking water and wastewater; (2) the challenges hindering its adoption by utilities; and (3) whether the federal government could play a role in encouraging more utilities to adopt asset management for infrastructure investments. The GAO report added the following benefits accruing from infrastructure planning with an asset management plan: "Among other things, comprehensive asset management allows utility managers to obtain better information on the age and condition of existing assets, determine the level of maintenance needed to optimize asset performance and useful life, asses the risks associated with the failure of various assets, understand the tradeoffs and implications of management decisions about the assets, and use better information to justify proposed rate increases or capital investments. Water industry officials agree that by making informed decisions about investing in capital assets, drinking water and wastewater utilities can better justify the rate increases associated with making needed improved improvements to their infrastructure". With asset life cycle planning, the interrelated set of four fundamental concepts are the heart of the approach: linking a strategy for addressing infrastructure needs; collecting and organizing information about the assets; analyzing the data to set priorities; and making fact-based decisions (Fig. 17.1).

Asset Management Plan Components

A first-time asset management plan does not have to be complicated. Plans for either a water system or a wastewater system, or a combined system, are similar in scope, the only significant differences are in the asset infrastructure needed for each operation. Hence the components of an asset management plan for either system will include these same basic categories. The five basic components to include in the plan are asset inventory, the LOS to provide,

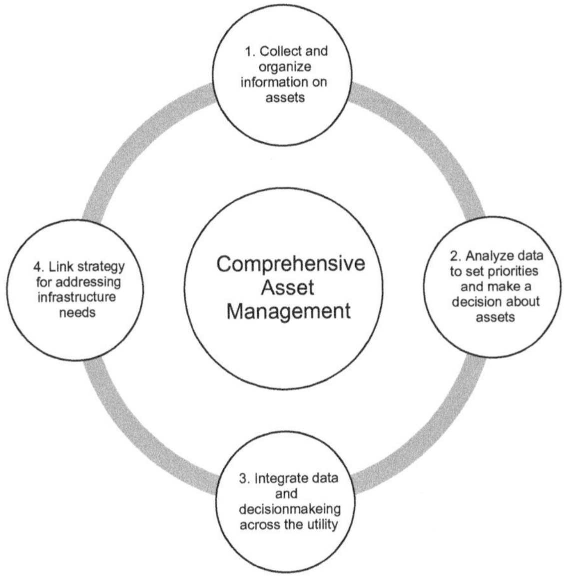

Fig. 17.1 Major elements of a comprehensive asset management system (*Source* The author)

the determination of which assets are critical for sustainable operations, life cycle costing, and a long-term funding strategy. Each of these components will have its own number of subcomponents. What sub-components are included in an asset management plan depends on what is needed for the type and size of the service, the staff needed to complete the plan, and the system's financial assets. The key to goals that are meaningful to the system and its community of ratepayers are:

1. Keep the objectives simple and achievable.
2. The quality of the objectives—not the quantity—is most important.
3. The objectives should be relevant, measurable and within the capabilities of the staff and the utility.
4. The objectives must be within federal, state and local regulations.
5. Engage staff in the objective identifying and quantifying processes.
6. Seek and heed customer input on the relevant objectives.
7. Review and update the goals regularly, letting everyone know of the changes.

Table 17.1 Form for core components of an asset management plan

Asset management plan component	Plan sub-component	Comments
Asset inventory	**Category and item list**	**Asset name and ID code**
	List of assets	
	Location of assets (map)	
	Asset condition rating	
	Remaining useful life	
	Remaining life value	
	Asset value if new	
Level of service	**Level of service statement for key services**	**Service provided to customers**
	Identify critical service items	Include measurement metrics
Critical status	**Criticality assessment score**	**Scale of necessity for continuous service**
	Identify most critical assets	Needed for sustainability
	Likelihood of failure	5-point scale
	Consequences of failure	In dollars and good will
Life cycle costing		**Cost if failure vs cost new**
	Detailed operation and maintenance program	
	Each asset location necessary	
	Repair and replacement schedule	
	Capital improvement plan (CIP)	5, 10 and 20 year plans
Long-term funding strategy	**Loan, grant or rate increase?**	**5-year financial plan (CIP)**
	CIP funding strategy	List possible funding options
	Rates structure	Flat or use-based?
	Repair and Replace (infrastructure) (R&R) funding schedule	Work from critical list

Table 17.1 is a version of the basic components included in a typical water or wastewater asset management plan, and Table 17.2 is an expanded outline for the steps to be taken in the completion of the first section of the plan. Table 17.3 is an asset inventory example, with the identification codes for the system's assets.

Table 17.2 Possible sub-topics that might be included in the asset inventory phase

Step	Data location	Responsibility
Collect and organize assets		
Locate and map all assets	All	General Manager (GM)
Develop identifiers for water assets	Plant	GM, Staff
Develop identifiers for sewer assets	Plant	Staff
Code current condition water	Plant and field	GM, Staff
Code current condition sewer	Plant and field	Staff
Enter asset IDs on spreadsheet	Plant	GM, Staff
Initial analysis of assets		
Identify all assets life-span	Supplier	GM
Identify current value of water assets	Field	GM, Staff
Identify current value of sewer assets	Field	Staff
Identify replacement value of water assets	Office	GM, Staff
Identify replacement value of sewer assets	Office	Staff
Integrate data for decision making		
Enter data on spreadsheet	Office	GM
Ensure water/sewer data compatibility	All	GM, Staff
Develop framework for R&R strategy	Office	GM

Table 17.3 Example of an asset inventory and asset categories for a small water system

Water plant assets	Asset ID	Classification ID	Item description
Water wells	WW-001-000		Well 1
		WA-001-001	Well 1 casing
		WA-001-002	Well 1 pump
		WA-001-003	Flow meter
		WA-001-003	Backflow meter
	WW-002-000		Well 2
		WA-002-001	Well 2 casing
		WA-002-002	Well 2 pump
		WA-002-003	Flow meter
		WA-002-004	Backflow meter
	WW-003-000		Well 3
		WA-003-001	Well 3 (casing)
		WA-003-002	Well 3 (pump)
	WA-004-000		Well 4
		WA-004-001	Well 4 (casing)

(continued)

Table 17.3 (continued)

Water plant assets	Asset ID	Classification ID	Item description
		WA-004-002	Well 4 (pump)
Water treatment	WT-001-001		Water treatment station 1
		WT-001-0001	Station 1 KMnO4 pump
		WT-001-0002	Station 1 Cl2 pump
		WT-001-0003	Filtration system
		WT-001-0004	Cl2 and PH analyzer
		WT-001-0005	
	WT-001-0000		Water treatment station 2
		WT-002-0001	Station 2 KMnO4 pump
		WT-002-0002	Station 2 Cl2 pump
		WT-002-0003	Station 2 FeCl3 pump
		WT-002-0004	Filtration system
		WT-002-0005	PH and Cl analyzer
Water storage	RS-001-000		
		RS-001-001	Reservoir
		RS-001-002	Tank mixer
		RS-001-003	Booster pump 1
		RS-001-003	Booster pump 2
Water distribution assets	**WD-001-0000**		
		WD-001-0001	Asbestos concrete mains
		WD-001-0002	PVC mains
		WD-001-0003	Iron mains
		WD-001-0004	¾ inch lateral poly
		WD-001-0005	1 inch lateral poly
		WD-001-0006	Pressure-reducing station 1
		WD-001-0007	Pressure-reducing station 2
		WD-001-0008	Water meters
Scada system	**SS-001-0000**		

Level of Service

A water or wastewater LOS are statements that establish the framework upon which the system will set fundamental operational goals for the reliable and sustainable delivery of to its customers. In short, they function as operational guideposts for the system's management and operations personnel. LOS objectives commonly include statements about how much water the water supply system will typically be able to supply, how often and for how long water restrictions might occur, how quickly water main breaks are repaired, how quickly ratepayer concerns are addressed, and related performance guides. LOS objectives may also provide a basis for water supply reliability planning. In arid areas, they might include the water system's clearly defined customer service responses for needed emergency water restriction in the possible occurrence of a prolonged drought or other emergency.

A guide for establishing LOS objectives and relevant metrics for measuring performance was published by the EPA. The guide was based on the EPA's 2008 list of *Ten Attributes of Effectively Managed Water Utilities*. The EPA supported by a team of representatives from professional organizations and 16 water and wastewater utilities identified these ten factors for utility performance measurement (EPA 2008; WRF 2014):

1. Product quality
2. Customer satisfaction
3. Employee and leadership development
4. Operational optimization
5. Financial viability
6. Infrastructure stability
7. Operational resiliency
8. Community sustainability
9. Water resource adequacy
10. Stakeholder understanding and support

Table 17.4 is an example of a possible scoring system for the annual assessment of the system's performance. Table 17.4 is an example of a set of LOS objectives for a small water and wastewater system. LOS statements should be measurable (Table 17.5).

Table 17.4 Asset management plan level of service scores

Asset management levels of service performance scores	
Objective performance score	**Description**
0	Does not meet objective
1	Meets the objective 10% of the time
2	Meets the objective 25% of the time
3	Meets the objective 50% of the time
4	Meets the objective 75% of the time
5	Meets the objective 100% of the time

Table 17.5 Sample list of level of service statements and performance scoring system

	Level of service goal performance score					
Water system	**0**	**1**	**2**	**3**	**4**	**5**
All federal and state water quality regulations will be met at all times						
Breaks will be repaired within 6 h of initiation of repair 95% of the time						
Customer complaints will be responded to within 24 h (Monday–Friday).						
Notice of scheduled service interruptions given to all customers no less than 5 days prior to the event						
Adequate water storage and treatment capacity will be maintained at all times						
The system will manage a minimum pressure of 40 psi						
Rates will be reviewed annually and raised as needed to ensure full cost recovery						
Source water pumps for 3 of four existing wells will be functional 99% of the time						
Water losses will be maintained below 7.5%						
Customers will receive written notice 24 h in advance of any planned interruption in service						
Service will be restored within 6 h of shutdown 90% of the time						
Wastewater system	**0**	**1**	**2**	**3**	**4**	**5**
The system will meet all state and federal regulatory standards 100% of the time						
Zero discharges of untreated or 'under-treated' wastewater at all times						
Reduce Inflow and Infiltration (I&I) by 10% annually						
Proper sludge management occurs 100% of the time						
Respond within 24 h to all complaints of customers' sewage blockages						

Identifying Critical Assets

The LOS is also a starting point for assessing overall system performance and for identifying critical assets. Not all assets are critical. For example, say that a LOS item is "sources of a leak in a water service lateral line will be identified and examined within 4 hours of receiving notification 95 percent of the time." The leak in the customer's water line is important to the customer, of course. But if the system is unable to respond within the four-hour objective, it is unlikely that the leak in the lateral line will not force the system to shut down. The lateral line is not a critical asset; still, it should be attended to as soon as possible. An example of a critical asset would be if the system has only one water source and no alternate source possible. The supply source and the infrastructure drawing water from that source are critical assets for the system.

Determining which assets are critical allows the system management to prioritize and justify its repair and replacement program. Thus, identifying which assets are critical is the first step in the critical asset assessment. The threat of a critical asset failing and compromising the ability of the system to supply water to customers should be known and weighed.

Determining the probability of the asset's failure is the second step in assessment. Informed decisions on replacing critical assets are based on the level of damage that could accrue from asset failure, the level of risk, the significance of a failure and whether replacement equipment or parts are readily available. If a critical pump at the bottom of a well breaks down and the cost of a new one is US$500, it may appear to be relatively inexpensive to fix. However, if the pumps are made in small quantities in the Black Forest of Germany and there is a 20-day wait before a new one is available, the real cost is not in fact US$500, but actually in the thousands of dollars in lost revenue and even more in lost customer relations. An asset management plan will give the system manager an estimate of the likelihood of failure; recognizing the importance of the pump, the manager can proactively plan by making sure a back-up replacement is readily available for replacing the pump before it fails.

The probability of failure is the sum of the following considerations: (1) the age and condition of the asset, the environmental conditions in which it operates, prior history of repairs to the asset, its maintenance history, and operational experience of other systems with similar assets and conditions; (2) the cost of operating and repairing the asset against the cost of its replacement; and (3) whether the asset is near or past its expected life span and whether major overhaul is scheduled. Table 17.6 is a rating schedule system that could be used or modified for use to record the probability of failure. All assets may be similarly rated with this scale. Table 17.7 is a rating scale for the potential consequences of an asset failure.

Table 17.6 Rating plan for probability of failure for a water system asset

Rating	Failure probability	Experience with this asset
5	Imminent	Asset (or component) fails continuously; very high likelihood of permanent failure at any time
4	Probable	Asset (or component) fails frequently; total failure expected to occur several times in life of unit
3	Sporadic	Asset (or component) has failure a few times, enough so that total failure could occur
2	Remote	Asset (or component) has never failed; however, expected life span suggests possibility of failure could fail
1	Improbable	Asset (or component) has never failed and is not likely to fail within remaining life span

Source From material in NJDOP (2016)

Table 17.7 Rating plan for quantifying potential consequences of an asset failure

Rating	Failure consequences	Level of effect on system
5	Catastrophic disruption of service	Massive system failure, severe economic, environmental, health and or social consequences; persistent and extensive damage; system unable to meet LOS objectives
4	Major disruption of service	Major system damage with loss of system capacity, major consequences or costs; LOS objectives compromised
3	Moderate disruption of service	Moderate effects from failure; loss of some system capacity; important LOS objectives still achieved
2	Minor disruption of service	Minor effects of failure; some loss of system capacity; minimal costs or impacts to system
1	Insignificant damage or disruption	Negligible effects on or cost to system; no significant consequences or impact on LOS

Source From material in NJDOP (2016)

A Combined Asset Criticality Rating

Establishing the critical state of an asset is possible by an analysis that compares the value determined for the probability of failure of an asset with a value for what are expected to be the consequences of failure. The assets with the highest probability of failure and the highest consequences from that failure are the most critical. Table 17.8 is a method for determining the combined score of the two ratings. The data gives the manager the information needed to place results of the analysis in the system's repair replacement schedule, and to include the cost of replacing the pump in the budget for the coming year or the year following. The consequences in terms of poor customer relations

Table 17.8 A multiplication method for establishing a hierarchy score for a critical asset

Consequence (cost) of failure							
			Low				High
			1	2	3	4	4
	Low	1	1	2	3	4	5
Probability of failure		2	2	4	6	8	10
		3	3	6	9	12	15
		4	4	8	12	16	20
	High	5	5	10	15	20	25

of a failure are too high to wait until the pump in well X fails before replacing it. Moreover, the well in question fills a need for redundancy in primary supply points for the water, although one pump alone cannot meet peak water demand during the high occupancy summer months. While proactively replacing the pump for this well, work on other related aspects of the pump can be scheduled. Similar criticality analyses should be done every year.

The following is an example of how this matrix can be used to determine a composite criticality score for an asset. The asset is the pump in well 2; the hypothetical data are from a small water and wastewater utility.

- Asset: the pump in well number 2
- Service history: the pump was replaced March 2006
- Probability of failure: pump has a supplier-provided life span of 15 years. It is now 12 years since it was replaced and will likely need to be replaced in the next 2–4 years.
- Consequence of failure: well 2 pump is heavily worked due to high water demand in summer months. Well 1 alone is unable to keep up with peak demand. Therefore, failure of pump in well 2 during coming summer season will require water restrictions resulting in poor customer relations.
- Criticality score = probability of failure × consequence of failure.

Life Cycle Costing

Life cycle costing (LCC) is considered by many to be the heart of asset management. The chief benefit to a water system is that it enables a schedule to be established for the maintenance, repair and replacement of system infrastructure. A schedule that depicts exactly what maintenance to do and when during the life cycle of an asset can extend the asset's life span and lower the risk of unexpected failure. At the same time, the prioritization

Table 17.9 Examples of major water system asset repair sand replacements, 15-year schedule

Year	Item	Brief description	Estimated cost (US$)	Estimate method	One-time or occurring	Time period of activity
2019	New meter electronics	For smart meter system	17,500	Supplier quote	One time	1–2 months
2020	Smart meters	Replace 1/5 of meters	15,000	Supplier quotes	Recurring	Every 3 years
2021	Water lateral pipes	Replace 500–600 ft/yr.	35,000	Past cost experience	Recurring	Every 4 years
2022	Emergency generator	Replace at well 2	25,000	Supplier estimates	One-time	6 months
2023	Smart meters	Replace 1/5 of meters	15,000	Supplier quotes	Recurring	Every 3 years
2024	Water lateral pipes	Replace 500–600 ft/yr.	35,000	Past cost experience	Recurring	Every 4 years
2025	Well 1 upgrade	Replace casing and pump	20,000	Well 2 experience	One-time	4 months
2026	Smart meters	Replace 1/5 of meters	18,000	Supplier estimates	Recurring	Every 3 years
2027	Water lateral pipes	Replace 500–600 ft/yr.	35,000	Past cost experience	Recurring	Every 4 years
2028	Water main upgrades	Refurbish 100–200 ft/yr.	100,000	Industry estimates	Recurring	Initial action
2029	Smart meters	Replace 1/5 of meters	18,000	Supplier estimates	Recurring	Every 3 years
2030	Water lateral pipes	Replace 500–600 ft/yr.	35,000	Past cost experience	Recurring	Every 4 years
2031	Water main upgrades	Refurbish 100–200 ft/yr.	100/000	Industry estimates	Recurring	Every 4 years
2032	Smart meters	Replace 1/5 of meters	20,000	Supplier estimates	Recurring	Every 3 years
2033	Water lateral pipes	Replace 500–600 ft/yr.	35,000	Past cost experience	recurring	Every 4 years

Source From concepts in NJDEP (2016)

of critical assets provides a means of scheduling and budgeting for capital improvements. There will always be competing demands for the water system's available resources. Cash flow management is simplified if the pattern of capital improvements is predictable over the long term.

The repair and replacement schedule process begins with an analysis of the estimated costs of a system or an important component over its entire live span. The operating costs that might be included in estimating an asset's costs include the cost of repairs, the cost of a failure, the costs of repairs, the cost of a new or replacement asset, and social cost of lost customer goodwill. Maintenance costs include the costs of maintaining the asset over its life span and the cost of preventative maintenance. It may also include the costs for disposal of a failed asset. Table 17.9 is an example of a basic maintenance, repair and replace schedule.

Long-Term Funding Strategy

With the need to repair, replace and expand, water and wastewater infrastructure has expanded greatly, resulting in the need for billions of dollars. A plan behind the development of asset management was to giver water and wastewater system managers a method for identifying and programing the funds needed for the repair and replacement of the deteriorating aged utility infrastructure most had installed 50–100 years earlier.

The developers of the asset management system recognized that a funding strategy would be necessary for supporting the management decisions that would make it possible for systems to achieve sustainability. The funding strategy should be based on estimates of the system's annual spending on operations and maintenance, for repairing and replacing existing infrastructure and for the infrastructure expansion needed to supply the water needs of a growing population, and for repairs and upgrades of water treatment facilities to keep up with safety and environmental standards. These estimates had to be balanced by the funding resources available to the system.

The funding resources available to water and wastewater systems fall into two major funding categories: external sources and internal sources. Loans and grants are the two major external sources. Most water utility loans funds are the Drinking Water State Revolving Fund (DWSRF) created by the 1996 amendments to the Safe Drinking Water Act (SDWA), a loan program funded by the federal government and administered by each state. The low interest rate revolving funds are managed by each individual state. Most of these state funds are only provided for capital projects.

The U.S. Department of Agriculture's rural development program administers a loan and grant program for funding clean and reliable drinking water systems, sanitary sewage disposal, sanitary solid waste disposal, and storm water drainage to households and businesses in eligible rural areas. Funding for many municipal utility infrastructure capital projects is provided by bonds. Most internal funding resources are from water utility rates and fees and these are usually enough for the operation and maintenance of the system. Rate structures based on full-cost pricing also include all anticipated costs for scheduled repair and infrastructure replacement and for debt service. An annual contribution to an emergency operating reserve is increasingly included in the budget structure.

Summary

Change efforts in organizations often achieve only partial results or fail to achieve desired results. An asset management program is a change that many small water suppliers have difficulty implementing. Involving all water-supplier personnel in the planning process has been shown to go far in overcoming this barrier. Implementing an asset management plan is fast becoming a required preliminary step for acceptance in a revolving capital improvement loan. Elements of the process were described in some detain in this chapter to illustrate the type of information needed for scheduling the repair and replacement of potentially failing infrastructure.

An asset management plan is designed to help a water system to reach the goal of transitioning to a point where it can sustainably provide the LOS it can afford to provide its ratepayers. Good asset management ensures that the infrastructure needs to meet those operational goals are met at the lowest possible life cycle cost commensurate with required sustainability.

Water systems that repair and replace their assets according to a schedule that omits costly unanticipated failures are more likely to provide safe and reliable drinking water for their customers, and save money doing so. Managing a water or a wastewater utility can be greatly improved simply by going through the steps required for producing a successful asset management plan.

The asset management plan is the first step on a pathway to sustainability. The chief benefit of operating a system according to asset management principles is achieving the ability to maintain a reliable service in the face of unexpected and uncontrollable emergencies. Successful implementation of asset management will go far in a water system achieving sustainability even when confronted with the greater number of and more severe natural disasters associated with climate change.

References

EPA (Environmental Protection Agency). 2008. Effective utility management: A primer for water and wastewater utilities. Accessed August 19, 2018 from https://nepis.epa.gov/Exe/ZyPDF.cgi/P10053BJ.PDF?Dockey=P10053BJ.PDF.

NJDEP (New Jersey Department of Environmental Protection). 2016. Asset management guidance and best practices. Accessed July 28, 2018 from https://www.nj.gov/dep/watersupply/pdf/guidance-amp.pdf.

NMEFC (New Mexico Environmental Finance Center). 2006. *Asset Management: A Guide for Water and Wastewater Systems*. Socorro, NM: New Mexico Institute of Mining and Technology.

WRF (Water Research Foundation). 2014. Performance benchmarking for effectively managed water utilities. Accessed August 19, 2018 from www.waterrf.org/publicreportlibrary/4313b.pdf.

WSDOH (Washington State Department of Health). 2009. Asset management for small water systems. Washington State Department of Health. Retrieved June 5, 2018 from www.doh.wa.gov/Portals/1/Documents/Pubs/331-445.pdf.

18

Planning for Achieving Sustainability

Achieving sustainability usually requires implementing a transformational change in the way in which the system is governed and managed. Making that change begins with planning. Water managers must know who are and what they want to be. They are suppliers of a public service and they should want to be the best supplier of clean, safe water at the most affordable price possible. However, eagerness to accept and pay for change is not a widespread characteristic among many water ratepayers. The persons and organizations that depend upon water and wastewater services can often have very different attitudes toward change than those held by elected commissioners. Also, systems facing the need for implementing and overseeing the changes may not have the resources needed for assessing the state of the current infrastructure, predicting future needs, evaluating existing repairs and programming regular repairs and replacement of assets, and reassigning the resources to meet the needs.

General Water Planning

The path to sustainability is long and difficult, strewn with barriers of increasing regulations, rising costs, aging infrastructure, growing demands for water, and the effects of climate change. If they are still focused only on day-to-day operations and are not already changing to this new way of functioning, water utility managers must get moving. Nobody and no outside intervention is going to remove the barriers for them. If they are going to get around these

D. E. McNabb, *Global Pathways to Water Sustainability*,
https://doi.org/10.1007/978-3-030-04085-7_18

barriers and move forward on the path toward the golden goal of sustainability, water system managers need to start planning the steps they are going to take to get there. It is not going to be easy, particularly for the thousands of small- and medium-sized water systems, wastewater systems or combined operations. Planning with sustainability as the paramount goal will help water system managers to identify what they need to do and which pathway to follow to get there. Water industry leaders have long known that comprehensive planning is one of the utility's most important responsibilities.

General Utility Planning (GUP) consists of two distinct but related areas: operations planning and capital improvement planning. An abridged model of the combined scope of the overall planning program is shown in Fig. 18.1. Operations planning is short-term planning of the tasks for achieving specific objectives. The operational plan includes the tactics that the utility will take in implementing the strategies included in the strategic plan. Examples of the work activities for which operational plans are prepared include billing meter reading and general infrastructure maintenance, testing and other laboratory procedures, and other operations and administrative functions.

The strategic plan is different to operational planning, mainly because it is a long-term plan that does not identify any specific guides for employing resources. Rather, it is an explanation of the strategies that the utility will follow to achieve its goals over the next five, ten or twenty years. The objectives in the operational plan are results-oriented, specific, time-framed and measurable. The goals in the strategic plan state the guiding strategies that are seldom changed.

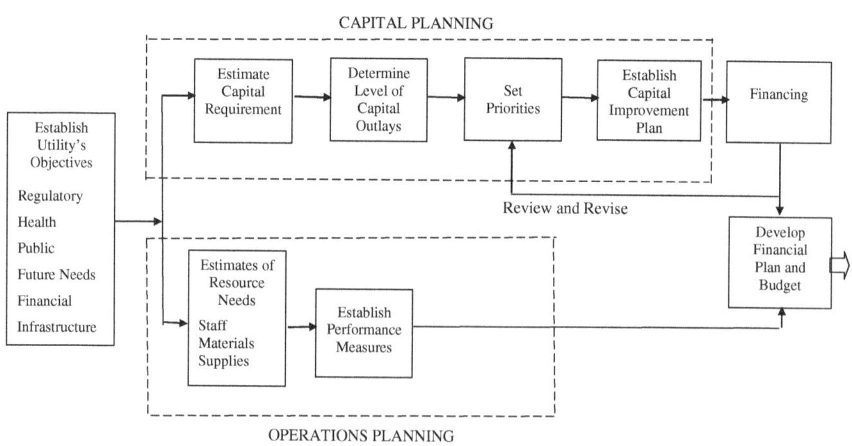

Fig. 18.1 Overview of utility general planning processes (*Source* After material in EPA 2016)

Capital planning focuses on identifying and planning the funding for the utility's capital needs over a five, ten or twenty-year period. It includes financial and facility planning for maintaining and expanding the physical facilities, including water mains, storage capacity, treatment equipment, buildings, vehicle needs and other major equipment.

Strategic Planning

The journey to water sustainability begins with a plan that reveals which barriers to avoid, which to go around, and which, if any, to follow. Strategic Sustainability Planning (SSP) is one such plan. SSP consists of a combination of elements from three planning models; the combination adopts elements of strategic planning, asset management planning and financial planning. There is no one best way to combine these tools, or which section or tool to emphasize. They all begin with making sure that you know who you are, where you want to go, and how you plan to get there. This preliminary work is known as situation analysis, other times it is referred to as SWOT analysis (for: strengths, weaknesses, opportunities and threats). It is defined as a systematic analysis of the internal and external factors that shape an organization, the capabilities, customers and competitors within the organization's current and projected environment and its financial viability.

Key Elements of a Strategic Plan

Regardless of how the strategic planning process is applied or the rationales for its implementation, there is general agreement of a set of from five to eight key elements that belong in every such plan. Strategic planning includes these key elements:

1. Adoption of the organization's mission, its vision, and its core values
2. Internal and external analysis (SWOT analysis)
3. Identify long-term goals
4. The annual or short-term objectives on the path to achieving its goals
5. Action plans for each objective
6. Schedule for progress measurements
7. Quarterly progress analysis and revisions where necessary
8. Annual progress report and preparation for the next year's plan.

Water systems large and small define their strategic plans in ways that are important to them. The following examples illustrate the diversity in the ways water utilities describe the strategic planning process. The Colorado Springs, Colorado utilities department describes its strategic plan this way: "Our strategic plan is an integral tool to guide us toward sustaining fiscal responsibility, planning for long term risks, and identifying opportunities to attract and satisfy our citizen owners and valued customers. A sound plan puts us all on the same page so that our work is focused, productive and allows us to deliver on our mission: 'to provide safe, reliable, competitively priced electric, natural gas, water and wastewater services to the citizen owners and customers of Colorado Springs Utilities'" (CSU 2018).

The Seattle, Washington public utilities describes the strategic plan for its water division in the following way: "Seattle Public Utilities provides reliable, high-quality drinking water to 1.4 million people who live in Seattle, King County and portions of southern Snohomish County. Thoughtful planning and strategic investment in our drinking water supplies and infrastructure will help ensure our region has enough water to meet the needs of our customers for decades … The 2019 Water System Plan … describes near- and long-term plans for the regional water system. It also focuses on updates to the water system and programs since the 2013 Water System Plan was completed. The Plan describes how Seattle Public Utilities: meets current and future water demands, ensures high-quality drinking water, enhances system reliability and resiliency, and cost-effectively invests in and maintains the water system" (SPU 2018).

The Columbus, Ohio Water Works (CWW) describes how it uses its strategic planning process and its vision this way: "Columbus Water Works uses the strategic planning process as a management tool to help focus our efforts and energy to meet our organizational performance goals. Our commitment to this endeavor not only ensures all members of the CWW team are working towards the same objectives, but it allows us to continuously evaluate and assess our progress in response to a volatile, uncertain, and ever-changing environment. For CWW, the emphasis placed on strategic planning at all levels in the organization allows us to provide the highest level of quality customer service while ensuring our operational activities support our purpose and vision … As a premier leader in the water and wastewater industry, our strategic focus in the upcoming years is providing exceptional customer service, cultivating a work environment that ensures employee engagement, implementation of new technologies to support operational cost-efficiencies, protecting the environment to ensure the sustainability of our natural resources for future generations, and strengthening our community and regional partnerships" (CWW 2018).

The 2013 AWWA Gold Award winner, City of Raleigh (NC) PUD describes how it uses the strategic plan and the planning process this way: "A strategic plan is developed to help define the future direction of an organization, creating a road map for making resource allocation decisions that take us where we want to go. With many challenges and limited resources, we make investment decisions for the utility every day. The strategic plan will help us decide how to optimize our investments in our people, work processes and the tools needed to do the job of providing our customers with world class service … The Public Utilities Department (PUD) is faced with many challenges. Ongoing reductions in per capita water consumption and the relatively high fixed cost of running the utility have resulted in challenges generating enough revenue to run the utility, even with a series of significant rate increases and more than 10% of our positions vacant. The economic downturn and the accompanying reduction in revenue from new development has also significantly affected our revenue. The bond rating agencies in turn have increased the creditworthiness requirements necessary to maintain the AAA credit rating, also adding higher revenue requirements. Finally, we will have even higher revenue needs in the future to address aging pipeline repair and replacement costs. In June 2010, the City Council proposed creating a Task Force … to advise on conforming the water utility's business model to reflect a sustainable Strategic Plan" (CRPUD 2013).

The Mission Statement

Planning for achieving sustainability and for overcoming the barriers that stand in the way begins with organizational agreement on statements pertaining to the utility's mission, its vision and its values. Only then will the assessment of the system's current strengths and weaknesses, opportunities and threats to sustainable service make sense. This is followed by development of a plan for making the water system strong enough and organizationally capable of overcoming the barriers of climate change, population growth-driven water demand and the need for repairing and replacing failing infrastructure. Adopting a sustainability-focused operation cannot begin until the organization's sustainability goals are identified, attainable objectives for achieving the goal or goals are defined, and the strategies designed for achieving the objectives are agreed upon. Only then can the necessary tasks be assigned and scheduled.

Public water suppliers in many parts of the world—including the USA—are confronted with a litany of formidable challenges as they maintain and expand their infrastructure systems for acquiring, treating and supplying such services as clean and safe freshwater to their customers. Similar barriers block sustainability in efforts to collect and treat all wastewater, and to manage stormwater runoff. These problems have intensified as pressure from population growth, urbanization and extreme weather events. System breakdowns have increased along with more and more devastating floods. New challenges to utility operations have emerged for coping with these system breakdowns. The effects of climate change and escalating costs of energy, labor and supplies are among the challenges utilities face in their commitment to providing safe and reliable water supplies and wastewater treatment services.

As the many barriers to sustainability discussed in earlier chapters converge, it has become clear to many utility managers that their operations must focus not just on meeting current needs, but instead now require planning in order to ensure that they have the resilience to overcome more and more severe extreme weather events. In planning for sustainability, they must also consider that changing operations must not curtail their responsibility for acting as stewards of the environment and communities they serve. To adjust their planning to include strategies to prevail over these barriers, many small- and medium-sized independent water and wastewater utilities must now engage in long-range strategic planning that goes beyond every system's five-year capital improvement plan. The previous chapter focused on the planning for asset management. This chapter incorporates asset management's contribution into the long-term strategic planning for implementing a strategy of resilient and sustainable operations for water and wastewater utilities.

Establishing a Vision

Establishing a vision for what you believe the water or wastewater utility will look like at some agreed upon date in the future is perhaps one of the most difficult tasks in the strategic planning process. For a commercial business, that future date could be any time from three to five years in the future. However, water utilities function on a longer time span. Infrastructure projects can take five or more years before they are implemented. Thus, strategic planning for a water utility is often for even longer time periods. A vision statement is a statement of what management wants the utility to be at a time from 10 to 20 years in the future.

Strategic planning is problematic because no one can predict with total accuracy the impact of all the future forces in the water environment. Too many both internal and external uncontrollable forces will have an influence on the outcome. A vision is what the utility's administrators and managers *want* the utility to be in 10 or 20 years in the future, not what it *will* be. The strategic plan includes the decisions that the leaders will make for that vision to come to pass. The long-term operational decisions for achieving that vision of the future are, in a sense, based on the best guess of what the administrators and managers believe will happen, and how the decisions they make now will be affected by what they believe that the future environment will be like. What they can be assured will happen is that changes will occur, and they must be ready to react positively to those changes. Changes will vary in many ways, including the following:

- The regulatory environment changes that utilities will have to do to comply with are influenced by unforeseen political pressures.
- The utility's service requirements in their community will either be greater as a result of population growth or far less because of outward migration.
- The population in most of the developed world is aging and along with this change in demographics are changes in water consumption patterns.
- Climate change-influenced changes in weather patterns will result in more water for some regions and less for others; weather extremes will certainly occur more often.
- The economy will either continue to grow or decline; whichever happens will have different impacts on operational costs and customers' willingness to absorb rate increases.
- Water infrastructure will continue to age, and more will have to be replaced. How much and when they will have to be replaced is hard to predict with any degree of certainty.

Core Values of the Utility

A water utility's core values are what the management believes in and how it behaves. They can be listed as a few or concepts or spelled out in some detail. The following abbreviated value statements are sections of the mission, vision and values statement for the Eagle River Water and Sanitation District in Vail, Colorado (ERWSD 2015).

- **Service**: Our core purpose is to provide our customers with safe and reliable water and wastewater services, in compliance with regulatory requirements.
- **Accountability**: Our organization is the steward of public funds collected to provide a defined service to our constituents. We manage these funds responsibly, fairly and with integrity and believe that our constituents are best served through continuous improvement and value enhancement.
- **Customer Confidence**: We foster customer confidence through clarity, honesty and transparency in all communications and work to educate our customers, community partners, and employees on the issues pertinent to responsible water resources management. **Community Partnership**: We forge community partnerships for more efficient use of public funds, to gain a greater understanding of our constituency, and foster the public's continued enjoyment and recreational uses of our rivers.
- **Personnel Dedication**: We respect and value the contributions of all our personnel. **Environmental Stewardship**: We are committed to safely complying with all applicable regulations to preserve the environment, esthetic beauty, and economy in and around our service area.
- **Leadership and Continuous Improvement**: We champion organizational initiatives that support our vision, set a higher standard for excellence, and are consistently aligned with our values.
- **Long Range Planning and Preparation**: We challenge ourselves to meet the evolving needs of the future through foresight, solid and specific planning efforts, and maintenance of organizational flexibility. We prepare for the unforeseen through commitment to safety, security and emergency preparedness.

Mission, Vision and Values for a Small Water Utility

The place to begin planning for sustainable operations is to ensure there is internal organization agreement with, and external stakeholder awareness of, what the organization does and what it stands for. An example of how a water utility must deal with change is the situation administrators of a nearly 60-year-old island vacation recreational community found themselves confronted with. Their community was slowly but steadily morphing into a retirement community with older permanent residents. A new elected board of three commissioners was faced with guiding the water and wastewater utility through the transformation. The board of water and wastewater decided it was time to change from a contracted management system to a full-time employed staff. With that change under its belt, it began preparing an asset management plan.

That plan would then become a major element in the district's first strategic sustainability plan. The process began with a consensus agreement on statements of district mission, vision and its values. A preliminary draft of these statements is shown in Box 18.1. Preliminary elements of the asset management plan were described in Chapter 17.

Box 18.1 Draft mission statement, vision and value statements of a small water system

Mission statement
The Island Water and Sewer District provides its customers a reliable supply of high quality, safe, drinking water as well as the essential collection, treatment and discharge services of wastewater for all its customers at the lowest reasonable rates consistent with economically sound and environmentally responsible practices while complying with all federal, state and local rules and regulations.

Vision statement
The district will achieve sustainability with all infrastructure replacements and improvements necessary for meeting the safe drinking water supply and wastewater service requirements required by this and future generations in a sustainable recreational and residential community.

Our core values

- Reliability: We are commitment to maintaining reliable infrastructure and cost-effective utility services for safe drinking water and solid waste collection, treatment and disposal.
- Integrity: We maintain financial stability and strength while continually improving performance.
- Sustainability: We always work towards achieving sustainability in all of our services.
- Open communication: We willingly share operational information, thoughts and concerns with all our customers.
- Customer service: We believe in providing the highest level of customer service commensurate with the available resources.
- Cooperation and collaboration: We are always open to suggestions on best practices information and innovation from our staff and other experienced water professionals.
- Valued workforce: We believe in supporting continual education and training for all workers.

Situation Analysis

With agreement on the utility's mission, its vision and core values guiding further planning, it is time to begin a thorough analysis of the current practices and operating environment of the water utility. This step is commonly referred to as a SWOT analysis, to ensure recognition of the utility's strengths, its weaknesses, and the opportunities and threats that allow or hinder the achievement of its sustainable operation goals. The analysis must not be unwilling to admit its weaknesses; actions for converting weaknesses to strengths is what objectives are for. Other ways of conducting the analysis are to focus on understanding the utility's political, economic, social and technological (PEST) environments and their ability to hinder or help in achieving the utility's goals and objectives. An abridged example of the key findings of a SWOT analysis for a small water and wastewater district is shown in Table 18.1. Once the SWOT factors are identified, leadership can establish goals that enable it to build on its strengths, overcome the weaknesses, and to use its strengths to take advantage of opportunities while mitigating threats.

Table 18.1 Results of a hypothetical SWOT analysis for a small water and sewer utility

Internal factors	
Strengths	**Weaknesses**
Existing staff is dedicated, supportive and committed to cross-training for continuous improvement of their water and wastewater knowledge and skills Utility has recently installed new US$1.3 million water filtration system, added a fourth well and refurbished another well Customer base is supportive and acceptant of the periodic need for rate increases for infrastructure repair and replacement Utility has a sound financial status with highly favorable state audits Utility has reputation for fast, friendly and helpful customer service with water and wastewater issues Prior leadership and staff have produced good capital improvement plans	Recent staff turnover has resulted in new leadership and office staff Organizational structure is lean, resulting with recurring need for part-time workers or outside contract assistance Utility has never produced a strategic plan Utility is only beginning its first asset management plan Maintenance is often more reactive than proactive Much of the existing infrastructure is at or nearing its expected lifespan Utility has high debt ratio, precluding low probability of future borrowing Dependence upon off-site laboratory service delays necessary results and reports
External environment	
Opportunities	**Threats**
Utility has a growing, financial stable ratebase, accustomed to the need for periodic rate increases Staff has good relationships with all regulatory agencies Asset management plan and maintenance schedule preparation underway New leadership and permanent staff with cross-training and dual certification reinforces operational reliability	Increases in rate of infrastructure failures expected New regulatory requirements for drinking water Heavy seasonal water demand periodically approaches authorized water rights High arsenic levels in some wells Off-site alternative water supply unavailable

Establishing Goals and Operational Objectives

In 2008, the EPA published a widely used guidebook to help water and wastewater utilities establish their long-term goals and the operational objects for achieving those goals: *Ten Attributes of Effectively Managed Water Sector Utilities*. There are still applicable and are presented below. The attributes were developed by the AWWA and water professionals and have been shown to be applicable to both large and small water and wastewater utilities. They provide a framework for planning for operations, infrastructure repair and replacement, customer service, community relations, environmental stewardship, and financial performance:

1. Product quality: Produces potable water, treated effluent, and process residuals in compliance with regulatory and reliability requirements and consistent with customer, public health, and ecological needs.
2. Customer satisfaction: Provides reliable, responsive, and affordable services in line with explicit, customer-accepted service levels.
3. Stakeholder understanding and support: Encourages understanding and support from regulatory and administrative agencies, community and watershed interests for service levels, rate structures, operating budgets, capital improvement programs, and risk-management decisions.
4. Employee and leadership development: Recruit and retain a workforce that is competent, motivated, adaptive, and safe-working, and provides a focus on and emphasizes opportunities for professional and leadership development and strives to create an integrated and well-coordinated senior leadership team.
5. Operational optimization: Ensures ongoing, timely, cost-effective, reliable and sustainable performance improvements in all facets of its operations; minimizes resource use, loss and impacts from day-to-day operations; and maintains awareness of information and operational technology developments to anticipate and support timely adoption of improvements.
6. Financial viability: Understands the full life-cycle cost of the utility and establishes and maintains an effective balance between long-term debt, asset values, Operations and Management (O&M) expenditures, and operating revenues.
7. Infrastructure stability: Understands the condition of and costs associated with critical infrastructure assets and maintains and enhances the condition of all assets over the long term at the lowest possible life-cycle cost and acceptable risk consistent with customer, community and

regulator-supported service levels, consistent with anticipated growth and system reliability goals.

8. Operational resiliency: Ensures utility leadership and staff work together to anticipate and avoid problems.
9. Community sustainability: Is explicitly cognizant of and attentive to the impact its decisions have on current and long-term future community and watershed health and welfare.
10. Water resource and disposal capacity adequacy: Ensures water availability consistent with current and future customer needs through long-term resource supply and demand analysis, conservation, and public education, and explicitly considers its role in water availability and manages operations to provide for long-term aquifer and surface water sustainability and replenishment and residuals and biosolids disposal capacity.

The guide recognizes that each utility will have what it believes are its own most important performance targets. The five-step process illustrated in Fig. 18.2 is a model of the steps for a utility to follow when making these decisions. The objectives to choose will reflect the utility's own strategic objectives, priorities, and the needs of the community it serves.

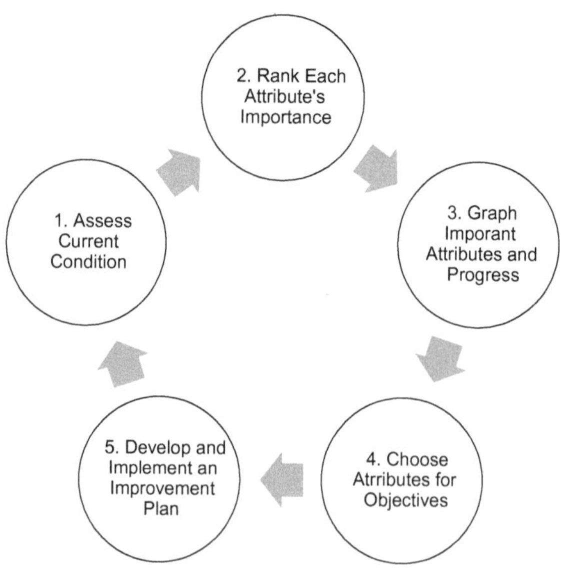

Fig. 18.2 Assessing and selecting which trait to focus on (*Source* EPA 2008)

Ranking an attribute's importance can be done using a five-point scale, beginning from the current level of achievement for each attribute. The most important of the attributes that are selected, together with the utility's strengths may then be chosen for inclusion in the annual improvement plan. An example of how certain items can be selected for specific action and follow-on performance measurements on achieving the goal are included in Fig. 18.2.

Formulate Key Strategic Goals

In 2011, water department staff and a water advisory board of the City of Boulder, Colorado published the final draft of its Water Quality Strategic Plan for 2009, in which it listed the following five strategic goals.

Goal 1. Provide safe and high-quality drinking water.
Goal 2. Manage pollutants from wastewater and other point sources.
Goal 3. Manage pollutants from stormwater and other nonpoint sources.
Goal 4. Protect, preserve and restore natural water systems.
Goal 5. Conserve water resources.

The recommended strategic goals for water quality are shown in Table 18.2.

Preparing the Finance Plan

The last step in the draft strategic plan is the preparation of a finance plan to ensure that the money is available to carry out the tasks identified in the work schedule for achieving the objectives. Financial management includes developing financial plans and budgets, plans for recovering costs, monitoring and updating accounting and financial information, and participating in general planning activities.

The financial plan is considered to be an important element in the strategic planning process because it functions as a roadmap for identifying the utility's financial requirements over the long term. The key to utility sustainability is for the utility to be self-sustaining and financially self-sufficient. Up to this point, the financial plan includes a revenue-generating system that ensures recovery of the total cost of operations. Funding for capital improvements are typically provided by either grants or long-term, low-interest loans.

Table 18.2 Strategic actions and performance measures for water quality

	Recommendation	Performance measures
1.	Evaluate and update policies in the Boulder Valley Comprehensive (BVC) Plan to incorporate water quality goals	Adoption of the five water quality goals into the BVC plan in 2010
2.	Perform analysis on city plans, policies and projects to identify gaps in meeting water quality goals	All policy and master plans reviewed and updated to incorporate water quality goals; development of a matrix of water quality indicators and evaluation of plan review based on criteria
3.	Develop annual work plans and water quality reports	Development of program work plans and annual reports
4.	Prepare for future water quality regulations	Develop and implement a program to conduct special studies, track new and proposed regulations and participate in rule-making

Source City of Boulder (2011)

The budget, a major component of the financial plan, functions as a controlling device. It gathers financial and performance data by all aspects of the operation, allowing for detailed budget reviews, comparing performance levels to the planning objectives, and providing the utility management with information for decisions about the cost and level of service provided.

Utility Sustainability Plan

The EPA has undertaken a series of studies in which the need for water utilities to plan for sustainability is encouraged. One such study, the 2012 report on Planning for Sustainability, was developed to serve as a step-by-step guide to lead water and wastewater utilities through the planning process. The importance of planning was clearly spelled out in the report's introduction:

> Water utilities typically have a long-term planning horizon and long-term infrastructure operation and maintenance commitments. The costs and potential benefits of investment decisions will be realized over a long period of time. Accordingly, the EPA's Sustainability Policy calls for drinking water and wastewater systems to undertake "robust and comprehensive" planning to

ensure that water infrastructure investments are cost-effective over their lifecycle, resource efficient, and consistent with other relevant community goals. Throughout this Policy, EPA emphasizes the important relationship between utility and community sustainability. The core mission of water sector utilities is to provide clean and safe water in compliance with all applicable standards and requirements at an affordable price in order to protect public health and enhance the economic, environmental, and social sustainability of the communities they serve. (EPA 2016)

Preparing for Sustainability Planning

Functioning as prelude to preparation of a strategic plan are a comprehensive asset management plan and a policy and practice of engagement with customers and the community served by the water system. The asset management plan described in the previous chapter provides an inventory, condition and criticality of the utility's infrastructure, a risk-based schedule for maintenance and repair or replacement, and access to a financial plan for ensuring the capital for these activities. Without the support of ratepayers and the community, the necessary rate increases are difficult to achieve—making sustainability nearly impossible to achieve. The many published success stories available on the Internet reveal that while large utilities are more likely to use the strategic planning process as a vehicle for engaging the community and their ratepayers, smaller and rural water systems with as few as 50 connections are have also begun to do so.

Elements of Sustainability Planning

The EPA identifies these four key components in a sustainability plan: goal setting, objective setting and attainment strategies, alternatives analysis, and a financial strategy. An example of a goal for a combined water and wastewater system might be: "The utility will reduce the stormwater I&I into the wastewater treatment system that results of discharge of untreated sewage." I&I—the Inflow and Infiltration of groundwater and stormwater into wastewater or sanitary sewer systems—can overload the ability of the utility to collect and treat normal sewage inflow, resulting in the release of untreated sewage into the environment. Sewage overflows have occurred in many locations during heavy rainfall, events which often result in costly fines for the utilities and damage to the environment.

Planning Step 1: Setting Sustainability Goals

The sustainability planning process begins with the utility setting the sustainability goals that emerge from an internal analysis of the barriers to sustainability in the community it serves and the strengths and weaknesses that emerge from a SWOT analysis. A selected list of typical sustainability goals included in the EPA's planning for sustainability guide is presented in Table 18.3.

The first step in sustainability goal setting is to conduct an internal analysis to identify the priorities and potential opportunities, infrastructure and operational strengths that can enable the utility to achieve that goal. Equally important is a vulnerability assessment. The second step is determination of the sustainability priorities of the community served by the utility. Once these are known, the third step—coordinating the utility's sustainability

Table 18.3 Potential water and/or wastewater utility sustainability goals

Goal	Examples of potential goals
Improve compliance	Work with regional laboratories and regulatory agencies to improve state-wide network of water supply test results
Reduce overall infrastructure costs	Investigate availability of innovative methods for finding and repairing leaks without pavement excavation
Extend projected adequacy of current water supplies	Design and implement water conservation program and investigate installation of new water meters
Reduce stormwater impacts	Implement green architecture test installations
Enhance community livability	Target water infrastructure projects to support community infrastructure and encourage redevelopment
Reduce long-term operation costs	Where possible, use natural treatment systems such as functioning wetlands
Improve operational resilience	Begin planning for sustainable operations during extreme weather events
Reduce vulnerability of water supply	Invest in real-time water quality monitoring, install new isolation shutoff valves
Ensure a safe, sustainable workforce	Implement safe workplace changes, knowledge retention and new knowledge through regular training
Reduce energy costs	Invest in more energy efficient equipment

Source After material in EPA (2008) with local content

activities with those of the community and the relevant stakeholders—can result in a synergistic effect that benefits both bodies. The fourth step is documenting sustainability goals so that they are available to guide following strategy and objective setting.

Planning Step 2: Setting Objectives and Strategies

Once a goal is identified, an analysis of the existing actions in that area should be evaluated and a baseline set. Objectives can then be set for the selected activities to be employed for moving forward from the baseline; for example, a utility might have identified a goal of forging a stable, reliable and knowledgeable long-term management team. The baseline would be analysis of the existing operating structure, which in this case might be operating under a contract management system with little say in planning for the future. An objective could be to conduct a search of other regional utilities of the same or similar size and experts at the state professional water management association. Once that data is collected, a second objective would be to select a management structure with a paid permanent general manager with the certifications and the skills necessary for achieving the utility's sustainability goal. Thus, achieving the goal of a more efficient and stronger management system includes sustainability considerations into the setting of the objective.

An operations example of a measurable objective and strategy for achieving the goal of improving the way the utility manages stormwater to lessen the likelihood of flooding from expected stronger storms might be stated this way: "The utility will reduce total I&I inflow to the sanitary system by 30 percent by 2025." The utility can adopt different strategies to achieve this goal. Inflow is stormwater that enters into sanitary sewer systems from drains, manholes and other access points; infiltration is groundwater that enters sewer systems through cracks and/or leaks in the sewer pipes (Global Water 2011). Sources of inflow include roof or foundation drains, downspouts, other architectural structure drains including outdoor basement stairwells, parking lots, roads and driveways, basement sump pumps, and even urban creeks and streams. Often, these sources are improperly or illegally connected to sanitary sewer systems. When stormwater combined with inflow from improper connections from sources other than household or business fixtures and drains enter the sewer system, it quickly contributes to the treatment plant overflow.

Planning Step 3: Alternatives Analysis

Analysis of alternatives to solutions to a strategy for achieving the objective calls for cost–benefit analysis and explicit and consistent strategy selection criteria. An initial action for identifying and measuring the status of the sewer mains could be the installation and monitoring of flowmeters that record limitations to increased flow in sewage lines during storms. Among the alternatives for achieving the objective is the increasingly important strategy of achieving this objective by green infrastructure. Since the 1960s, water and wastewater utilities have been expected to contribute to environmental protection in their communities and, since the 1990s, they have been encouraged to prepare strategic plans for ensuring sustainability. Environmental protection options include catch basins, urban green areas and permeable roads and parking lots, and facilities for collecting stormwater that allow it soak into the ground and recharge an underlying aquifer. Another strategy might be to use a video camera to inspect sewer mains for breaks and for intrusions of root blockages that restrict the higher flow during storms.

Analysis of the cost of a sustainability action for correcting the problem immediately or at a later period should include the social and economic costs that could occur from an extreme weather event. The two main sources of funds for operations and infrastructure repair, replacement and expansion are utility rate payments and borrowing. The traditional approach is for repairs and replacement to be paid for by current users through rate revenues; costs for expansion are paid for by loans that are paid for by new users. Repairing less critical I&I (inflow and infiltration) points from breaks and removing root blockages could be scheduled for repair at regular periods. Large repairs and replacement programs such as changing combined sewer and storm drainage systems into separate infrastructure must be included in the capital improvement plan.

A scorecard approach is often used when analyzing alternative actions. Two alternatives are scored using the same criteria and numeric scale. The alternative with the most favorable score for benefits and for costs is the preferred choice.

Summary

The combination of increasing scarcity of freshwater supplies in many parts of the world, the need for greater food supplies, changing precipitation patterns, pollution of much of the world's dwindling surface water resources, and the growing flood of wastewater resulting from an urbanized society has brought about greater emphasis on planning for sustainability. To get to scarcity planning, water managers must expand the context of their GUP. What had for a long time been a local problem of acquiring, treating and distributing drinking water and providing sanitation service has now become a global challenge that cannot be ignored. The combined effects of these barriers to sustainability have made it a global water-scarcity problem. Finding additional and alternative water supplies is a salient sustainability goal—and a global necessity.

Planning for sustainability begins with establishing a framework for decision making. GUP provides this framework. A comprehensive strategic plan can then be used to build on the information gathered in the general planning elements by helping water managers identify what they will do in the future, who will do it and how it will be paid for. Strategic planning is the critical intermediate step in sustainability planning. Sustainability planning then allows the water manager to focus efforts and resources on actions that contribute most to achieving the goal of a sustainable water supply, treatment and delivery system.

References

City of Boulder. 2011. Water quality strategic plan 2009. Accessed August 29, 2018 from https://bouldercolorado.gov/water/water-utility-master-plan.

CRPUD (City of Raleigh Public Utilities Department). 2013. Strategic plan and change. Accessed August 24, 2018 from file:///C:/Users/David/Downloads/StrategicPlanforWeb%20(1).pdf.

CSU (Colorado Springs Utilities). 2018. Strategic planning. Accessed August 24, 2018 from www.csu.org/pages/strategic-planning-r.aspx.

CWW (Columbus Water Works). Strategic plan. Accessed August 24, 2018 from www.cwwga.org/plaintext/aboutus/strategicplan.aspx.

EPA (Environmental Protection Agency). 2008. Effective utility management: A primer for water and wastewater utilities. Accessed August 19, 2018 from https://nepis.epa.gov/Exe/ZyPDF.cgi/P10053BJ.PDF?Dockey=P10053BJ.PDF.

EPA (Environmental Protection Agency). 2016. Planning for sustainability: A handbook for water and wastewater utilities. Accessed November 7, 2017 from www.epa.gov/sites/production/files/2016-01/documents/planning-for-sustainability-a-handbook-for-water-and-wastewater-utilities.pdf.

ERWSD (Eagle River Water and Sanitation District). 2015. Mission, vision and values. Accessed August 30, 2018 from www.erwsd.org/about/mission-vision-values/.

Global Water. 2011. *Inflow and Infiltration*. New York: Xylem Inc. Accessed August 25, 2018 from www.globalw.com/support/inflow.html.

SPU (Seattle Public Utilities). 2018. 2019 Strategic plan. Accessed August 24, 2018 from www.seattle.gov/util/Documents/Plans/Water/WaterSystemPlan/index.htm.

Glossary

Aquifer Large body of permeable or porous material situated below the water table that contains or transmits groundwater.

Biofuels Solid and liquid fuels from a biomass source, such as wood, waste, hydrogen gas and alcohols.

Biomass Plant matter such as trees, scrap wood, grasses, agricultural crops or other biological material; a renewable fuel source for generating electrical power, a transportation fuel, or as chemical feedstocks.

Catchment area A hydrological space in which all precipitation that falls and does not evaporate eventually ends up in the same river or lake. Catchment areas are separated from one another by watersheds.

City The UN does not recognize any universally accepted definition for city. Different countries, regions and scientists use different definitions. For this application, a city is an urban concentration with a population of more than 10,000 residents.

Climate change Significant changes in measurable climate factors such as rain, temperature, wind, or precipitation, that extends for decades or longer.

Cubic meters One cubic meter equals 264.17 US gallons; 1 billion cubic meters equals 264,172,051,241.56 US gallons.

Desalination Reducing the contents of total dissolved solids or salt and minerals in sea or brackish water.

Deforestation Converting forested lands to other uses, such as agriculture, settlement, or other non-forested uses. Deforestation is considered a major cause of climate change.

Evapotranspiration The process by which surface soil loses moisture by evaporation or by absorption by plants then lost from transpiration.

D. E. McNabb, *Global Pathways to Water Sustainability*,
https://doi.org/10.1007/978-3-030-04085-7

Floods A hydrological event that occurs when excessive water from any source that temporarily covers surface areas that are normally dry. When floods occur in urban areas, they can cause enormous damage and loss of life.

Freshwater Water with less than 1000–2000 parts per million (ppm) of dissolved salts.

Geothermal energy Electric power generation using naturally heated gases, steam or water extracted from within the earth.

Global warming A long-term trend of a continuing average increase in the atmospheric temperature measured near the surface of the earth and the troposphere.

Green roofs Building roofs that are all or partially covered with vegetation planted over a waterproof layer. Green roofs may be used to reduce surface runoff or for rainfall harvesting.

Greenhouse gases A group of gases that include but are not limited to water vapor, carbon dioxide, nitrous oxide, chlorofluorocarbons, ozone, hydrocarbons, sulfur dioxide and others. These gases absorb atmospheric infrared radiation.

Groundwater Water that is below the land surface in pores or crevices of soil, sand and rock, contained in an aquifer.

Groundwater Water below the surface of the earth as in an aquifer.

Infiltration A method of storing water by natural soaking into the ground; a method of recharging aquifers.

Megacity Definitions vary from cities with more than 5 million to those with more than 20 million inhabitants. In this work, the UN measure of more than 20 million is used.

Nexus A connection or series of connections linking two or more things; such as the nexus between water and energy, or the nexus between water, population growth and agriculture.

Non-freshwater resources Includes brackish or saltwater; urban or industrial wastewater (with or without treatment); and agricultural drainage water. Also referred to as alternative or non-conventional water resources.

Non-renewable water resources Deep aquifers that have a minimal rate of recharge during an average human lifetime.

Operational objectives Explicit, measurable and attainable statements of what the utility will do within a specific period of time to achieve its long-term goals.

Ponding or pluvial floods Excessive amounts of rainwater that would normally be absorbed into the ground or collected in a stream or lake but instead floods surface lands rather than flowing into a stream, lake or reservoir.

Renewable energy Electric energy generated from renewable sources such as wind, hydropower, solar power, geothermal power and biomass (vegetation).

Renewable water resources Total amount of surface and groundwater resources generated via the hydrological cycle.

Re-used water treatment Processes related to re-using or recycling the treated wastewater effluent that is not discharged (conventional tertiary treatment, membrane treatment).

Runoff Surface water flow, such as rainwater over the ground to a drainage system. It often refers to water flow when the earth is saturated or during periods of extreme weather events.

Slum A residence in which a group of individuals live under the same roof where one or more of the following conditions do not apply: access to improved (treated) water; access to improved sanitation facilities, no more than three adults sharing one room; improved or durable dwelling; security of tenure (UN).

Stewardship The careful and responsible management of a designated resource, region or public asset.

Strategies The methods a utility will use to accomplish its objectives.

Surface water Natural water in lakes, rivers, streams or reservoirs.

Sustainability The use of natural, community and utility resources in a manner that satisfies current needs without compromising future needs or options. The EPA adds this definition: "Sustainability is based on a simple principle: Everything that we need for our survival and well-being depends, either directly or indirectly, on our natural environment. To pursue sustainability is to create and maintain the conditions under which humans and nature can exist in productive harmony to support present and future generations."

Sustainability goals Broad qualitative statements of what the water system desires to achieve over its long-term operations.

Sustainable Activities involving resource use for meeting the present needs of society without lessening the availability for following generations to meet future needs.

Wastewater treatment Involves collection (pumping, transporting sewage), treatment (primary, secondary, tertiary) and discharge.

Water consumption The volume withdrawn that is not returned to the source (i.e. it is evaporated or transported to another location) and is no longer available for other uses.

Watercourse A surface water system, such as a stream, river, ditch, culvert, canal, or other path over which water flows.

Water distribution Delivery of treated water to the customers via distribution networks (pumping, pressurizing, storing and distributing).

Water scarcity When renewable annual freshwater water supplies fall below 1,000 m^3 per person; **absolute scarcity** is when supplies fall below 500 m3 per person. (see water shortage).

Water sector Includes all processes whose main purpose is to treat/process or move water to or from the end-use, including: groundwater and surface water extraction, long-distance water transport, water treatment, desalination, water distribution, wastewater collection, wastewater treatment and water re-use.

Water shortage The situation within all or part of [a public or private supply] district when insufficient water is available to meet the present and anticipated needs of the [water rights] permit system, or when conditions are such as to require temporary reduction in the total use within a particular area to protect water resources from serious harm (Ecology dictionary 2014).

Watershed A specific surface area or a natural dividing line such as a ridge or high points of land that separates waters flowing into separate basins, rivers, lakes or seas.

Watershed health The ability of ecosystems to provide the functions needed by plants, wildlife and humans, including the quality and quantity of land and aquatic resources.

Water stress When renewable annual freshwater water supplies fall below 1700 cubic meters (m3) per person.

Water treatment Process of removing contaminants from water or wastewater in order to bring it up to water quality standards and for storage in freshwater reservoirs.

Water withdrawal The volume of water removed from a source; withdrawals are always equal to or greater than consumption.

Sources EIA **(2016) and USGS (2017)**

Bibliography

Abramson, Mark A., Jonathan D. Breul, and John Kamensky. 2006. *Six Trends Transforming Government*. Washington, DC: IBM Center for the Business of Government.

Adikari, Yogamath, and Junichi Yoshiami. 2009. Global trends in water related disasters: An insight for policymakers. United Nations Educational, Scientific and Cultural Organization (UNESCO). Accessed December 19, 2017 from http://unesdoc.unesco.org/images/0018/001817/181793e.pdf.

AFBF (American Farm Bureau Federation). 2017. Clean Water Act, WOTUS. 2017. Accessed November 21, 2017 from https://www.fb.org/issues/regulatory-reform/clean-water-act/.

Agrihome Expressions. 2016. Conservation agriculture (CA): Increase yields and improve farmlands. Accessed August 14, 2018 from http://agrihomegh.com/conservation-agriculture-benefits/.

Agyeman, Julian. 2007. Environmental justice and sustainability. In *Handbook of Sustainable Development*, ed. Giles Atkinson, Simon Dietz, and Eric Neumayer, 171–188. Cheltenham: Edward Elgar.

Ali, Babkir, and Amil Kuman. 2017. Development of water demand coefficients for power generation from renewable energy technologies. *Energy Conversion and Management* 143: 470–481.

Allan, J. David, Margaret Palmer, and N. Leroy Poff. 2005. Climate change and freshwater ecosystems. In *Climate Change and Biodiversity*, ed. Thomas E. Lovejoy and Lee J. Hannah, 274–290. New Haven: Yale University Press.

American Rivers. 2017. What is integrated water management? Accessed June 3, 2018 from www.americanrivers.org/conservation-resources/integrated-water-management/what-is-integrated-water-management/.

D. E. McNabb, *Global Pathways to Water Sustainability*,
https://doi.org/10.1007/978-3-030-04085-7

AP (Associated Press). 2018. Extreme heat and wildfires made worse by climate change, say scientists. 3:40 am, July 28. Accessed July 31, 2018 from www.nbcnews.com/news/weather/extreme-heat-wildfires-worsened-climate-change-say-scientists-n895496.

Appel, Todd, Bethany Bezak, and John Lisle. 2017. DC water green infrastructure financing: Pay for success can help utilities pursue innovate solution. *AWWA Journal* 109 (10): 28–31.

Armstrong, Adrian. 2006. Ethical issues in water use sustainability. *Area* 38 (1): 9–15.

ASCE (American Society of Civil Engineers). 2017. Infrastructure report card: Drinking water. Accessed January 23, 2017 from www.infrastructurereportcard.org/wp-content/uploads/2017/01/Drinking-Water-Report-Final.pdf.

ASCE (American Society of Civil Engineers). 2018. 2017 infrastructure report card: Drinking water. American Society of Civil Engineers. Accessed May 21, 2018 from www.infrastructurereportcard.org/wp-content/uploads/2017/01/Drinking-Water-Final.pdf.

ATS Irrigation, Inc. 2018. What are the different types of farm irrigation systems? Accessed August 15, 2018 from www.atsirrigation.com/what-are-the-different-types-of-farm-irrigation-systems/.

AWWA (American Water Works Association). 2012. Buried no longer: Confronting America's water infrastructure challenge. Accessed May 21, 2018 from www.awwa.org/portals/0/files/legreg/documents/buriednolonger.pdf.

Ayala, Edgardo. 2010. El Salvador: Most water-stressed country in Central America. International Press Service (IPS). Accessed January 25, 2018 from www.ipsnews.net/2010/12/el-salvador-most-water-stressed-country-in-central-america/.

Babbitt, Christina H., Mark Burbach, and Lisa Pennisi. 2015. A mixed-methods approach to assessing success in transitioning water management institutions: A case study of the Platte River Basin, Nebraska. *Ecology and Society* 20 (1): 54–64.

Baea, Jinwon, and Sandy Dall'erba. 2018. Crop production, export of virtual water and water-saving strategies in Arizona. *Ecological Economics* 146: 148–156.

Bahri, Akiça. 2012. Integrated urban water management. TEC Background Papers No. 16, Global Water Partnership, Technical Committee. Accessed October 24, 2017 from www.gwp.org/globalassets/global/toolbox/publications/background-papers-integrated-urban-water-management-2012.pdf.

Bai, Xuemei, Timon McPhearson, Helen Cleugh, Harini Nagendra, Xin Tong, Tong Zhu, and Yong-Guan Zhu. 2017. Linking urbanization and the environment: Conceptual and empirical advances. *Annual Review of Environment and Resources* 42: 215–240.

Baird, Gregory M. 2010. The infrastructure investment gap. *American Water Works Journal* 102 (6): 26–33.

Baker, C.J., K.E. Saxton, and R.R. Ritchie. 2002. *No-Till Seeding: Science and Practice*, 2nd ed. Oxford, UK: CAB International.

Barbosa, Marina Costa, Khorshed Alam, and Shahbaz Mushtaq. 2016. Water policy implementation in the state of São Palo Brazil: Key challenges and opportunities. *Environmental Science and Policy* 60: 11–18.

Barkin, David. 2014. Mexico City's water crisis. North American Congress on Latin America (NACLA). Accessed January 29, 2018 from https://nacla.org/article/mexico-citys-water-crisis.

Barlow, Maude, and Tony Clarke. 2004. The struggle for Latin America's water. *Global Policy Forum.* Accessed January 26, 2018 from www.globalpolicy.org/social-and-economic-policy/global-public-goods-1-101/46052-the-struggle-for-latin-americas-water.html.

Barrett, Oliver-Leighton. 2014. Brazil: Water woes, climate change and security. The Center for Climate and Security, Washington, DC. Accessed July 15, 2018 from climateandsecurity.org/2014/09/03/brazil-water-woes-climate-change-and-security/.

Beecham, S., and R. Fallahzadadeh. 2011. Innovative approaches to urban water management in developing countries. *Sustainable Development and Planning V* 150: 1743–3541.

Bel, Roland, Vladimir Smirnov, and Andrew Wait. 2018. Managing change: Communication, management style and change in organizations. *Economic Modelling* 69: 1–12.

Belloti, William, and John F. Rochecouste. 2014. The development of conservation agriculture in Australia—Farmers as innovators. *Industrial Soil and Water Conservation Research* 2 (1): 21–34.

Berger, Amélie, Theodor Friedrich, and Josef Kienzle. 2010. Conservation agriculture. In *Soils, Plant Growth and Crop Production*, ed. Willy H. Verheye, vol. 1, 79–105. Oxford: EOLSS Publishing.

Bernados, Brian. 2018. Reverse osmosis for direct portable reuse in California. *American Water Works Association Journal* 110 (1): 28–36.

Bhalage, Pradeep, B.B. Jadia, and S.T. Sangale. 2015. Case studies of innovative irrigation management techniques. *Aquatic Procedia* 4: 1197–1202.

Bichai, Françoise, Arani Kajenthira Grindle, and Sharmila L. Murphy. 2018. Addressing barriers in the water-recycling innovation system to reach water security in arid countries. *Journal of Cleaner Production* 171: S97–S109.

Biswas, Asit K., Cecilia Tortajada, and Udisha Saklan. 2017. India is facing the worst water crisis in generations. *Quartz India.* Accessed December 13, 2017 from http://qz.com/931878/india-is-facing-the-worst-water-crisis-in-generations.

Blackstock, K.L., and C.E. Carteri. 2007. Operationalizing sustainability science for a sustainable directive? Reflecting on three pilot projects. *The Geographical Journal* 173 (4): 343–357.

Boccaletti, Guilio, Martin Studhtey, and Marc van Olst. 2010. Confronting South Africa's water challenge. Accessed July 8, 2018 from www.mckinsey.com/business-functions/sustainability-and-resource-productivity/our-insights/confronting-south-africas-water-challenge.

Borins, Sandford. 2014. *The Persistance of Innovation in Government.* Washington, DC: Brookings Institution.

Buckley, Patricia, Lester Gunnion, and Will Sarni. 2016. The aging water infrastructure: Out of sight, out of mind? Accessed November 8, 2017 from https://dupress.deloitte.com/dup-us-en/economy/issues-by-the-numbers/us-aging-water-infrastructure-investment-opportunities.html.

Busby, Joshua. 2018. Warming world. *Foreign Affairs* 97 (4): 49–55.

Carter, Nicole. 2013. Energy-water nexus: The energy sector's water use. Congressional Research Service (CRS), Washington, DC. Accessed June 22, 2018 from https://fas.org/sgp/crs/misc/R43199.pdf.

Carvalho, Anne E., Luciano Menezes Costa, and Bezerra Sampaio. 2015. Paths to universalize water and sewage services in Brazil: The role of regulatory authorities in promoting efficient service. *Utilities Policy* 34: 1–10.

CASQA (California Stormwater Quality Association). 2018. The Los Angeles stormwater capture master plan: Harvesting stormwater for municipal supply. Accessed July 6, 2018 from www.casqa.org/asca/los-angeles-stormwater-capture-master-plan-harvesting-local-stormwater-municipal-supply.

CCAFS (Climate Change, Agriculture and Food Security). 2014. Climate smart agriculture. Wageningen University and Research, The Netherlands. Accessed August 11, 2018 from https://ccafs.cgiar.org/climate-smart-agriculture-0#.W28Vfc5KiUk.

CDC (Center for Disease Control). 2016. Types of agriculture and water use. Accessed January 6, 2017 from https://www.cdc.gov/healthywater/other/agricultural/types.html.

Ceratti, Mariana. 2016. Brazil may be the owner of 20% of the world's water supply but it is still very thirsty. *The World Bank*, August 5. Accessed April 15, 2018 from www.worldbank.org/en/news/feature/2016/07/27/how-brazil-managing-water-resources-new-report-scd.

CFR (Council on Foreign Relations). 2017. Water stress in sub-Saharan Africa. Accessed January 12, 2018, from www.cfr.org/backgrounder/water-stress-sub-saharan-africa.

Chartzoulakis, Konstantinos, and Maria Bertaki. 2015. Sustainable water management in agriculture under climate change. *Aquatic Prodedia* 4: 88–98.

Chesbrough, Henry W. 2003. The era of open innovation. *MIT Sloan Management Review* 44 (3): 35–41.

Chesbrough, Henry W. 2011a. *Open Services Innovation*. San Francisco: Jossey-Bass.

Chesbrough, Henry W. 2011b. Bringing open innovation to services. *MIT Sloan Management Review* 52 (2): 85–90.

Chi, Linli, and Jun Shi. 2012. Urbanization and its environmental effects in Shanghai, China. *Urban Climate* 2: 1–15.

Chidammodzi, C.L., and V.S. Muhandiki. 2017. Water resources management and integrated water resources management implementation in Malawi: Status and implications for lake basin management. *Lakes and Reservoirs: Research and Management* 22: 101–114.

Chini, Christopher, and Ashlynn S. Stillwell. 2018. The state of U.S. urban water: Data and the energy-water nexus. *Nexus*, February 26. Accessed March 15, 2018 from www.water-energy-food.org/news/2018-02-26-the-state-of-us-urban-water-data-and-the-energy-water-nexus/.

Cho, Renee. 2011. How china is dealing with its water crisis. *State of the Planet* blog. Columbia University Earth Institute. Accessed April 14, 2014 from https://blogs.ei.columbia.edu/2011/05/05/how-china-is-dealing-with-its-water-crisis/.

CIA (Central Intelligence Agency). 2013. Map of India. Central Intelligence Agency. Accessed December 9, 2017 from https://www.cia.gov/library/publications/resources/cia-maps-publications/India.html.

City of Boulder. 2011. Water quality strategic plan 2009. Accessed August 29, 2018 from https://bouldercolorado.gov/water/water-utility-master-plan.

Cohen, Steven. 2011. *Sustainability Management*. New York: Columbia University Press.

Conte, Giulio, Andrea Bolognesi, Cristina Bragalli, Sara Branchine, Asessandro De Carli, Chiara lenzi, Fabio masi, Antonio Massarutto, Mardo Pollastri, and Haria Prindipi. 2012. Innovative urban water management as a climate change adaptation strategy: Results from the implementation of the project "Water Against Climate Change (WATACLIC)". *Water* 4 (4): 1025–1038.

Copeland, Claudia. 2012. Water infrastructure financing: History of EPA appropriations. Congressional Research Service (CRS), Washington, DC. Accessed August 2, 2018 from https://fas.org/sgp/crs/misc/96-647.pdf.

Crase, Lin, Bethany Cooper, Brian Dollery, and Rui Marques. 2018. One person's drain is another's water supply: Why property rights, scope, measurement and hydrology matter when it comes to integrated water resources management. *Ecological Economics* 147: 436–441.

Crow, Patrick. 2017. New report highlights staggering costs ahead for water infrastructure. *WaterWorld* 28 (4). Accessed November 8, 2017 from www.waterworld.com/articles/print/volume-28/issue-4/departments/washington-update/new-report-highlights-staggering-costs-ahead-for-water-infrastructure-by-patrick-crow-washington-correspondent.html.

CRPUD (City of Raleigh Public Utilities Department). 2013. Strategic plan and change. Accessed August 24, 2018 from file:///C:/Users/David/Downloads/StrategicPlanforWeb%20(1).pdf.

CRWUA (Colorado River Water Users Association). n.d. Urban uses. Accessed December 29, 2017 from www.crwua.org/colorado-river/uses/urban-uses.

CSU (Colorado Springs Utilities). 2018. Strategic planning. Accessed August 24, 2018 from www.csu.org/pages/strategic-planning-r.aspx.

Cui, Lin li, and Jun Shi. 2012. Urbanization and its environmental effects in Shanghai, China. *Urban Climate* 2 (2012): 1–15.

Currell, Mathew J., and Dongmei Han. 2017. The global drain: Why China's water pollution problems should matter to the rest of the world. *Environment: Science and Policy for Sustainable Development* 59 (1): 16–29.

Custodio, Emilio, Vicki Kretsinger, and M. Ramóm Llamas. 2005. Intensive development of groundwater: Concept, facts and suggestions. *Water Policy* 7 (2): 151–162.

CWW (Columbus Water Works). Strategic plan. Accessed August 24, 2018 from www.cwwga.org/plaintext/aboutus/strategicplan.aspx.

Dallman, Suzanne. 2010. Living within our means: Contemplating a sustainable urban water budget. *The International Journal of Interdisciplinary Social Sciences* 58: 1833–1882.

Dascher, Erin D., Jiyun Kang, and Gwendolyn Husteydt. 2014. Water sustainability: Environmental attitude, drought attitude and motivation. *International Journal of Consumer Studies* 38: 457–474.

De Lima, Gabriela N., Magda A. Lombardo, and V. Magaña. 2018. Urban water supply and the changes in the precipitation patterns in the metropolitan area of São Paulo, Brazil. *Applied Geography* 94: 223–229.

De Sam Lazaro, Fred. 2010. Delhi's dilemma: Rising numbers, limited water. Pulitzer Center on Crisis Reporting, Washington, DC. Accessed December 9, 2017 from https://pulitzercenter.org/projects/population-growth-water-scarcity-new-delhi.

Devine, Jon. 2017. Trump's attack on clean water: What you need to know. Natural Resources Defense Council (NRDC). Accessed November 21, 2017 from https://www.nrdc.org/experts/trumps-attack-clean-water-what-you-need-know.

Dilekli, Naci, Faye Duchin, and Ignacio Cazcarro. 2018. Restricting water withdrawals of the thermal power sector: An input-output analysis for the northeast of the United States. *Journal of Cleaner Production* 198: 258–268.

Dimick, Dennis. 2014. If you think the water crisis can't get any worse, wait until the aquifers are drained. Accessed November 13, 2017 from https://news.nationalgeographic.com/news/2014/08/140819-groundwater-california-drought-aquifers-hidden-crisis/.

Dobermann, Achim, and Rebecca Nelson. 2013. Opportunities and solutions for sustainable food production. Sustainable Development Solutions Network (United Nations). Accessed November 22, 2017 from http://unsdsn.org/wp-content/uploads/2014/02/130112-HLP-TG7-Solutions-for-sustainable-food-production.pdf.

Dos Santos, Sylvana M., and Maria Mariah M.W.E.C. de Farias. 2017. Potential for rainwater harvesting in a dry climate: Assessments in a semiarid region in northeast Brazil. *Journal of Cleaner Production* 164: 1007–1015.

Dour, M.K., Muhammad Nafees, Shafaqat Ali, Muhammad Rizwan, Raees Ahmad Bajwa, Muhammad Bilal Shakoor, Muhammad Umair Arshad, Shahzad Ali, Shahid Catha, Farah Deeba, Waheed Murad, Ijaz Malook, and Shui Jin Zhu. 2017. Drinking water quality status and contamination in Pakistan. *BioMed Research International* 2017: 1–18.

Duesterhous, Richard. 1990. Sustainability's Promise. *Journal of Soil and Water Conservation* 45 (1): 4.

Duke Staff. 2015. How much water does U.S. fracking really use? Duke University Campus Research. Accessed January 10, 2018 from https://today.duke.edu/2015/09/frackfoot.

Dunn, A.D., and E. Derrington. 2010. Investment in water and wastewater infrastructure: An environmental justice challenge, a governance solution. *Natural Resources & The Environment* 24 (3): 3–8.

EcoWatch. 2015. Eight major cities funning out of water. Accessed July 14, 2018 from www.ecowatch.com/8-major-cities-running-out-of-water-1882061458.html.

EEA (European Environment Agency). 2012. Climate change—A threat to aquatic ecosystems. Accessed November 30, 2017 from http://www.climate-and-freshwater.info/climate_change/.

EIA (U.S. Energy Information Administration). 2014. Many newer power plants have cooling systems that reuse water. Accessed July 20, 2018 from www.eia.gov/todayinenergy/detail.php?id=14971.

EIA (U.S. Energy Information Administration). 2016a. What is U.S. electricity generation by energy source? Accessed February 8, 2018 from www.eia.gov/tools/faqs/faq.php?id=425&t=3.

EIA (Energy Information Administration). 2016b. World Energy Outlook 2016. Accessed December 7, 2018 from www.eia.gov/outlooks/ieo/pdg/0484)2016).pdf.

EIA (U.S. Energy Information Administration). 2017. Hydroelectric generators are among the United States' oldest power plants. Accessed July 25, 2018 from www.eia.gov/todayinenergy/detail.php?id=30312.

EIA (U.S. Energy Information Administration). 2018. Consumption of fuels for electricity generation, 2005–2017. Accessed June 22, 2018 from www.eia.gov/totalenergy/data/monthly/pdf/sec7_9.pdf.

EPA (Environmental Protection Agency). 2008. Effective utility management: A primer for water and wastewater utilities. Accessed August 19, 2018 from https://nepis.epa.gov/Exe/ZyPDF.cgi/P10053BJ.PDF?Dockey=P10053BJ.PDF.

EPA (Environmental Protection Agency). 2013. Rural and small systems guidebook to sustainable utility management. Cooperative Program of the Environmental Protection Agency's Office of Wastewater Management, and the U.S. Department of Agriculture. Accessed March 14, 2018 from http://dnr.mo.gov/env/wpp/pdwd/docs/sustainable-management.102413.pdf.

EPA (Environmental Protection Agency). 2015. Drinking water infrastructure needs survey and assessment: Sixth Report to Congress. Accessed August 10, 2018 from www.epa.gov/sites/production/files/2018-03/documents/sixth_drinking_water_infrastructure_needs_survey_and_assessment.pdf.

EPA (Environmental Protection Agency). 2016a. Climate impacts on water resources. US Environmental Protection Agency. Accessed November 24, 2017 from https://19january2017snapshot.epa.gov/climate-impacts/climate-impacts-water-resources_.html#main-content.

EPA (Environmental Protection Agency). 2016b. Polluted runoff: Nonpoint source pollution. US Environmental Protection Agency. Accessed November 7, 2017 from https://www.epa.gov/nps/what-nonpoint-source.

EPA (Environmental Protection Agency). 2016c. Planning for sustainability: A handbook for water and wastewater utilities. Accessed November 7, 2017 from www.epa.gov/sites/production/files/2016-01/documents/planning-for-sustainability-a-handbook-for-water-and-wastewater-utilities.pdf.

EPA (Environmental Protection Agency). 2017. Climate impacts on coastal areas. Environmental Protection Agency. Accessed November 28, 2017 from https://19january2017snapshot.epa.gov/climate-impacts/climate-impacts-coastal-areas_.html.

EPA (Environmental Protection Agency). 2018. Green infrastructure. U.S. Environmental Protection Agency. Accessed July 3, 2018 from https://www.epa.gov/green-infrastructure/what-green-infrastructure#main-content.

ERWSD (Eagle River Water and Sanitation District). 2015. Mission, vision and values. Accessed August 30, 2018 from www.erwsd.org/about/mission-vision-values/.

Espey, Jessica, Hayden Dahmm, and Laurie Manderino. 2018. Leaving no U.S. city behind: The U.S. cities sustainable development goals index. United Nations Sustainable Development Solutions Network. Accessed August 5, 2018. Available at http://unsdsn.org/wp-content/uploads/2018/06/US-Cities-Index-Report.pdf.

Falkenmark, Malin. 1990. Water issues confronting humanity. *Journal of Peace Research* 27 (2): 177–190.

Fant, Charles, C. Adam Schlosser, Xiang Gao, Kenneth Strzepek, and John Reilly. 2016. Projections of water stress based on an ensemble of socioeconomic growth and climate change scenarios: A case study in Asia. MIT Joint Program on the Science and Policy of Global Change. Accessed January 13, 2018 from https://globalchange.mit.edu/sites/default/files/MITJPSPGC_Reprint_16-6.pdf.

FAO (Food and Agriculture Organization of the United Nations). 2012. China: Geography, climate and population: Water report. Aquastat, Food and Agriculture Organization of the UN. Accessed December 9, 2017 from www.fao.org/nr/water/aquastat/countries_regions/CHN/.

FAO (Food and Agriculture Organization of the United Nations). 2013. FAO success stories on climate-smart agriculture. Accessed August 13, 2018 from www.fao.org/3/a-i3817e.pdf.

FAO (Food and Agriculture Organization of the United Nations). 2015. India: Geography, climate and population: Water report. Food and Agriculture Organization of the United Nations, Aquastat. Accessed December 9, 2017 from http://www.fao.org/nr/water/aquastat/countries_regions/IND/print1.stm.

FAO (Food and Agriculture Organization of the United Nations). 2016a. Nigeria: Geography climate and population water report. Food and Agriculture Organization of the UN, Aquastat. Accessed December 9, 2017 from http://www.fao.org/nr/water/aquastat/countries_regions/NGA/.

FAO (Food and Agriculture Organization of the United Nations). 2016b. Russian Federation: Geography, climate and population. Food and Agriculture Organization of the UN, Aquastat. Accessed January 22, 2018 from www.fao.org/nr/water/aquastat/countries_regions/RUS/print1.stm.

FAO (Food and Agriculture Organization of the United Nations). 2017a. Agriculture and water quality interactions: A global review. Food and Agriculture Organization of the UN, Aquastat. Accessed November 21, 2017 from http://www.fao.org/fileadmin/templates/solaw/files/thematic_reports/TR_08.pdf.

FAO (Food and Agriculture Organization of the United Nations). 2017b. Water pollution from agriculture: A global review (executive summary). Food and Agriculture Organization of the UN. Accessed November 20, 2017 from http://www.fao.org/3/a-i7754e.pdf.

Farming First. 2013. Enhancing sustainable development through agriculture: Applying the Farming First principles to increase water use efficiency. Accessed January 21, 2018 from https://farmingfirst.org/wordpress/wp-content/uploads/2010/10/FarmingFirst-Water-Paper.pdf.

Fernandez, Sergio, and Hal G. Rainey. 2006. Managing successful organizational change in the public sector. *Public Administration Review* 66 (2): 168–176.

Fishman, Ram, Naresh Devineni, and Swaminathan Raman. 2015. Can improved agricultural water use efficiency save India's groundwater? Accessed April 12, 2018 from iopscience.iop.org/article/10.1088/1748-9326/10/8/084022.pdf.

Fitzgerald, Joan, and Joshua Laufer. 2017. Governing green stormwater infrastructure: The Philadelphia experience. *Local Environment* 22 (2): 256–268.

Furlong, Casey, Saman De Silva, Lachlan Guthrie, and Robert Considine. 2016. Developing a water infrastructure planning network for the complex modern planning environment. *Utilities Policy* 38: 1–10.

Furlong, Casey, Ryan Brotchie, Robert Considine, Greg Finlayson, and Lachian Guthrie. 2017. Key concepts for integrated urban water management infrastructure planning: Lessons from Melbourne. *Utilities Policy* 45: 84–96.

Gallegos, Tanya J., Brian A. Varela, Seth S. Haines, and Mark A. Engles. 2015. Hydraulic fracturing water use variability in the United States and potential environmental implications. *Water Resources Journal* 51 (7): 5839–5845.

GAO (U.S. Government Accountability Office). 2016. *Water Infrastructure*. U.S. Government Accountability Office. Washington, DC: GAO Reports.

Geer, Bobbi Watt, Jill K. Meher, and Michele T. Cole. 2008. Managing nonprofit organizations: The importance of transformational leadership and commitment to operating standards for nonprofit accountability. *Public Performance and Management Review* 32 (1): 51–75.

Geospatial World. 2010. Middle East, North Africa vulnerable to water stress. Accessed January 16, 2017 from www.geospatialworld.net/news/middle-east-north-africa-vulnerable-to-water-stress/.

Geri, Laurance and David E. McNabb. 2011. *Energy Policy in the U.S.* Boca Raton, FL: CRC Press.

Gjorgiev, Blaže, and Giovanni Sansavini. 2017. Water-energy nexus: Impact on electrical energy conversion and mitigation by smart water resources management. *Energy Conversion and Management* 148: 1114–1126.

Global Environment Facility. 2015. The importance of water sustainability. Accessed October 11, 2017 from www.thegef.org/news/importance-water-sustainability.

Global Water. 2011. *Inflow and Infiltration*. New York: Xylem Inc. Accessed August 25, 2018 from www.globalw.com/support/inflow.html.

Gray, Regina C. 2015. Guest Editor's introduction: Building a research agenda for creating sustainable and inclusive communities for all. *Cityscape: A Journal of Policy Development and Research* 17(2): 2–15.

Green, Alison, and Jerry Hauser. 2012. *Managing to change the world: The nonprofit manager's guide to getting results*. San Francisco: Wiley.

Grimaldos, Angel Simón. 2015. Water for a sustainable future: New paradigm, new vision. In *Water and Sustainable Development*, ed. Ramiro A. Lopera, 68–73. United Nations Water for Life. Accessed December 7, 2017 from www.un.org/waterforlifedecade/pdf/wm-iii-eng.pdf.

Grubert, Emily A., Fred C. Beach, and Michael E. Webber. 2012. Can switching fuels save water? A life cycle quantification of freshwater consumption for Texas coal- and natural gas-fired electricity. *Environmental Research Letters* 7 (4): 1–11.

GWP (Global Water Partnership). 2000. Integrated water resources management. TAC Background Papers (4). Accessed August 4, 2018 from www.gwp.org/globalassets/global/toolbox/publications/background-papers/04-integrated-water-resources-management-2000-english.pdf.

Hadadin, Nidal, Qaqish Maher, Emad Adawwi, and Ahmed Bdour. 2010. Water shortage in Jordan: Sustainable solutions. *Desalination* 250: 197–202.

Haider, Husnain, Rehan Sadiq, and Soloman Tessfamariam. 2015. Multilevel performance management framework for small to medium sized water utilities in Canada. *Canadian Journal of Civil Engineering* 42 (11): 899–900.

Hall, Richard H. *Organizations: Structures, Processes and Outcomes*, 7th ed. Upper Saddle River, NJ: Palgrave.

Hamman, Philippe, Virginie Anquetin, and Céline Monicolle. 2017. Contemporary meanings of the 'Sustainable City': A comparative review of the French and English language literature. *Sustainable Development* 25 (4): 336–355.

Han, Dongmei, Matthew J. Currell, and Guoliang Cao. 2016. Deep challenges for China's war on water pollution. *Environmental Pollution* 218: 1222–1233.

Hays, Jeffrey. 2008. Water in China. Facts and Details. Accessed December 9, 2017 from http://factsanddetails.com/china/cat13/sub85/item317.html.

Herrick, Charles, and Joanna Pratt. 2012. Sustainability in the water sector: Enabling lasting change through leadership and cultural transformation. *Nature and Culture* 7 (3): 285–313.

Hill, Christopher. 2017. Sustainable cities water index: Harnessing water for long-term success. Accessed September 3, 2018 from www.arcadis.com/en/united-states/our-perspectives/2016/sustainable-cities-water-indexharnessing-water-for-long-term-success/.

Hobbs, Peter R. 2006. Conservation agriculture: What it is and why is it important for future sustainable food production? Paper presented at the international workshop on Increasing Wheat Yield Potential, Obregon, Mexico, March 20–24.

Hoekstra, Arjen, and Mestin M. Mekonnen. The water footprint of humanity. In *Proceedings of the National Academy of Sciences of the United States of America* 109 (9): 3232–3237.

Hui, Rui, Jonathan Herman, Jay Lund, and Kaveh Madani. 2018. Adaptive water infrastructure planning for nonstationary hydrology. *Advances in Water Resources* (Forthcoming): 1–12. Accessed July 2, 2018 from https://doi.org/10.1016/j.advwatres.2018.05.009.

Hukka, Jarmo J., and Tapio S. Katko. 2015. Resilient asset management and governance for deteriorating water service infrastructure. *Procedia of Economics and Finance* 21: 112–119.

Huskova, Ivana, Evgenii S. Matrosov, Julien J. Harou, Joseph R. Kasprzyk, and Cris Lambert. 2016. Screening robust water infrastructure investments and their trade-offs under global change. *Global Environmental Change* 41: 216–227.

ICWE (International Conference on Water and the Environment). 1992. The Dublin statement on water and the environment. Accessed August 4, 2018 from https://www.gdrc.org/uem/water/dublin-statement.html.

Iglesias, Ana, and Luis Garrote. 2015. Adaptation strategies for agricultural water management under climate change in Europe. *Agricultural Water Management* 155: 113–123.

Immigroup. 2004. Public domain map of China. Accessed December 9, 2017 from https://www.immigroup.com/china-visa.

Industrial Waterworld. 2018. Produced water treatment plant to double capacity. *Water Briefs* 10 (3): 10.

IPCC (Intergovernmental Panel on Climate Change). 2014. Summary for policymakers. In *Climate Change 2014: Impacts, Adaptation, and Vulnerability. Part A: Global and Sectoral Aspects*, 1–32. Contribution of Working Group II to the Fifth Assessment Report. Cambridge, UK: Cambridge University Press.

IPS (Inter Press Service). 2017. World water week: "Water and waste: Reduce and reuse", August 22. Accessed February 24, 2018 from http://www.ipsnews.net/2017/08/2017-world-water-week-water-waste-reduce-reuse/.

IWA (International Water Association). 2017. Sustainability in water supply. Accessed October 11, 2017 from www.iwapublishing.com/news/sustainability-water-supply.

Jackson, Tim. 2007. Sustainable consumption. In *Handbook of Sustainable Development*, ed. Giles Atkinson, Simon Dietz, and Eric Neumayer, 254–268. Cheltenham: Edward Elgar.

Jain, Sharad K. 2011. Population rise and growing water scarcity in India—Revised estimates and required initiatives. *Current Science* 101 (3): 271–276. National Commission for Integrated Water Resources Development (NCIWRD), India Ministry of Water Resources.

Jarmillo, Pablo, and Ali Nazemi. 2018. Assessing urban water security under changing climate: Challenges and ways forward. *Sustainable Cities and Society* (Forthcoming): 1–12.

Jiménez Cisneros, Blanca E., Taikan Oki, Nigel W. Arnell, Gerado Benito, J. Graham Cogley, Petra Döll, Tong Jiang, and Shadrack S. Mwakalila. 2014. Freshwater resources. In *Climate Change 2014: Impacts, Adaptation, and Vulnerability. Part A: Global and Sectoral Aspects*, ed. C.B. Field, V.R. Barros, D.J. Dokken, K.J. Mach, M.D. Mastrandrea, T.E. Bilir, M. Chatterjee, K.L. Ebi, Y.O. Estrada, R.C. Genova, B. Girma, E.S. Kissel, A.N. Levy, S. MacCracken, P.R. Mastrandrea, and L.L. White, 229–269. Cambridge: Cambridge University Press.

Jones, J.A.A. 1999. Climate change and sustainable water resources: Placing the threat of global warming in perspective. *Hydrological Sciences Journal* 44 (4): 541–557.

Karl, Thomas R., and Kevin E. Trenberth. 2005. What is climate change? In *Climate Change and Biodiversity*, ed. Thomas E. Lovejoy and Lee J. Hannah, 15–28. New Haven: Yale University Press.

Kassam, Amir, Rolf Derpsch, and Theodor Friedrich. 2014. Global achievements in soil and water conservation: The case of Conservation Agriculture. *International Soil and Water Conservation Research* 2 (1): 5–13.

Khalid, Iram, and Mazhar Abbas Khan. 2016. Water scarcity: QA major human security challenge to Pakistan. *South Asia Studies* 31 (2): 143–156.

Khalkhali, Masoumeh, Kirk Westphal, and Weiwei Mo. 2018. The water-energy nexus at water supply and its implications on the integrated water and energy management. *Science of the Total Environment* 636: 1257–1267.

Kimmilman, Michael. 2017. Can Jakarts save itself from sinking in 10 years? *The Seattle Times*, December 27, A1, 9.

Kirshen, Paul, Semra Aytur, Jory Hecht, Andrew Walker, David Burdick, Stephen Jones, Neil Fennessey, Renee Boudeau, and Lorilee Mather. 2018. Integrated urban water management applied to adaptation to climate change. *Urban Climate* 24: 246–263.

Kopocia, Ken. 2015. The facts about the Clean Water Rule and agriculture. Environmental Protection Agency blog. Accessed November 21, 2017 from https://blog.epa.gov/blog/2015/06/clean-water-rule-and-agriculture/.

Král, Pavel, and Věra Králová. 2016. Approaches to changing organizational structure: The effects of drivers and communication. *Journal of business Research* 69: 5169–5174.

Lall, Upmanu, Tanya Heikkila, Casey Brown, and Tobias Siegried. 2008. Water in the 21st century: Defining the elements of global crises and potential solutions. *Journal of International Affairs* 61 (2): 1–17.

Laloux, Frederic, and Etienne Appert. 2016. *Reinventing Organizations*. Brussels: Nelson Parker.

Laquinta, David L., and Axel S. Drescher. 2000. Defining the peri-urban: Rural-urban linkages and institutional connections. Research paper funded by the UN Food and Agricultural Organization (FAO). Accessed August 9, 2018 from www.fao.org/docrep/003/X8050T/x8050t02.htm.

Lee, Junesbok, and Tamin Younos. 2018. Sustainability strategies at the water-energy nexus: Renewable energy and decentralized infrastructure. *Journal of the American Water Works Association* 10 (2): 32–39.

Leech, M., I. Scoones, and A. Stirling. 2007. Pathways to sustainability: An overview of the STEPS Center approach. Concept paper, University of Sussex STEPS Centre, Brighton, UK. Accessed May 2, 2018 from http://steps-centre.org/wp-content/uploads/final_steps_overview.pdf.

Levidow, Les, Daniele Zaccaria, Rodrigo Maia, Eduardo Vivas, Mladen Todorovic, and Alessendra Scardigno. 2014. Improving water-efficient irrigation: Prospects and difficulties of innovative practices. *Agricultural Water Management* 46: 4184–4194.

Li, Liu, and Jensen M. Bergen. 2018. Green infrastructure for sustainable urban water management: Practices of five forerunner cities. *Cities* 74: 126–133.

Liao, Xiawei, Jim W. Hall, and Nick Eyre. 2016. Water use in China's thermoelectric power sector. *Global Climate Change* 41: 142–152.

Liao, Xiawei, Xu Zhaob, Jim W. Hall, and Dabo Guan. 2018. Categorizing virtual water transfers through China's electric power sector. *Applied Energy* 226: 252–260.

Likhotal, Alexander. 2013. The future of water: Strategies to meet the challenge. *Cadmus* 2 (1): 85–92.

Lindahl, Karin Beland, Susan Baker, Lucy Rist, and Anna Zachrisson. 2016. Theorising pathways to sustainability. *International Journal of Sustainable Development & World Ecology* 23 (5): 399–411.

Llanilla, Marc. 2013. China's top 6 environmental problems. *Live Science*, March 15. Accessed January 21, 2018 from www.livescience.com/27862-china-environmental-problems.htm.

Lovejoy, Thomas E. 2005. Conservation with a changing climate. In *Climate Change and Biodiversity*, ed. Thomas E. Lovejoy and Lee Hannah, 325–328. New Haven: Yale University Press.

Lovejoy, Thomas E., and Lee Hannah. 2005. Global greenhouse gas levels and the future of biodiversity. In *Climate Change and Biodiversity*, ed. Thomas E. Lovejoy and Lee Hannah, 387–395. New Haven: Yale University Press.

Luthra, Sonia, and Amrita Kundu. 2013. India's water crisis and cures: An interview with Kirit S. Parikha. The National Bureau of Asian Research (NBR). Accessed June 28, 2018 from http://www.nbr.org/research/activity.aspx?id=356.

Maddocks, Andrew, Robert S. Young, and Paul Reig. 2015. Ranking the world's most water stressed countries in 2040. World Resources Institute, Washington, DC. Accessed November 4, 2017 from http://www.wri.org/blog/2015/08/ranking-world%E2%80%99s-most-water-stressed-countries-2040.

Maestu, Josephina. 2015. Water and sustainable development: Improving the water related sustainable development goals. The relevance of infrastructure. In *Water and Sustainable Development*, ed. Ramiro A. Lopera, 4–11. United Nations Water for Life. Accessed December 7, 2017 from www.un.org/waterforlifedecade/pdf/wm-iii-eng.pdf.

Magill, Bobby. 2015. Study: Water use skyrockets as fracking expands. Climate Central. Accessed January 10, 2018 from www.climatecentral.org/news/fracking-water-use-skyrockets-19177.

Mahto, Shankar. 2013. Present status of water use efficiency on irrigation projects in India and action taken for its improvement including role of role of farmers. Training Program on Increasing Water Use Efficiency (WUE) in irrigation sector, NWA, Pune, January 21–February 1, 9–19.

Manocha, Nishtha, and Vladan Babovic. 2017. Development and valuation of adaptive pathways fore storm water management infrastructure. *Environmental Science and Policy* 77: 86–97.

Manokar, A. Muthu, D. Prince Winston, A.E. Kabeel, and Ravishankar Sathyamurthy. 2018. Substantial fresh water and power production by integrating PV panel to inclined solar still. *Journal of Cleaner Production* 172: 2711–2719.

Marco, Tomás A. Sancho. 2015. Water and sustainable development. In *Water and Sustainable Development*, ed. Ramiro A. Lopera, 54–67. United Nations Water for Life. Accessed December 7, 2017 from www.un.org/waterforlifedecade/pdf/wm-iii-eng.pdf.

Marlow, David R., Magnus Moglia, David J. Beale, and Anna Stenstromer. 2012. Embedding sustainability into a utility's business culture. *Journal of the American Water Works Association* 104 (2): E121–E126.

Marques, Rui Cunha, and João Miranda. 2016. Redrafting water governance: Guiding the way to improve the status quo. *Utilities Policy* 43: 1–3.

Matthews, John H., and Guillermo Mendoza. 2015. Designing for climate confidence: Moving beyond uncertainty in sustainable water management. In *Water and Sustainable Development*, ed. Ramiro A. Lopera, 44–53. United Nations Water for Life. Accessed December 7, 2017 from www.un.org/waterforlifedecade/pdf/wm-iii-eng.pdf.

Maxwell-Gaines, Chris. 2018. What is rainwater harvesting? Innovative Water Solutions. Accessed July 6, 2018 from www.watercache.com/faqs/rainwater-harvesting-defined.

Maya, Anna. 2017. Securing Southern California's water future. *Water World* 33 (11): 10–13, 39–40.

Mayo, Martin. 2013. Conservation agriculture: Manual for implementation. International Fund for Agricultural Development. Accessed August 12, 2018 from https://images.agri-profocus.nl/upload/post/Conservation_Agriculture_manual_for_implementation1418995455.pdf.

Mayor, Beatriz, R. Rodríguez Casado, J. Landeta, E. López-Gunn, and F. Villarroya. 2016. An expert outlook on water security and water for energy trends to 2030–2050. *Water Policy* 18 (1): 1–18.

McDonald, Robert J., Katherine Weber, Julie Padowski, Martina Flöke, Christof Schneider, Pamela A. Green, Thomas Gleeson, Stephanie Eckman, Bernard Lehner, Deborah Balk, Timothy Boucher, Günther Grill, and Mark Montgomery. 2014. Water on an urban planet: Urbanization and the reach of urban water infrastructure. *Global Climate Change* 27: 96–105.

McKinsey & Co. 2009. Charting our water future. 2030 Water Resources Group. *Sustainability & Resource Productivity* 18 (1): 1–18. Accessed December 15, 2017 from www.mckinsey.com/~/media/mckinsey/dotcom/client_service/sustainability/pdfs/charting%20our%20water%20future/charting_our_water_future_full_report_.ashx.

McNabb, David E. 2016. *Public Utilities: Meeting 21st Century Management Challenges*, 2nd ed. London: Edward Elgar.

McNabb, David E. 2017. *Managing the Water Resource in an Era of Climate Change*. New York: Palgrave Macmillan.

McNicoll, Geoffrey. 2007. Population and sustainability. In *Handbook of Sustainable Development*, ed. Giles Atkinson, Simon Dietz, and Eric Neumayer, 125–139. Cheltenham: Edward Elgar.

Mees, Hannelore, Cathy Suykens, and Ann Crabbe. 2017. Evaluating conditions for integrated water resource management at sub-basin scale: A comparison of the Flemish Sub-Basin boards and Walloon River contracts. *Environmental Policy and Governance* 27: 59–73.

Meindl, Cristopher F. 2011. Water, water, everywhere? *Southeastern Geographer* 51 (4): 615–640.

Mejia-Dugand, Santiago, Olof Hjelm, and Leo Baas. 2017. Public utility companies in liberalized markets—The impact of management models on local and regional sustainability. *Utilities Policy* 49: 137–144.

Melo dos Santos, Sylvana, and Maria Mariah M.W.E.C. de Farias. 2017. Potential for rainwater harvesting in a dry climate: Assessments in a semiarid region in northeast Brazil. *Journal of Cleaner Production* 164: 1007–1015.

Michell, Nick. 2016. How to plug the gap in water investments. Accessed November 9, 2017 from http://news.devfinance.net/how-to-plug-the-gap-in-water-investments.

Mikovits, Christian, Wolfgang Rauch, and Manfred Kleidorfer. 2018. Importance of scenario analysis in urban development for urban water infrastructure planning and management. *Computers, Environment and Urban Systems* 68: 9–16.

Miralles-Wilhelm, Fernando, Leon Clarke, Mohamad Hejazi, Sonny Kim, Kelly Gustafson, Raul Muñoz-Castillo, and Neal Graham. 2017. *Physical Impacts of Climate Change on Water Resources*. Washington, DC: Global Water Practice, World Bank Group. Accessed October 16, 2017 from file:///C:/Users/David/

Downloads/112726-WP-PUBLIC-Physical-impacts-of-climate-change-on-water-resources.pdf.

Molden, David. 2012. Water for a food-secure world. Presentation at the Achieving Sustainable Food Security: Challenges and Opportunities workshop of the National Research Council. Part II, 142–144. The National Archives Press, Washington, DC.

Morris, John C. 2017. Planning for water infrastructure: Challenges and opportunities. *Public Works Management and Policy* 22 (1): 24–30.

Mosey, Helen, and Erin Nealer. 2016. *Barriers to Bankable Infrastructure: Incentivizing Private Investment to Fill the Global Infrastructure Gap*. Washington, DC: Center for Strategic and International Studies. Accessed November 14, 2017 from https://csis-prod.s3.amazonaws.com/s3fs-public/publication/160308_Moser_BarriersBankableInfrastructure_Web.pdf.

Moss, Brian. 2008. Water pollution by agriculture. *Philosophical Transactions: Biological Sciences* 363 (1491): 659–666.

Mugabi, Josses, Sam Kayaga, and Cyrus Njiru. 2007. Strategic planning for water utilities in developing countries. *Water Utilities* 15: 1–8.

Mung, Alex. 2013. 7 reasons why water is vital for a sustainable future. World Economic Forum. Accessed October 23, 1017 from www.weforum.org/agenda/2015/03/7-reasons-why-water-is-vital-for-a-sustainable-future/.

Murphy, Maripat. 2018. 2018 state of the water industry: The challenge of building resilience. *Journal of the American Water Works Association* 110 (8): 61–71.

Mustafa, Daanish, Majed Akhter, and Natilie Nasrallah. 2013. Understanding Pakistan's water-security nexus. United States Institute of Peace. Accessed January 18, 2018 from www.usip.org/sites/default/files/PW88_Understanding-Pakistan%27s-Water-Security-Nexus.pdf.

Muys, Jerome, Van P. Hildebrand, and the Firm of Sullivan and Worcester. 2011. Water and renewables: What are the options? Accessed November 3, 2017 from http://www.renewableenergyworld.com/articles/print/volume-14/issue-5/other/the-last-word-water-and-renewables.html.

Narcizo de Lima, Gabriela, Magda A. Lombardo, and V. Magaña. 2018. Urban water supply and the changes in the precipitation patterns in the metropolitan area of São Paulo, Brazil. *Applied Geography* 94: 223–229.

NASA (National Aeronautics and Space Administration). 2015. San Joaquin Valley is still sinking. Accessed November 13, 2017 from https://earthobservatory.nasa.gov/IOTD/view.php?id=89761.

NASA (National Aeronautics and Space Administration). 2016. Study maps hidden water pollution in the U.S. National Aeronautics and Space Administration. Accessed November 13, 2017 from www.nasa.gov/feature/jpl/study-maps-hidden-water-pollution-in-us-coastal-areas.

Niemczynowicz, Janusz. 1993. New aspects of sewerage and water technology. *Ambio* 22 (7): 449–455.

NJDEP (New Jersey Department of Environmental Protection). 2016. Asset management guidance and best practices. Accessed July 28, 2018 from https://www.nj.gov/dep/watersupply/pdf/guidance-amp.pdf.

Nkhonjera, German K. 2017. Understanding the impact of climate change on the dwindling water resources of South Africa, focusing mainly on Olifants River Basin. *Environmental Science & Policy* 71: 19–29.

NMEFC (New Mexico Environmental Finance Center). 2006. *Asset Management: A Guide for Water and Wastewater Systems*. Socorro, NM: New Mexico Institute of Mining and Technology.

NOAA (National Oceanic and Atmospheric Administration). 2017. Billion-dollar water and climate disasters: National centers for environmental information. Accessed December 19, 2017 from www.ncdc.noaa.gov/billions/events/US/1980-2017.

NRCS (Natural Resources Conservation Services). 2012. Inefficient use of irrigation water. U.S. Department of Agriculture. Accessed December 4, 2012 from https://www.nrcs.usda.gov/Internet/FSE_DOCUMENTS/stelprdb1187280.pdf.

OHCHR (Office of the High Commissioner). 2003. *General Comment No. 15*: The right to water. United Nations, January 20. Accessed October 23, 2017 from http://www.refworld.org/pdfid/4538838d11.pdf.

OHCHR (Office of the High Commissioner). 2017. Vienna declaration and program of action. Vienna World Conference on Human Rights, United Nations. Accessed October 23, 2017 from http://www.ohchr.org/EN/ProfessionalInterest/Pages/Vienna.aspx.

Oppenheimer, Joan, Allegra Da Silva, Zita L.T. Yu, Mark Hanna, and Ken Susilo. 2017. Capturing alternative water sources to supplement drinking water supply. *Journal of the American Water Works Association* 109 (1): 18–25.

Pacific Institute. 2014. Defining water scarcity, water stress and water risk: It's not just semantics. *Pacific Instituter Insights*, February 4. Accessed June 30, 2018 from http://pacinst.org/water-definitions/.

Padowski, Julie C., and James W. Jawitz. 2009. The future of global water scarcity: Policy and management challenges and opportunities. *The Whitehead Journal of Diplomacy and International Relations* 10 (2): 99–114.

PAI (Population Action International). 2011. Why population matters to water resources. Accessed December 4, 2017 from https://pai.org/wp-content/uploads/2012/04/PAI-1293-WATER-4PG.pdf.

Passioura, John B. 1977. Grain yield harvest index and water-use of wheat. *Journal of the Australian Institute of Agricultural Science* 43 (3/4): 117–120.

Payet-Burin, Raphael, Federica Bertoni, Claus Davidsen, and Peter Bauer-Gottwein. 2018. Optimization of regional water-power systems under cooling constraints and climate change. *Energy* 155: 484–494.

Pedraza, Lisdey Espinoza, and Markus Heinrich. 2016. Water scarcity: Cooperation or conflict in the Middle East and North Africa? *Foreign Policy Journal*, September 2. Accessed January 25, 2018 from https://www.

foreignpolicyjournal.com/2016/09/02/water-scarcity-cooperation-or-conflict-in-the-middle-east-and-north-africa/.

Peer, Rebecca A.M., and Kelly T. Sanders. 2018. The water consequence of a transitioning US power sector. *Applied Energy* 210: 613–622.

Petelet-Giraud, Emmanuelle, Lisa Cary, Paul Cary, Guillaume Bertrand, Armelle Goglio-Jacqumot, Ricardo Hirata, Luc Aquilina, Lincoln Muniz Alves, Veridiana Martins, Ana Maria Melo, Suzana Montenegro, Eliot Chatton, Melissa Franzen, and Axel Aurouet. 2018. Multi-playered water resources, management, and uses under the impacts of global changes in a southern coastal metropolis: When will it be already too late? Crossed analysis in Recife, NE Brazil. *Science of the Total Environment* 618: 645–657.

Peters, Robert L., and Thomas E. Lovejoy. 1992. *Global Warming and Biodiversity*. London: Yale University Press.

Petit, Oliver. 2016. Paradise lost? The difficulties in defining and monitoring integrated water resources management indicators. *Environmental Sustainability* 21: 58–64.

Piesse, Mervyn. 2016. Indonesian water security: Improving but still subject to shocks. Future Directions International strategic analysis paper. Accessed January 19, 2018 from http://www.futuredirections.org.au/publication/indonesian-water-security-improving-but-still-subject-to-shocks/.

Postel, Sandra L. 2007. Aquatic ecosystem protection and drinking water utilities. *Journal of the American Water Works Association* 99 (2): 52–63.

Pradhan, Aliza, Catherine Chan, Pravat Kumar Roul, Jacqueline, and Brent Sipes. 2018. Potential of conservation agriculture (CA) for climate change adaptation and food security under rainfed uplands of India: A transdisciplinary approach. *Agricultural Systems* 163: 27–35.

Pradhan, Piyush, Ajay Verma, and Vithlesh Kumar. 2018. Need of conservation agriculture in India: Sustainability. *International Journal of Current Microbilogy* 7 (1): 308–314.

Proskuryakova, Liliana N., Oxcan Saritas, and Sergey Sivaev. 2018. Global water trends and future scenarios for sustainable development. *Journal of Cleaner Production* 170: 867–879.

Pushard, Doug. 2018. Gray water: The new green? *Green Fire News*, April 4. Accessed July 7, 2018 from https://greenfiretimes.com/2018/04/gray-water-the-new-green/.

PWD (Philadelphia Water Department). 2018. Green stormwater management programs. Accessed August 8, 2018 from www.phillywatersheds.org/what_were_doing/green_infrastructure/programs.

Rahman, Ashig Ur. 2016. Urban sustainability through strategic planning: A case of metropolitan planning in Khulna city, Bangladesh. *Journal of Urban Management* 5: 16–22.

Randall, Alan. 2007. Benefit cost analysis and a save minimum standard of conservation. In *Handbook of Sustainable Development*, ed. Giles Atkinson, Simon Dietz, and Eric Neumayer, 91–105. Cheltenham: Edward Elgar.

RCT (Railroad Commission of Texas). 2018. Water use in association with oil and gas activities. Retrieved June 21, 2018 from www.rrc.state.tx.us/about-us/resource-center/faqs/oil-gas-faqs/faq-water-use-in-association-with-oil-and-gas-activities/.

Reclamation (U.S. Bureau of Reclamation). 2016. SECURE Water Act Section 9503(c): Reclamation climate change and water. Prepared for United States Congress. Bureau of Reclamation, Policy and Administration, Denver, CO.

Reguckia, Pawel, Marek Lewkowicza, Renata Krzyżyńskab, and Hussam Jouharac. 2018. Numerical study of water flow rates in power plant cooling systems. *Thermal Science and Engineering Progress* 7: 27–32.

Reig, Paul, Andrew Maddocks, and Francis Gassert. 2013. *World's most water stressed countries*. Washington, DC: World Resources Institute. Accessed November 4, 2017 from www.wri.org/blog/2013/12/world%E2%80%99s-36-most-water-stressed-countries.

Ribeiro, João M.P., Stephane L. Bocasanta, Risa O. Ávila, Mica Magoto, Ana V. Jonck, George M. Gabriel, José B.S.O. de Andrade Guerra. 2018. The adoption of strategies for sustainable cities: A comparative study between Seattle and Florianópolis legislation for energy and water efficiency in buildings. *Journal of Cleaner Production* 197: 366–378.

Rogers, John, Kristen Averyt, Steve Clemmer, Michelle Davis, Francisco Flores-Lopez, Peter Frumhoff, Doug Kenney, Jordan Macknick, Nadia Madden, James Meldrum, Jonothan Overpeck, Sandra Sattler, Erika Spanger-Siegfried, and David Yates. 2013. Water-smart power: Strengthening the U.S. electricity system in a warming world. Union of Concerned Scientists, Cambridge, MA. Available at www.ucsusa.org/sites/default/files/legacy/assets/documents/clean_energy/Water-Smart-Power-Full-Report.pdf.

Rogers, Kyle. 2018. Day Zero: City braces for water supply drying up. *Municipal Sewer and Water*, January 31. Accessed February 3, 2018 from www.mswmag.com/online_exclusives/2018/01/day-zero-city-braces-for-water-supply-drying-up.

Rosenberg, Matt. 2017. The largest cities in the world. Accessed October 29, 2017 from www.thoughtco.com/a-list-of-the-largest-cities-in-the-world-1435767.

Roser, Max, and Esteban Ortiz-Ospina. 2017. World population growth. Published online at OurWorldInData.org. Accessed December 3, 2017 from https://ourworldindata.org/world-population-growth/.

Roudi-Fahimi, Farzaneh, Liz Creel, and Roger-Mark de Souza. 2002. Finding the balance: Population and water scarcity in the Middle East and North Africa. MENA policy brief. Accessed January 1, 2018 from www.prb.org/pdf/FindingTheBalance_Eng.pdf.

Ruiz, Gerardo. 2015. Estimation of the groundwater recharge in the aquifer of Mexico City. *Procedia Environmental Sciences* 25: 220–226.

Rydin, Yvonne. 2007. Sustainable cities and local sustainability. In *Handbook of Sustainable Development*, ed. Giles Atkinson, Simon Dietz, and Eric Neumayer, 347–361. Cheltenham: Edward Elgar.

Salman, Salman M.A., and Siobhán McInerney-Lankford. 2004. *The Human Right to Water: Legal and Political Dimensions*. Washington, DC: The World Bank.

Savenije, H.H.G., and A.Y. Hoekstra. 2007. Water resources management. Vol. I: *Water Resources Management*. Encyclopedia of Life Support Systems. Accessed July 29, 2018 from www.eolss.net/sample-chapters/c07/e2-16-00-00.pdf.

Savenije, H.H.G., and P. Van der Zaag. 2008. Integrated water resources management: Concepts and issues. *Physics and Chemistry of the Earth* 33: 290–297.

Scherer, Laura, and Stephan Pfister. 2016. Global water footprint assessment of hydropower. *Renewable Energy* 99: 711–720.

Schnoor, Jerald L. 2008. Living with a changing water environment. *The Bridge* 38 (2): 46–54.

Schnoor, Jerald L. 2010. Water sustainability in a changing world. 2010 Clarke Prize Lecture, National Water Research Institute. Accessed October 11, 2017 from nwri-usa.org/pdf/2010/ClarkPrizelecture.pdf.

Schultz, Sebastian. 2012. Here to stay: The daily challenges facing Shanghai's float population. *The Global Urbanist*, October 23. Accessed January 21, 2018 from http://globalurbanist.com/2012/10/23/shanghais-floating-population.

Schuster-Wallace, C.J., M. Qadir, Z. Adeel, F. Ranaud, and S.K. Dickin. 2015. Putting water and energy at the heart of sustainable development. United Nations University. Accessed July 19, 2018 from http://inweh.unu.edu/wp-content/uploads/2015/03/Putting-Water-and-Energy-at-the-Heart-of-Sustainable-Development-Web.pdf.

Schwab. Donald P. 1999. *Research Methods for Organizational Studies*. Mahwah, NJ: Lawrence Erlbaum.

Senn, Dominique, and Dorothee Spuhler. 2015. Water, sanitation and urbanization. Sustainable Sanitation and Water Management (SWWM). Accessed October 25, 2017 from https://www.sswm.info/content/water-sanitation-and-urbanisation.

Senn, Dominique, and Dorthee Spuhler. 2018. Water, sanitation and urbanization. Accessed December 6, 2018 from https://sswm.info/node/7722.

Shaukh, Hina, and Ijaz Nabil. 2017. The six biggest challenges facing Pakistan's urban future. International Growth Center (IGI). Accessed January 17, 2018 from www.theigc.org/blog/the-six-biggest-challenges-facing-pakistans-urban-future/.

Shen, Yanjun, Changyuan Tang, Jieying Xiao, Taikan Oki, and Shinjiro Kanae. 2005. Effects of urbanization on water resource development and its problems in Shijiazhuang, China. In *Proceedings of the S2 symposium of the Seventh IAHS Scientific Assembly at Foz do Iguaçu, Brazin*, April. Accessed December 27, 2017 from http://hydro.sjziam.ac.cn/resources/Effects%20of%20urbanization%20on%20water%20resources%20development%20and%20its%20problems%20in%20Shijiazhuang,%20China.pdf.

Shevchuk, Anatoly. 2013. Water resource management, water ecosystem problems. In *Sustainable Development in Russia*, ed. Sergei Bobylev and Renat Perelet, 144–153. Berlin: Russian-German Exchange.

Shuster-Wallace, C.J., M. Qadir, Z. Adeel, F. Renaud, and S.K. Dickin. 2015. Putting water and energy at the heart of sustainable development. United Nations University (UNU). Accessed October 24, 2017 from http://inweh.unu.edu/wp-content/uploads/2015/03/Putting-Water-and-Energy-at-the-Heart-of-Sustainable-Development-Web.pdf.

Sinimbu, Fabiola, and Liria Jade. 2017. Over 850 Brazil cities face major water shortage issues. *Agenciea Brazil*, March 22. Accessed April 15, 2018 from http://agenciabrasil.ebc.com.br/en/geral/noticia/2017-03/over-850-brazil-cities-face-major-water-shortage-issues.

Slavikova, Sara. 2018. 10 benefits of crop rotation in agriculture. Accessed August 14, 2018 from https://greentumble.com/10-benefits-of-crop-rotation/.

Smart Cities Dive! (formerly Sustainable Cities Collective). 2018. Three cities innovate solutions for tackling water scarcity. Accessed July 24, 2018 from www.smartcitiesdive.com/ex/sustainablecitiescollective/friday-fun-three-cities-innovate-solutions-tackling-water-scarcity/1099086/.

Smidt, Samuel J., Erin M.K. Haacker, Anthony D. Kendall, Jillian M. Deines, Lisi Pei, Kayla A. Corterman, Haoyang Li, Xiao Liu, Bruno Basso, and David W. Hyndman. 2016. Complex water management in modern agriculture: Trends in the water-energy-food nexus over the High Plains Aquifer. *Science of the Environment* 566–567: 988–1001.

Snyder, Shannyn. 2017. Water in crisis: India. The Water Project. Concord, New Hampshire. Accessed December 9, 2017 from https://thewaterproject.org/water-crisis/water-in-crisis-India.

Soeder, Daniel J., and William M. Keppel. 2009. Water resources and natural gas production from the Marcellus shale. United States Geological Survey (USGS) Report. Accessed January 10, 2018 from https://pubs.usgs.gov/fs/2009/3032/pdf/FS2009-3032.pdf.

Sonneland, Holly K. 2016. Population trends in Latin America. Americas Society/Council of the Americas. Accessed January 14, 2018 from www.as-coa.org/articles/weekly-chart-population-trends-latin-america.

Soubbotina, Tatyana P. 2004. *Beyond economic growth: An introduction to sustainable development*, 2nd ed. World Bank. Accessed December 4, 2018 from http://documents.worldbank.org/curated/en/454041468780615049/pdf/2489402nd0edition0Beyond0economic0growth.pdf.

Sprang, Edward, W.R. Moomaw, K.S. Gallagher, P.H. Kirshen, and D.H. Marks. 2014. The water consumption of energy production: An international comparison. *Environmental Research Papers* 9 (10): 1–14. Accessed June 6, 2018 from iopscience.iop.org/article/10.1988/1748-9326/9/10/105002/pdf.

Spring, Ursula Oswald. 2011. *Water Resources in Mexico: Scarcity, Degradation, Stress, Conflicts, Management, and Policy*. New York: Springer.

SPU (Seattle Public Utilities). 2018a. City of Seattle 2018 NPDES Phase I Municipal Stormwater Permit. Stormwater Management Program. Accessed June 5, 2018 from www.seattle.gov/util/cs/groups/public/@spu/@drainsew/documents/webcontent/1_074774.pdf.

SPU (Seattle Public Utilities). 2018b. 2019 Strategic plan. Accessed August 24, 2018 from www.seattle.gov/util/Documents/Plans/Water/WaterSystemPlan/index.htm.

Sroufe, Robert. 2017. Integration and organizational change towards sustainability. *Journal of Cleaner Production* 162 (217): 315–329.

SSWM (Sustainable Sanitation and Water Management). 2014. Integrated water resources management (IWRM). Accessed December 20, 2017 from www.sswm.info/category/concept/iwrm.

Statista. 2018. The twenty countries with the largest population growth rates in 2017 (compared to the previous year). Accessed January 7, 2018 from www.statista.com/statistics/264687/countries-with-the-highest-population-growth-rate/.

Stillwell, Ashlynn S., Mary E. Clayton, and Michael E. Webber. 2011. Technical analysis of a river basin-based model of advanced power plant cooling technologies for mitigating water management challenges. *Environmental Research Letters* 6 (3): 1–11.

Stillwell, Ashlynn S., and Michael E. Webber. 2013. Evaluation of power generation operations in response to changes in surface water reservoir storage. *Environmental Research Letters* 7 (2): 1–11.

Strucki, Andrea. 2009. Open innovation models in public utilities as an option to react on reform process: The case of the postal sector. Paper presented at the second annual conference on Competition and Regulation in Networked Industries, Brussels, Belgium. Accessed August 9, 2015 from infoscience.epfl.ch/record/151576/files/astruckipaperCRN109.pdf.

Studart, Rogerio, and Kevin Gallagher. 2018. Guaranteeing sustainable infrastructure. *International Economics* 155 (2018): 84–91.

Swain, Ashok. 2017. Here's why the idea to link India's rivers will remain a pipe dream. *Outlook*, September 3. Accessed April 12, 2018 from https://www.outlookindia.com/website/story/heres-why-the-idea-to-link-indias-rivers-will-remain-a-pipe-dream/301157.

SWITCH (Sustainable Waster Management Improves Tomorrows Cities Health). 2006. Sustainable water management in the city of the future. Accessed October 11, 2017 from www.switchurbanwater.eu/outputs/pdfs/SWITCH-Final-Report.pdf.

SWITCH (Sustainable Waster Management Improves Tomorrows Cities Health). 2011. Switch in the city. International Water and Sanitation Center (IRC), The Hague. Accessed August 4, 2018 from www.switchurbanwater.eu/outputs/pdfs/SWITCH_in_the_City.pdf.

SWP (Sustainable Water Partnership). 2018. What is water security? Accessed August 7, 2018 from www.swpwater.org/what-is-water-security/.

Tabibian, M., and S. Movahed. 2016. Towards resilient and sustainable cities: A conceptual framework. *Scientia Iranica* 23 (5): 2081–2093.

Tao, Tao, and Kunlun Xin. 2014. A sustainable plan for China's drinking water. *Nature* 511: 527–529.

Tidwell, Vincent C., Jordan Macknick, Katie Zemlick, Jasmine Sanchez, and Tibebe Woldeyesus. 2014. Can switching fuels save water? A life cycle quantification of freshwater consumption for Texas coal- and natural gas-fired electricity. *Applied Energy* 131: 508–516.

Tisdell, Clement A. 2009. Sustainable agriculture. In *Handbook of Sustainable Development*, ed. Giles Atkinson, Simon Dietz, and Eric Neumayer, 362–375. Cheltenham: Edward Elgar.

Tortajada, Cecilia. 2008. Challenges and realities of water management of megacities: The case of Mexico City metropolitan area. *Journal of International Affairs* 61 (2): 147–166.

Tortajada, Cecilia. 2016. Policy dimensions of development and financing of sater infrastructure: The cases of China and India. *Environmental Science & Policy* 64: 177–187.

Tropp, Håkan. 2007. Water governance: Trends and needs for new capacity development. *Water Policy* 9 (S2): 19–30.

TSP (The Shift Project). 2015. Breakdown of electricity generation by energy source. Accessed February 8, 2018 from http://www.tsp-data-portal.org/Breakdown-of-Electricity-Generation-by-Energy-Source#tspQvChart.

Turner, Neil C. 2004. Agronomic options for improving rainfall-use efficiency of crops in dryland farming systems. *Journal of Experimental Botany* 55 (407): 2413–2425.

TVA (Tennessee Valley Authority). n.d. Caledonia combined cycle plant. Accessed July 24, 2018 from www.tva.gov/Energy/Our-Power-System/Natural-Gas/Caledonia-Combined-Cycle-Plant.

UACOG (Utility Advisors and Collaborating Organization Group). 2008. Effective utility management: A primer for water and wastewater utilities. Accessed August 12, 2015 from 2008-06eumprimer.pdf.

UACOG (Utility Advisors and Collaborating Organization Group). 2013. Rural and small systems guidebook to sustainable utility management. Accessed August 12, 2015 from Sustainable-Management-of-Rural-and-Small-Systems-Guide-Final-10-24-13.pdf.

UCLA (University of California at Los Angeles). n.d. What is sustainability? University of California at Los Angeles Sustainability Committee. Accessed October 19, 2017 from https://www.sustain.ucla.edu/about-us/what-is-sustainability/.

UCS (Union of Concerned Scientists). 2011. Nuclear power and water: Quick facts on nuclear power generation and water use. Cambridge, MA. Accessed November 1, 2017 from http://www.ucsusa.org/sites/default/files/legacy/assets/documents/nuclear_power/fact-sheet-water-use.pdf.

UCS (Union of Concerned Scientists). 2016. Energy and water use. Union of Concerned Scientists. Accessed June 19, 2018 from www.ucsusa.org/clean-energy/energy-water-use#.WylAIFVKiUk.

UCS (Union of Concerned Scientists). 2017a. Energy and water use: Our energy choices have a direct effect on our water resources. Union of Concerned Scientists. Accessed December 6, 2017 from www.ucsusa.org/clean-energy/energy-water-use#.WigZ0FWnGUk.

UCS (Union of Concerned Scientists). 2017b. How do we know that humans are the major cause of global warming? Union of Concerned Scientists. Accessed December 2, 2017 from www.ucsusa.org.global-warming/science-and-impacts/science/human-contriution-to-gw-faq.html#.WiLROnm1vGg.

Udias, Angel, Marco Pastori, Anna Malago, Olga Vigiak, Nikolaos P. Nikolaidis, and Faycal Bouraoui. 2018. Identifying efficient agricultural irrigation strategies in Crete. *Science of the Total Environment* 633: 271–284.

Udimal, Thomas B., Zhuang Jincai, Emmanuel C. Ayamba, and Samuel M. Owusu. 2017. China's water situation: The supply of water and the pattern of it uses. *International Journal of Sustainable Built Environment* 6 (2017): 491–500.

UN (United Nations). 1992. The Dublin Statement on water and sustainable development. Accessed December 7, 2018 from www.un-documents.net/h2o-dub.htm.

UN (United Nations). 1997. Earth summit. United Nations. Accessed October 22, 2017 from http://www.un.org/geninfo/bp/envirp2.html.

UN (United Nations). 2006. Gender, water and sanitation: A polity brief. United Nations Water and the Interagency Task Force on Gender and Water. Accessed November 8, 2017 from http://www.un.org/waterforlifedecade/pdf/un_water_policy_brief_2_gender.pdf.

UN (United Nations). 2010. Sustainable water management in cities: Engaging stakeholders for effective change and action. Final report. Accessed August 4, 2018 from http://www.un.org/waterforlifedecade/swm_cities_zaragoza_2010/pdf/final_report_swm_cities.pdf.

UN (United Nations). 2014a. International decade for action 'Water for Life' 2005–2015. Accessed November 5, 2017 from http://www.un.org/waterforlifedecade/food_security.shtml.

UN (United Nations). 2014b. World's population increasingly urban with more than half living in urban areas. United Nations. Accessed October 20, 2017 from http://www.un.org/en/development/desa/news/population/world-urbanization-prospects-2014.html.

UN (United Nations). 2014c. Integrated water resources management (IWRM). United Nations. Accessed June 3, 2018 from http://www.un.org/waterforlifedecade/iwrm.shtml.

UN (United Nations). 2014d. World water development report 2014: Water and energy. Accessed July 2018 from http://www.unesco.org/new/en/natural-sciences/environment/water/wwap/wwdr/2014-water-and-energy/.

UN (United Nations). 2015. Water for a sustainable world: Facts and figures. United Nations. Accessed October 11, 2017 from http://www.unesco.org/new/en/natural-sciences/environment/water/wwap/wwdr/2015-water-for-a-sustainable-world/.

UN (United Nations). 2017. World population prospects. Accessed October 20, 2017 from https://esa.un.org/unpd/wpp/Publications/Files/WPP2017_KeyFindings.pdf.

UN (United Nations). 2018. 2018 World Water Development Report. Accessed July 23, 2018 from http://unesdoc.unesco.org/images/0026/002614/261424e.pdf.

UNDESA (United Nations Department of Economic and Social Affairs). 2014. Water and cities. Accessed January 8, 2017 from www.un.org/waterforlifedecade/water_cities.shtml.

UNDP (United Nations Development Program). 2017a. Goal six targets. Accessed December 7, 2017 from www.undp.org/content/undp/en/home/sustainable-development-goals/goal-6-clean-water-and-sanitation/targets/.

UNDP (United Nations Development Program). 2017b. Human Development Report. Accessed January 14, 2018 from http://hdr.undp.org/sites/default/files/2016_human_development_report.pdf.

UNESCO (United Nations Educational, Scientific and Cultural Organization). 2017. *Climate change and water.* Overview from the World Water Development Report 3: Water in a changing world. Accessed October 16, 2017 from unesdoc.unesco.org/images/0018/001863/186318e.pdf.

UNISDR (United Nations Office for Disaster Risk Reduction). 2015. The economic and human impact of disasters in the last 10 years. Accessed December 19, 2017 from https://www.unisdr.org/we/inform/disaster-statistics.

UNU-INWEH (United Nations University—Institute for Water, Environment and Health). 2013. Catalyzing water for sustainable development and growth. Framing water within the post 2015 development agenda: Options and considerations. Accessed December 16, 2017 from http://www.zaragoza.es/contenidos/medioambiente/onu/1093-eng_Catalyzing_water_for_sustainable_development_and_growth.pdf.

UN Water. 2014a. A post-2015 global goal for water: Synthesis of key findings and recommendations from UN-Water. Accessed December 16, 2017 from www.un.org/waterforlifedecade/pdf/27_01_2014_UN-Water_paper_on_a_post2015_global_goal_for_water.pdf.

UN Water. 2014b. Water for life: Integrated water resources management (IWRM). Accessed December 20, 2017 from www.un.org/waterforlifedecade/iwrm.shtml.

UN Water. 2017. Water and urbanization. Accessed January 21, 2018 from www.unwater.org/water-facts/urbanization/.

UN Water. 2018. Nature-based solutions for water. United Nations World Water Development Report 2018. Accessed July 16, 2018 from unesdoc.unesco.org/images/0026/002614/261424e.pdf.

USAID (United States Agency for International Development). 2017. Water and wastewater project. Accessed November 23, 2017 from www.usaid.gov/jordan/fact-sheets/water-and-wastewater-infrastructure-project.

USDA (United States Department of Agriculture). 2007. Sustainable agriculture: Definition and terms. Accessed November 10, 2017 from https://www.nal.usda.gov/afsic/sustainable-agriculture-definitions-and-terms.

USDA (United States Department of Agriculture). 2012. Improving irrigation water use efficiency. Natural Resources Conservation Service, U.S. Department of Agriculture. Accessed November 13, 2017 from https://www.nrcs.usda.gov/wps/portal/nrcs/detail/az/soils/?cid=nrcs144p2_065174.

USDA (United States Department of Agriculture). 2017. Irrigation and water use: Overview. Economic Research Service, U.S. Department of Agriculture. Accessed April 7, 2018 from www.ers.usda.gov/topics/farm-practices-management/irrigation-water-use.aspx.

USGS (United States Geological Survey). 2010. Canal providing for crop flood irrigation. Accessed December 8, 2017 from www.usgs.gov/media/images/canal-providing-water-crop-flood-irrigation.

USGS (United States Geological Survey). 2014. Estimated use of water in the United States in 2010. U.S. Geological Survey. U.S. Department of the Interior Circular 1405. Accessed September 24, 2015 from pubs.usgs.gov/circ/1405/pdf/circ1405.pdf.

USGS (United States Geological Survey). 2015. 2011–2014 hydraulic fracturing water use United States Geological Survey. Accessed January 11, 2018 from https://tse1.mm.bing.net/th?id=OIP.jWq9ymU_Kl8LqVB3hkDWhgHaFc&pid=Api&rs=1&c=1&qlt=95&w=141&h=104.

USGS (United States Geological Survey). 2016a. Water use in the United States: Thermoelectric power water use. Department of the Interior. Accessed October 28, 2017 from https://water.usgs.gov/watuse/wupt.html.

USGS (United States Geological Survey). 2016b. Total water use. Department of the Interior. Accessed November 10, 2017 from https://water.usgs.gov/wateruse/wuto.html.

USGS (United States Geological Survey). 2016c. Total water use in the United States in 2010. Department of the Interior. Accessed November 10, 2017 from https://water.usgs.gov/edu/wateruse-total.html.

USGS (United States Geological Survey). 2017a. The water cycle. Department of the Interior. Accessed November 5, 2017 from https://water.usgs.gov/edu/watercycle.html.

USGS (United States Geological Survey). 2017b. Water science: Glossary of terms. Department of the Interior. Accessed October 30, 2017 from https://water.usgs.gov/edu/dictionary.html.

Van den Berg, Jeoen C. 2007. Sustainable development in ecological economics. In *Handbook of Sustainable Development*, ed. Giles Atkinson, Simon Dietz, and Eric Neumayer, 63–77. Cheltenham: Edward Elgar.

Verbruggen, Aviel. 2018. Renewable and nuclear power: A common future? *Energy Policy* 36: 4036–4037.

Vidal, John. 2017. As water scarcity deepens across Latin America, political instability grows. *The Guardian*, March 1. Accessed January 26, 2018 from www.theguardian.com/global-development-professionals-network/2017/mar/01/water-scarcity-latin-america-political-instability.

Vigerstol, Kari. 2011a. Eco logic: Sustainability risk assessment, part 1: defining the area of influence and sustainability boundaries. *Journal of the American Water Works Association* 101 (1): 107–109.

Vigerstol, Kari. 2011b. Eco logic: Water sustainability risk assessment, part 2: primary and secondary effects. *Journal of the American Water Works Association* 103 (3): 20–23.

Vos, Jeroen, and Rutgerd Boelens. 2014. Sustainability standards and the water question. *Development and Change* 45 (2): 205–230.

Vucetich, John A., and Michael P. Nelson. 2010. Sustainability: Virtuous or vulgar? *Bioscience* 60 (7): 539–544.

Wagena, Moges B., and Zachary M. Easton. 2018. Agricultural conservation practices can help mitigate the impact of climate change. *Science of the Environment* 635: 132–143.

Wang, Jiao, Linjun Zhong, and Charles Iceland. 2017. China's water stress in on the rise. World Resources Institute, Washington, DC. Accessed December 8, 2017 from www.wri.org/blog/2017/01/chinas-water-stress-rise.

Wang, Sicong, and Shifeng Wang. 2017. Implications of improving energy efficiency for water Huiresources. *Energy* 140: 922–928.

Ward, Frank A. 2007. Decision support for water policy: A review of economic concepts and tools. *Water Policy* 9 (1): 1–31.

Water Deeply. 2017. Oroville dam incident explained: What happened, why and what's next. Accessed July 3, 2018 from www.newsdeeply.com/water/articles/2017/06/14/oroville-dam-incident-explained-what-happened-why-and-whats-next.

WCED (World Commission on Environment and Development). 1987. *Our Common Future*. New York: Oxford University Press.

WEF (World Economic Forum). 2014. The global infrastructure gap. Accessed November 14, 2017 from http://reports.weforum.org/strategic-infrastructure-2014/introduction-the-operations-and-maintenance-om-imperative/the-global-infrastructure-gap/.

WHO (World Health Organization). 2011. Water, sanitation and hygiene interventions and the prevention of diarrhea. Accessed October 24, 2017 from http://www.who.int/elena/titles/bbc/wsh_diarrhoea/en/.

WHO (World Health Organization). 2016. Urban population growth. World Health Organization. Accessed October 20, 2017 from www.who.int/gho/urban_health/situation_trends/urban_population_growth/en.html.

WHO (World Health Organization). 2017a. Progress on drinking water, sanitation, and hygiene, update and sustainable development goals. World Health Organization (WHO) and the United Nations Children's Fund (UNICEF) Joint Monitoring Program (JMP). Accessed November 7, 2017 from http://apps.who.int/iris/bitstream/10665/258617/1/9789241512893-eng.pdf?ua=1.

WHO (World Health Organization). 2017b. Drinking water fact sheet. World Health Organization. Accessed October 23, 2017 from www.who.int/mediacentre/factsheets/fs391/en/.

Wikipedia Commons. 2015. Map of Nigeria. Accessed December 7, 2017 from https://commons.wikimedia.org/wiki/File:Nigeria_blank_with_rivers.svg.

WNA (World Nuclear Association). 2017. Cooling power plants. World Nuclear Association, London. Accessed November 3, 2017 from http://www.world-nuclear.org/information-library/current-and-future-generation/cooling-power-plants.aspx.

WNA (World Nuclear Association). 2018. Nuclear power in the European Union. Accessed July 24, 2018 from www.world-nuclear.org/information-library/country-profiles/others/european-union.aspx.

Woodward, Guy, Danial M. Perkins, and Lee E. Brown. 2010. Climate change and freshwater ecosystems: Impacts across multiple levels of organization. *Physical Transactions of the Royal Society B* 365 (1549): 2093–2136.

World Atlas. 2017. Highest usage of freshwater by country. Accessed December 8, 2017 from www.worldatlas.com/articles/highest-use-of-freshwater-by-country.html.

World Atlas. 2018. Countries with the biggest population declines. Accessed January 8, 2018 from www.worldatlas.com/articles/countries-with-the-biggest-population-decline.html.

World Bank. 2012. Gender equality and development. World Development Report 2012. Accessed October 24, 2017 from http://documents.worldbank.org/curated/en/492221468136792185/pdf/646650WDR0201200Box364543B-00PUBLIC0.pdf.

World Bank. 2013. Thirsty energy: Securing energy in a water-constrained world. Accessed October 24, 2017 from http://www.worldbank.org/en/topic/sustainabledevelopment/brief/water-energy-nexus.

World Bank. 2014. Water resources management: Sector results profile. Accessed November 13, 2017 from http://www.worldbank.org/en/results/2013/04/15/water-resources-management-results-profile.

World Bank. 2015. Thirsty energy update. Accessed October 24, 2017 from www.un.org/waterforlifedecade/images/newsletter/Thirsty_Energy_general_update_February_2015.pdf.

World Bank. 2016. Brazil may be the owner of 20% of the world's water supply but it is still very thirsty. Accessed July 9, 2018 from http://www.worldbank.org/en/news/feature/2016/07/27/how-brazil-managing-water-resources-new-report-scd.

World Bank. 2017a. Urban population (% of total). Accessed October 22, 2017 from https://data.worldbank.org/indicator/SP.URB.TOTL.IN.ZS.

World Bank. 2017b. Water resource management. Accessed March 29, 2018 from http://www.worldbank.org/en/topic/waterresourcesmanagement.

World Water Council. 2006. Water problems in Latin America. Accessed January 13, 2018 from http://www.worldwatercouncil.org/fileadmin/wwc/News/WWC_News/water_problems_22.03.04.pdf.

Worldmeters.info. 2018. Countries in the world by population. Accessed January 7, 2018 from http://www.worldometers.info/world-population/population-by-country/.

WPB. 2017. Sub-Saharan population. *World Population Review*. Accessed January 12, 2018 from http://worldpopulationreview.com/continents/sub-saharan-africa-population/.

WRC (World Resource Council). 2015. Ranking the world's most water-stressed countries in 2040. Accessed January 14, 2018 from www.wri.org/blog/2015/08/ranking-world%E2%80%99s-most-water-stressed-countries-2040.

WRF (Water Research Foundation). 2014. Performance benchmarking for effectively managed water utilities. Accessed August 19, 2018 from www.waterrf.org/publicreportlibrary/4313b.pdf.

WRI (World Resources Index). 2015. Environmental democracy index. Accessed December 6, 2018 from www.wri.org/publication/environmental-index.

Wright, Lawrence. 2018. The glut economy. *The New Yorker* 93 (42): 42–53.

WSDOE (Washington State Department of Ecology). 2017. Landowners fined for illegally pumping 500 million gallons from Odessa Aquifer. Washington State Department of Ecology. Accessed December 14, 2017 from https://www.ecology.wa.gov/About-us/Get-to-know-us/News/2017/Dec-19-Landowners-fined-for-illegally-pumping-500.

WSDOH (Washington State Department of Health). 2009. Asset management for small water systems. Washington State Department of Health. Accessed June 5, 2018 from www.doh.wa.gov/Portals/1/Documents/Pubs/331-445.pdf.

WSDOH (Washington State Department of Health). 2011. *Water Use Efficiency Guidebook*, 3rd ed. Washington State Department of Health. Accessed August 21, 2015 from www.doh.wa.gov/Portals/1/Documents/Pubs/331-375.pdf.

WSP (Water and Sanitation Program). 2017. Clean water and sanitation reduce childhood malnutrition and diarrhea. Water and Sanitation Program, World Bank Group. Accessed October 23, 2017 from https://www.wsp.org/featuresevents/features/clean-water-and-sanitation-reduce-childhood-malnutrituion-and-diarrhea.

WSP (Water and Sanitation Program). 2018. Urban water and sanitation: The challenges. Water and Sanitation Program, World Bank. Accessed January 8, 2017 from www.wsp.org/topics/urban%20water%20supply%20and%20sanitation.

Wu, May, Michael Wang, and Salil Arora. 2009. Consumptive water use in the production of bioethanol and petroleum gasoline. Center for Transportation Research, Argonne National Laboratory (ANL). Accessed June 19, 2018 from www.great.es.anl.gov/publication-consumptive-water.

Wu, X.D., and C.Q. Chen. 2017. Energy and water nexus in power generation: The surprisingly high amount of industrial water use induced by solar power infrastructure in China. *Applied energy* 195: 125–136.

WWAP (United Nations World Water Assessment Programme). 2015. *The United Nations World Water Development Report 2015: Water for a Sustainable*

World. Paris: UNESCO. Available at http://unesdoc.unesco.org/images/0023/002318/231823E.pdf.

WWAP (United Nations World Water Assessment Programme). 2017. *The United Nations World Water Development Report 2017. Wastewater: The Untapped Resource.* Paris: UNESCO. Available at http://unesdoc.unesco.org/images/0024/002471/247153e.pdf.

WWF (World Wildlife Fund). 2017. Water scarcity. Accessed November 7, 2017 from https://www.worldwildlife.org/threats/water-scarcity.

Yates, Julian S., and Leila M. Harris. 2018. Hybrid regulatory landscapes: The human right to water, variegated neoliberal water governance, and policy transfer in Cape Town, South Africa, and Accra Ghana. *World Development* 110: 75–87.

Yigzaw, Wondmagegn, and Faisal Hossain. 2016. Water sustainability of large cities in the United States from the perspectives of population increase, anthropogenic activities, and climate change. *Earth's Future* 4 (12): 603–617.

Yu, Min, Chaoran Wang, Yi Liu, Gustaf Olsson, and Chunyan Wang. 2018. Sustainability of mega water diversion projects: Experience and lessons from China. *Science of the Total Environment* 619–620: 721–731.

Zaveri, Esha, Danielle S. Grogan, Karen Fisher-Vanden, Steve Frolking, Richard B. Lammers, Douglas H. Wrenn, Alexander Prusevich, and Robert E. Nicholas. 2016. Invisible water, visible impact: Groundwater use and Indian agriculture under climate change. Accessed April 12, 2018 from iopscience.iop.org/article/10.1088/1748-9326/4/8/084005.pdf.

Zhao, Kiki. 2014. Water from China's South-North transfer project flows to Beijing. *The New York Times, Sinosphere Blog,* December 25. Accessed June 27, 2018 from sinosphere.blogs.nytimes.com/2014/12/25/water-from-chinas-south-north-transfer-project-flows-to-beijing/.

Zhao, Zhen-Yu., Jian Zuo, and George Zillante. 2017. Transformation of water resource management: A case study of the South-to-North Water Diversion Project. *Journal of Cleaner Production* 163: 136–145.

Index

A
Agriculture irrigation 186, 196, 225
Agriculture sprinkler systems 186
Alternative water supplies 218, 219, 303
Asset management planning 265, 266, 287

B
Biofuel feedstocks 121
Brazil 9, 58, 79, 153–155, 162–164, 166, 193, 219, 230, 235, 236, 254, 255
Brundtland Commission Report 4

C
Capital planning 144, 287
Central America 53, 74–76, 112
China 9, 47, 54, 58, 60, 70, 77, 79, 83, 85, 90–93, 109, 123, 124, 127, 128, 135, 154, 155, 159–161, 166, 202, 206, 211, 217, 221–224, 226, 230, 243, 254–256, 259–261
Clean Water Act (CWA) 111, 138, 139, 248
Clean Water Rule of 2015 111
Climate change 4–6, 10, 15, 24, 28, 29, 31, 32, 35, 47–49, 53–63, 70, 72, 74, 76–79, 88, 95, 96, 104, 113, 127, 128, 133, 143, 148, 161–165, 175, 177, 182, 185, 188–190, 195, 197, 202, 205, 206, 208, 213, 217, 221, 223, 230, 233, 236–238, 247, 253, 261, 267, 283, 285, 289–291, 305
Climate Smart Agriculture (CSA) 187–190, 197
Coastal flooding 56, 57
Collected and stored stormwater 233
Community water utilities 136, 174
Conservation agriculture (CA) 187, 188, 190–195, 197, 198
Core values 287, 291, 293
Critical assets 266, 276, 277, 281

D. E. McNabb, *Global Pathways to Water Sustainability*,
https://doi.org/10.1007/978-3-030-04085-7

D

Desalination 13, 18, 27, 79, 113, 143, 160, 161, 218, 222, 253, 254, 261
Drip irrigation 103, 186

E

Establishing a vision 290
Exeter, New Hampshire 179
Extreme water stress 46, 68, 77, 78, 217, 221, 223, 230

F

Financial plan 266, 297–299
Flint, Michigan 84, 135
Flood irrigation 92, 103, 104, 186
Floods, droughts and windstorms 47
Fossil fuels 5, 28, 118, 119, 123, 128, 129, 207, 209, 213
Fracking 24, 125, 126
Funding strategy 271, 281
Furrow irrigation 186, 197

G

General utility planning (GUP) 286, 303
Global Change Assessment Model (GCAM) 59
Global warming 5, 10, 17, 19, 35, 47, 54, 57, 306
Goals and Operational Objectives 295
Greenhouse gasses 54, 203
Green infrastructure 170, 177, 179, 233, 236, 237, 240–243, 248, 249, 302
Gro Harlem Brundtland 19
Groundwater depletion 108

H

Harvested rainwater 233, 236
Hydraulic fracturing 119, 120, 125, 212
Hydrologic cycle 18, 28, 55, 160, 260
Hydropower 24, 60, 121, 122, 129, 163, 209, 210, 306

I

India 9, 54, 58, 60, 70, 77, 79, 83, 90, 124, 128, 153, 154, 156–158, 166, 195, 217, 225–227, 230, 235, 255, 257, 258
Indian National Perspective 158, 259
Indonesia 70, 77, 79, 83, 90, 94, 95
Integrated urban water management (IUWM) 174, 175, 178, 180, 266
Integrated water resources management (IWRM) 7, 13, 174, 175, 223
Iran 58, 72, 83, 90, 94
Irrigated agriculture 58–60, 100–102, 112, 223, 230, 257

J

Jordan 217, 228–230

L

Level of service (LOS) 41, 266, 267, 270, 274–276, 282, 298
Life cycle costing (LCC) 177, 271, 279

M

Malnutrition 11, 15, 20, 22, 23
McKinsey Group 156, 257
Mexico 54, 57, 58, 74, 75, 96, 163–165, 240, 242

Micro irrigation 187
Middle East 14, 53, 54, 58, 71, 72, 101, 106, 192, 228–230, 239, 253
Millennium Declaration 11, 12
Mission Statement 289
Municipal graywater 233, 234

N

Nigeria 70, 73, 79, 163
Non-point pollution 8
North Africa 53, 54, 58, 71, 72, 192
North America 58, 60, 71, 74, 88
No-till agriculture 193
Nuclear-generated electricity 127

O

Ogallala Aquifer 100
Openness to innovation 37

P

Pakistan 60, 70, 77, 90, 93, 94
Peri-urban areas 143
Point source pollution 7, 159, 259
Poor sanitation 10, 22, 23, 95, 225
Population growth 3–5, 9, 10, 22, 24–26, 28, 29, 31, 32, 35, 41, 49, 58, 67, 70, 71, 73–79, 83, 88, 93–95, 97, 109, 113, 138, 143, 144, 148, 154, 156, 175, 185, 189, 190, 195, 197, 207, 213, 217, 221, 224, 226, 236, 253, 255, 257, 268, 290, 291, 306
Public utilities 34, 265, 269, 288

R

Recycled water 18, 27, 70, 79, 113, 147, 181, 218, 247, 253, 261
Repair and replacement schedule 281
Right of access 20

S

Salinization 109
Solar power 121, 209, 210, 213, 306
South Africa 11, 73, 74, 127, 153–155, 161, 162, 165, 166, 217, 254–256
South America 60, 74, 76, 77, 106, 162, 192, 194
South-to-North Water Diversion Project (SNWDP) 161, 260
Strategic Sustainability Planning (SSP) 287
Sub-Saharan Africa 71, 73, 74, 79, 88, 193
Supplemental water solutions 233
Sustainability 3–6, 8–10, 14, 15, 17, 24–27, 29, 31, 32, 35, 41, 44, 48, 49, 53, 64, 67, 69, 79, 83–85, 89, 95, 97, 100, 112, 113, 117, 124, 128, 136, 142, 143, 148, 153, 154, 156, 158, 160, 161, 166, 170–172, 174, 177–182, 187, 190, 197, 213, 217–219, 221, 229, 230, 237, 247, 255, 261, 265–269, 275, 281–283, 285–290, 293, 296–300, 302, 303, 307
Sustainability goals 6, 12, 33–35, 43, 49, 182, 191, 269, 289, 300, 301, 303, 307
Sustainability management 26, 113, 173

Sustainability pathways 46, 154, 156, 158, 187, 255, 258
Sustainability planning 174, 175, 177, 299, 300, 303
Sustainable agriculture 4, 26, 99, 100, 102, 185, 187, 188, 190, 198
Sustainable cites 169, 170, 179, 182
Sustainable development 4, 7, 10–12, 14, 15, 19, 20, 22, 26, 27, 29, 43, 100, 144, 172, 190, 234
Sustainable Water Management Improves Tomorrows Cities Health (SWITCH) 178
Sustainable Water Partnership (SWP) 153

T

Texas 18, 96, 100, 119, 125, 172, 193, 212
Thermal electric generation 7, 101, 123, 127, 129, 201
Thermoelectric generating plants 123, 202, 213
Three water crises 6, 15

U

Union of Concerned Scientists (UCS) 119, 123, 127, 130, 206
United States of America (USA) 23, 71, 96, 100, 103, 105, 106, 109, 110, 117, 118, 122–125, 127, 136, 138–140, 143, 146, 163, 165, 171, 172, 174, 176, 191–193, 201–203, 205, 207, 209–213, 235, 237, 240, 245, 265, 266, 290
Urbanization 4, 5, 9, 10, 17, 22, 24, 25, 28, 29, 31, 35, 47, 49, 54, 58, 59, 69, 73, 75–77, 79, 83–85, 88–95, 97, 100, 104, 106, 109, 113, 128, 129, 138, 139, 144, 148, 154, 156, 163, 171, 175, 217, 221, 223, 224, 226, 227, 230, 251, 253, 255, 257, 290
Urban management challenges 88, 178
Urban water runoff 89, 234, 238, 242
U.S. Geological Survey (USGS) 124, 125, 202, 308

W

Water and energy 24, 117, 118, 123, 124, 129, 130, 181, 201, 202, 205, 306
Water efficiency 13, 42, 104, 106, 158, 258
Water efficiency rules 42
Water–energy nexus 117, 201, 202
Water for life 4, 12, 69
Water governance 44, 45
Water infrastructure 32, 33, 57, 133, 135, 136, 138, 139, 142, 144, 145, 148, 162, 174, 175, 182, 219, 224, 230, 266, 291, 299
Water pollution 6, 8, 13, 15, 19, 22, 23, 46, 56, 57, 60, 72, 88, 109, 110, 156, 159, 160, 181, 185, 223, 257, 259, 260
Water-related disasters 43, 44, 46–48, 69
Water-related diseases 22, 32, 73
Water-related infrastructure 10, 88, 138
Water Resources Group (WRG) 154, 161, 255
Water Resources Management (WRM) 174, 292
Water scarcity 4, 5, 7, 9, 10, 13, 15, 26, 27, 49, 53, 54, 56–59, 63, 64, 74, 76, 77, 79, 92, 102, 104, 113, 114, 138, 143, 147, 148, 153, 161–166, 171, 175, 185, 189, 202, 204, 205, 207, 218, 223, 227, 229, 230, 236, 239, 249, 253, 254, 307

Water security 5, 6, 153, 154, 156, 161, 218, 256
Water shortages 55, 72, 76, 91, 127, 158, 165, 223, 259
Water-smart Power 201, 209
Water stress 4, 9, 13, 14, 68, 71–78, 96, 156, 207, 212, 218, 223, 228–230, 251, 257, 308
Water-stressed environments 22, 73
Weather 5, 10, 24, 31, 35, 38, 47, 53–56, 60, 63, 75, 85, 95, 96, 127, 133, 134, 141, 148, 163, 164, 180, 181, 189, 190, 193, 195, 202, 226, 237, 242, 243, 245, 290, 291, 302, 307
Wind turbines 210
World Bank 20, 23, 24, 34, 59, 71, 84, 90, 102, 139, 154, 163, 255
World Development Report on Gender Equality and Development 23
World population growth 67

FSC
www.fsc.org
MIX
Papier | Fördert
gute Waldnutzung
FSC® C083411

Kolibri 360 GmbH
Ferdinand-Jühlke-Straße 7
99095 Erfurt, Deutschland
produktsicherheit@kolibri360.de